PCR Mutation Detection Protocols

METHODS IN MOLECULAR BIOLOGY™

John M. Walker, SERIES EDITOR

METHODS IN MOLECULAR BIOLOGY™

PCR Mutation Detection Protocols

Edited by

Bimal D. M. Theophilus

Department of Haematology, Birmingham Children's Hospital NHS Trust, Birmingham, UK

and

Ralph Rapley

Department of Biosciences, University of Hertfordshire, Hatfield, UK

Humana Press   Totowa, New Jersey

© 2002 Humana Press Inc.
999 Riverview Drive, Suite 208
Totowa, New Jersey 07512

www.humanapress.com

This publication is printed on acid-free paper. ∞
ANSI Z39.48-1984 (American Standards Institute) Permanence of Paper for Printed Library Materials.

Production Editor: Jessica Jannicelli.

Cover design by Patricia F. Cleary.

For additional copies, pricing for bulk purchases, and/or information about other Humana titles, contact Humana at the above address or at any of the following numbers: Tel.: 973-256-1699; Fax: 973-256-8341; E-mail: humana@humanapr.com; or visit our Website: www.humanapress.com

Printed in the United States of America. 10 9 8 7 6 5 4 3 2 1

Library of Congress Cataloging in Publication Data

PCR mutation detection protocols / edited by Bimal d. Theophilus and Ralph Rapley.
    p. cm. -- (Methods in molecular biology ; v. 187)
  Includes bibliographical references and index.
  ISBN 0-89603-617-0 (alk. paper)
    1. Mutation (Biology)--Laboratory manuals. 2. Polymerase chain reaction--Laboratory manuals. I. Theophilus, Bimal D. II. Rapley, Ralph. III. Methods in molecular biology (Totowa, N.J.) ; v. 187.

    QH462.A1 P37 2002
    576.5'49--dc21

                                                                                      2002020563

Preface

As we enter the new millennium, it is tempting to speculate what may lie ahead in future years, decades, and even centuries. In the area of the medical and life sciences at least, we can speculate with perhaps more certainty than may be possible in other areas. The exciting stage at which we find ourselves in the field of molecular genetics means that we can be in no doubt that the application of DNA technology will underlie many major advances in medicine in the coming decades. While international research efforts seek to demonstrate the viability of gene therapy, a major present application of human molecular genetics is the identification of disease-causing mutations. This information may be used for prenatal and carrier diagnoses, or to aid early detection and determine appropriate treatment of various disease states. While, traditionally, progress has been in diseases caused by mutations in single genes, present research is unraveling the underlying molecular basis of multigene disorders such as cancers, as well as identifying increasing numbers of disease-associated single nucleotide polymorphisms (SNPs). In addition, the completion of the human genome project will no doubt advance the pace of discovery even further, and also provide new possibilities for diagnosis and treatment.

The rapidly increasing applications of DNA technology to disease diagnosis has spawned numerous molecular diagnostic laboratories with an interest in mutation detection methodology. Such laboratories would like the availability of a single mutation method that is cheap, fast, with 100% detection in kilobase lengths of DNA, and does not require specialized equipment or harmful reagents. However, because no such universally applicable method exists, the present state of play is a plethora of methodology, from which the user makes a choice based on facilities, expertise, frequency of use, detection rate demanded, and whether the application purpose is diagnostic (detection of the presence or absence of a known mutation) or involves screening a candidate gene for a new unidentified mutation.

PCR Mutation Detection Protocols comprises a comprehensive step-by-step guide that brings together the large number of PCR-based mutation detection methods described to date. Many of the earlier chapters describe the basic technology and techniques, e.g., the principles and methodology of PCR, labeling DNA probes, restriction fragment length polymorphism analysis, and Southern blotting. Further techniques are then presented covering both categories of

v

mutation detection: detection of the presence of a known mutation and screening for new mutations. The techniques presented in each involve different approaches appropriate to different mutation types: point mutations (e.g., ASO-PCR, SSCP, DGGE, chemical cleavage), deletions (multiplex PCR, FISH, blotting), nonsense mutations (PTT), etc. The new and exciting techniques of DNA array analysis are also presented. The final chapters deal with different approaches to DNA sequencing as a detection method in its own right, or for characterizing mutations previously located by one of the other screening techniques. Recently developed and experimental methods, such as conformation sensitive gel electrophoresis, are presented in addition to the more established methods.

Each chapter includes the underlying basis of the techniques, and enables the reader to select the optimum method to use in relation to the above criteria. Particularly useful are the Notes sections containing the small details necessary for the successful execution of the technique. *PCR Mutation Detection Protocols* is aimed at postgraduate scientists and researchers in diagnostic and research laboratories. In addition, the basic techniques covered in the introductory chapters will ensure the book constitutes a fitting initiation to molecular techniques for individuals in related medical and scientific fields.

Bimal D. M. Theophilus
Ralph Rapley

Contents

Contributors

RICHARD BAGNALL • *Division of Medical and Molecular Genetics, GKT School of Medicine, Guy's Hospital, London, UK*

HENRY BRZESKI • *Windber Research Institute, Windber, PA*

CHARLES R. CANTOR • *Sequenom Inc., San Diego, CA*

PETER A. DAVIES • *Institute of Medical Genetics, University of Wales College of Medicine, Cardiff, UK*

SARA A. DYER • *Regional Genetics Laboratory, Birmingham Women's Hospital, Birmingham, UK*

MOHAMMAD S. ENAYAT • *Department of Haematology, Birmingham Children's Hospital NHS Trust, Birmingham, UK*

FRANCESCO GIANNELLI • *Division of Medical and Molecular Genetics, GKT School of Medicine, Guy's Hospital, London, UK*

ANNE C. GOODEVE • *Division of Genomic Medicine, Royal Hallamshire Hospital, Sheffield, UK*

GEORGE GRAY • *Department of Clinical Chemistry, Birmingham Children's Hospital NHS Trust, Birmingham, UK*

ELAINE K. GREEN • *Department of Psychiatry, Queen Elizabeth Psychiatric Hospital, University of Birmingham, Birmingham, UK*

PETER M. GREEN • *Division of Medical and Molecular Genetics, GKT School of Medicine, Guy's Hospital, London, UK*

ANDREA M. GUILLIATT • *Department of Haematology, Birmingham Children's Hospital NHS Trust, Birmingham, UK*

CAROL A. HARDY • *Molecular Genetics Laboratory, Regional Genetics Service, Birmingham Women's Hospital, Birmingham, UK*

ADRIAN J. HARWOOD • *MRC Laboratory for Molecular Cell Biology and Department of Biology, University College London, London, UK*

LAURA HEISLER • *Third Wave Technologies, Inc., Madison, WI*

NICOLA LOUISE JONES • *Department of Haematology, Birmingham Children's Hospital NHS Trust, Birmingham, UK*

CHRISTIAN JURINKE • *Sequenom Inc., San Diego, CA*

HUBERT KÖSTER • *Sequenom Inc., San Diego, CA*

CHAO-HUNG LEE • *Department of Pathology and Laboratory Medicine, Indiana University School of Medicine, Indianapolis, IN*

RALPH RAPLEY • *Department of Biosciences, University of Hertfordshire, Hatfield, UK*

G. K. SURDHAR • *Department of Haematology, Birmingham Children's Hospital NHS Trust, Birmingham, UK*

ANN-CHRISTINE SYVÄNEN • *Department of Medical Sciences, Uppsala University, Sweden*

BIMAL D. M. THEOPHILUS • *Department of Haematology, Birmingham Children's Hospital NHS Trust, Birmingham, UK*

DIRK VAN DEN BOOM • *Sequenom Inc., San Diego, CA*

ANDREW J. WALLACE • *DNA Laboratory, Department of Medical Genetics, St. Mary's Hospital, Manchester, UK*

YVONNE WALLIS • *Regional Genetics Laboratory, Birmingham Women's Hospital, Birmingham, UK*

ANU WARTIOVAARA • *Department of Neurology, University of Helsinki, Helsinki, Finland*

NAUSHIN H. WASEEM • *Division of Medical and Molecular Genetics, GKT School of Medicine, Guy's Hospital, London*

IAN J. WILLIAMS • *Department of Haematology, Birmingham Children's Hospital NHS Trust, Birmingham, UK*

1

Agarose and Polyacrylamide Gel Electrophoresis

Andrea M. Guilliatt

1. Introduction

Electrophoresis through agarose or polyacrylamide gels is a standard method used to separate, identify, and purify nucleic acids. The technique is simple, rapid to perform and capable of resolving fragments that differ by as little as 0.2% in size. Electrophoresis occurs under the influence of an electric field: Charged molecules such as nucleic acids migrate in the direction of the electrode having the opposite charge (anode). The electrophoretic mobility of nucleic acids is determined by a number of parameters, but molecules of linear double-stranded DNA migrate through gel matrices at rates that are inversely proportional to the $\log_{10}$ of the number of base pairs *(1)* and therefore larger molecules migrate more slowly because of the greater frictional drag (*see* **Note 1**). Other factors affecting electrophoretic mobility include the pK value, base composition, concentration of gel matrix, composition and ionic strength of the electrophoresis buffer, temperature and the use of intercalating dyes such as ethidium bromide.

The matrix used for electrophoresis should have adjustable but regular pore sizes and be chemically inert, and the choice of which gel matrix to use depends primarily on the sizes of fragments being separated. Agarose gels are the most popular medium for the separation of moderate and large-sized nucleic acids and have a wide range of separation but a relatively low resolving power. Polyacrylamide gels are most effective for separating smaller fragments, and although the gels are generally more difficult to prepare and handle, they have three major advantages over agarose gels. They have a greater resolving power, can accommodate larger quantities of DNA without significant loss in resolution, and the DNA recovered from polyacrylamide gels is extremely pure.

From: *Methods in Molecular Biology, vol. 187: PCR Mutation Detection Protocols*
Edited by: B. D. M. Theophilus and R. Rapley © Humana Press Inc., Totowa, NJ

Two electrophoresis buffers are commonly used and contain EDTA and Tris-acetate (TAE) or Tris-borate (TBE) at a concentration of approx 50 mM. For historical reasons, TAE is the most commonly used buffer for agarose gel electrophoresis, but its buffering capacity is low and may become exhausted during extended electrophoresis. TBE is slightly more expensive, but it offers significantly higher buffering capacity. Although the resolving power of the buffers is almost identical, double-stranded linear DNA migrates approx 10% faster in TAE than in TBE. Electrophoresis buffers are routinely prepared as concentrated solutions and stored at room temperature (*see* **Note 2**).

The most convenient method for visualizing DNA in agarose and polyacrylamide gels is by staining with the fluorescent dye ethidium bromide (3,8-diamino-6-ethyl-5-phenyl-phenanthridium bromide), which contains a fixed planar group that intercalates between the stacked bases of the DNA *(2)*. The fixed position and the close proximity to the bases causes the bound dye to display an increased fluorescent yield compared to that of the free dye in solution. Ultraviolet (UV) radiation at a range of 260–360 nm is absorbed by the DNA and transmitted to the dye, and the energy is re-emitted at 590 nm in the red–orange region of the visible spectrum. Because the fluorescent yield of ethidium bromide: DNA complexes is greater than that of unbound dye, small amounts of DNA can be detected in the presence of free ethidium bromide in the gel. Ethidium bromide promotes damage of the nucleic acids when viewed under UV light (photonicking); therefore, if the nucleic acid is to be used in reactions following visualization, the gel should be viewed using long-wavelength UV light (300 nm).

1.1. Agarose Gel Electrophoresis

Agarose is a linear polymer extracted from seaweed that forms a gel matrix by hydrogen-bonding when heated in a buffer and allowed to cool. Many chemically modified forms of agarose are available commercially that gel or melt at different temperatures without any significant loss of mechanical strength. Although these different forms of agarose can be useful in both the qualitative and preparative electrophoresis of DNA, the resolving power is still not comparable to that of polyacrylamide gels.

The density and porosity of the gel matrix is determined by the concentration of agarose used, referred to as the percentage of agarose (w/v) in buffer (*see* **Note 3**). Typical agarose gel concentrations fall within the range of 0.3 to 2.5% (w/v), depending on the size of DNA fragments to be separated (**Table 1**). For most applications, only a single-component agarose is needed and no polymerization catalysts are required and they are, therefore, quick and easy to prepare. This coupled, with the lack of toxicity (unless in the buffers), is largely responsible for the popularity of agarose gel electrophoresis.

Table 1
Range of Separation of Linear DNA Molecules
in Different Agarose Gel Concentrations

Concentration of agarose (% [w/v])	Efficient range of separation of linear DNA molecules (kb)
0.3	5–60
0.6	1–20
0.7	0.8–10
0.9	0.5–7
1.2	0.4–6
1.5	0.2–3
2.0	0.1–2

Many configurations and sizes of agarose gel electrophoresis tanks are available, of which the most common is the horizontal slab gel. Because of their relatively poor mechanical strength, agarose gels are cast in clear plastic UV-transparent trays allowing handling and transfer of the gel once set (*see* **Note 4**). Electrophoresis is carried out with the gel submerged just beneath the surface of the buffer, and as the resistance of the gel is similar to that of the buffer, a current passes through the gel. The principle advantage of submarine gel electrophoresis is that the thin layer of buffer prevents the gel from drying out and provides some degree of cooling.

The electrophoretic behavior of DNA in agarose gels is not significantly affected by temperature or the base composition of the DNA *(3)*; therefore, agarose gels are generally run at room temperature unless low-melting-temperature agarose is used or the agarose concentration is less than 0.5% (w/v), when the mechanical strength can be improved by running at 4°C.

1.2. Polyacrylamide Gel Electrophoresis

Polyacrylamide gels are formed by the vinyl polymerization of acrylamide monomers, ($CH_2=CH-CO-NH_2$) crosslinked by the bifunctional co-monomer N,N'-methylene-bis-acrylamide ($CH_2=CH-CO-NH-CH_2-NH-CO-CH=CH_2$). The resulting crosslinked chains form a gel structure whose pore size is determined by the initial concentrations of both acrylamide and the crosslinker. The nomenclature introduced by Hjertén et al. *(4)* is now widely used to describe gel composition, the term T being the total monomer concentration (acrylamide and Bis) in grams/100 mL and C being the percentage (by weight) of total monomer T that is contributed by the crosslinker (Bis). The pore size of the gel can be altered in an easy and controllable fashion by changing the concentrations of the two monomers. The polymerization proceeds by a free-radi-

Table 2
Range of Separation of Linear DNA Molecules
in Agarose and Polyacrylamide Gels and the
Position of Migration of Bromophenol Blue and Xylene Cyanol

Agarose gel concentration (%[w/v])	Effective range of resolution (bp)	Xylene cyanol migration (bp)	Bromophenol blue migration (bp)
0.5–1.5	1000–3000	4000–5000	400–500
Acrylamide gel concentration (%[w/v])			
3.5	1000–2000	460	100
5.0	80–500	260	65
8.0	60–400	160	45
12.0	40–200	70	20
15.0	25–150	60	15
20.0	6–100	45	12

cal mechanism and the most common method of initiation is with ammonium persulfate, which produces oxygen free radicals by a base-catalyzed mechanism, typically tertiary aliphatic amines such as $N,N,N'N'$-tetramethylethylene-diamine (TEMED) (*see* **Note 5**). The length of the chains is determined by the concentration of the acrylamide in the polymerization reaction (between 3.5% and 20%). One molecule of crosslinker is included for every 29 monomers of acrylamide. The effective range of separation in nondenaturing gels containing different concentrations of acrylamide is shown in **Table 2**.

Polyacrylamide gels are usually run between two glass plates, ensuring uniform electrical conditions across the slab so that comparison between different sample zones is far more accurate and a large number of samples may be run on the gel.

Polyacrylamide gels are poured and run in 1X TBE at low voltages to prevent denaturation of small fragments of DNA by heat generated by passage of the electric current. Most species of double-stranded DNA migrate through the gel at a rate approximately inversely proportion to the log_{10} of their size, however, their electrophoretic mobility is affected by their base composition and sequence, so that two DNAs of exactly the same size can differ in mobility by up to 10%, as a result of secondary structures that may form at specific sequences in the double-stranded DNA *(5)*.

Denaturing polyacrylamide gels are used for the separation and purification of single-stranded fragments of DNA and are polymerized in the presence of an agent that suppresses base-pairing in nucleic acids, usually urea. Denatured DNA migrates though these gels at a rate that is almost completely dependent on its base composition and sequence and is discussed elsewhere in this volume (*see* Chapters 14–16).

2. Materials
2.1. Agarose Gel Electrophoresis

All of the chemicals used are of molecular biology grade, and solutions are prepared with double-distilled water unless otherwise stated.

1. Agarose gel apparatus, comprising:
   a. Gel tank and safety lid
   b. Gel tray
   c. Comb
   d. Gel caster (optional)
2. Power supply capable of at least 100 V, 100 mA.
3. Powdered agarose.
4. Electrophoresis buffer (*see* **Note 2** for formulations).
5. 10X Gel loading buffer: The loading buffer for sample application should contain 0.25% bromophenol blue (BPB) and 0.25% xylene cyanol as tracking dyes and 30% sucrose, glycerol, or Ficoll to increase the sample solution density (*see* **Note 6**).
6. Ethidium bromide solution is generally prepared as a stock solution at a concentration of 10 mg/mL in water and stored at room temperature protected from light. Ethidium bromide is toxic and a powerful mutagen; therefore, gloves should always be worn. Solutions containing ethidium bromide should be disposed of appropriately as discussed in the Material Safety Data Sheets.
7. Microwave oven or hot plate.
8. UV transilluminator and gel documentation system.

2.2. Polyacrylamide Gel Electrophoresis

All of the chemicals used are of molecular biology grade and solutions are prepared with double-distilled water unless otherwise stated.

1. Polyacrylamide gel apparatus, comprising:
   a. Gel tank and safety lid
   b. Glass plates
   c. Spacers and combs of the same thickness
   d. Clamps or gel caster assembly (optional)
2. 30% Acrylamide stock, prepared by the addition of 29 g of acrylamide and 1 g *N,N'*-methylene-bis-acrylamide to 100 mL water (*see* **Note 7**).
3. 10X TBE (*see* **Note 2** for formulation).
4. 10% Ammonium persulfate, prepared by adding 1 g ammonium persulfate to 10 mL water. This solution may be kept at 4°C for several weeks.
5. TEMED.

6. Power supply.
7. Siliconizing solution (dimethyl dichlorosilane [e.g., Sigmacote®]).

3. Methods

3.1. Agarose Gel Electrophoresis

3.1.1. Assembly and Pouring of the Gel

1. Seal the edges of the UV-transparent plastic casting tray with strong masking tape or use a commercial gel casting system (*see* **Note 8**).
2. Place the tray/gel caster onto a horizontal section of bench, using a glass leveling plate if necessary, and place the comb(s) in the appropriate position(s) so that wells are formed at the cathode end of the gel.
3. Add the desired amount of powdered agarose to a measured quantity of 1X electrophoresis buffer in an Erlenmeyer flask or beaker and cover with Saran-Wrap. Heat the mixture in a microwave oven swirling every 30 s until the agarose is visibly seen to have dissolved. Alternatively, the agarose can be heated using a hot plate. Any undissolved agarose appears as small translucent particles (*see* **Note 9**).
4. Allow the solution to cool to 50°C, unless a high concentration of agarose or high-gelling-temperature agarose is used where gelation will occur more rapidly. A low level (0.5 µg/mL) of ethidium bromide can be added at this stage, allowing the progression of the electrophoresis to be analyzed during electrophoresis by illuminating the gel with UV light (*see* **Note 10**).
5. Pour the agarose into the gel mold, ensuring that no air bubbles form between the teeth of the comb, and allow the gel to set at room temperature for 30–40 min.

3.1.2. Running the Gel

1. Carefully remove the comb and place the gel and tray into the gel tank oriented with the wells at the cathode end, and add sufficient 1X electrophoresis buffer to cover the gel to a depth of approx 1 mm (*see* **Note 11**).
2. Mix the DNA samples with gel loading buffer to produce a 1X concentration of buffer and load into the wells through the thin layer of running buffer. Placing a black piece of paper behind the wells may facilitate in the loading process by making the wells more visible (*see* **Note 12**).
3. Load a DNA size standard to allow the determination of the sizes of the DNA fragments, because although the tracking dyes in the loading buffer give a rough estimate of the migration of the DNA, they do not give the exact size. Size standards can be purchased commercially or prepared by restriction enzyme digestion of plasmid DNA, producing DNA fragments of known sizes.
4. Place the lid onto the gel tank, being careful not to disturb the samples, and begin electrophoresis (*see* **Note 13**).
5. When the dyes have migrated the appropriate distance on the gel as shown in **Table 2**, turn off the power supply and proceed with visualization of the DNA.

3.1.3. Staining and Visualization of the Nucleic Acids

The gel can be stained during electrophoresis by the addition of ethidium bromide as described in **Subheading 3.1.1.** or following electrophoresis by immersion in a solution of 0.5 µg/mL ethidium bromide.

1. If ethidium bromide has been incorporated in the gel, the DNA can be visualized progressively during the run. If post-electrophoretic staining is necessary, place the gel in an appropriate volume of 0.5 µg/mL ethidium bromide for 30 min and then destain in water for 10 min (*see* **Note 14**).
2. Following staining, place the gel on a UV transilluminator and photograph the gel using standard cameras and film such as a Polaroid camera with a red filter and 667 black and white film, or a charged-coupled devise (CCD)-based digital analysis system (*see* **Note 15**).

3.2. Polyacrylamide Gel Electrophoresis

There are many types of commercially available electrophoresis equipment with differing arrangements of glass plates and spacers. In all cases, the aim is to form an airtight seal between the plates and the spacers so that any unpolymerized gel solution does not leak.

Most vertical electrophoresis tanks are constructed to hold glass plates. Spacers vary in thickness from 0.5 to 2 mm, but the thicker the gel, the hotter it will become during electrophoresis and overheating may occur.

3.2.1. Assembly and Pouring of the Gel

1. Prepare the glass plates and spacers by washing with warm detergent and rinsing with water. The plates should only be held by the edges so that oils from hands do not become deposited on the working surface of the plates and lead to the formation of bubbles in the gel. Rinse the plates with ethanol and allow to dry. One surface of the glass plate should be periodically treated with silicone solution to prevent the gel from sticking to both plates and therefore reduce the possibility that the gel will tear when it is removed from the plates following electrophoresis.
2. Most modern commercial gel systems provide gel casting units for the preparation of polyacrylamide gels; therefore, the manufacturers guidelines should be followed. Generally, lay one plate on the bench siliconized side upward and position the spacers on the plate. Place the inner glass plate onto the spacers and seal the edges of the gel with electrical tape or a clamping unit (*see* **Note 16**).
3. Calculate and prepare the desired quantities of reagents needed to make sufficient solution to fill the gel mould. For example, to pour a 5% acrylamide gel in a total volume of 100 mL, add 10 mL 10X TBE and 16.67 mL 30% acrylamide to 72.23 mL water (*see* **Note 17**).
4. Immediately before pouring, add the ammonium persulfate solution and TEMED and mix. Quickly fill the mould with the solution, trying not to trap any air bubbles in the mould. Apply a comb to the top of the gel and then flush out the syringe and needle (*see* **Note 18**).

3.2.2. Assembly and Running of the Gel

1. When the gel has polymerized, assemble the gel tank apparatus as recommended by the manufacturers.
2. Fill the tank with 1X buffer, remove the comb, and wash out the wells with buffer. Remove the tape from the bottom of the plates or cut with a sharp blade (*see* **Note 19**).
3. Mix the DNA samples with appropriate gel loading buffer and apply to the wells (*see* **Note 20**).
4. Run the gel at a voltage between 1 and 8 V/cm. If electrophoresis is carried out at a higher voltage, differential heating in the center of the gel may cause bowing of the DNA bands or even melting of small strands of DNA.
5. When the marker dyes have migrated the desired distance, turn off the power supply and disconnect the leads. Remove and detach the glass plates and pry apart using a spatula.

3.2.3. Staining and Visualization of Nucleic Acids

1. Because polyacrylamide quenches the fluorescence of ethidium bromide, it is not possible to detect bands that contain less than 10 ng of DNA using this method. To stain the gel, gently submerge the gel and its attached glass plate in 0.5 µg/mL ethidium bromide in 1X TBE buffer for 10–30 min at room temperature.
2. Destain for 10 min in water and following removal from the glass plate view the gel as described in **Subheading 3.1.3.**

4. Notes

1. At low voltages, the rate of migration of linear DNA fragments is proportional to the voltage applied, but as the electric field strength is increased, the mobility of high-molecular-weight DNA fragments increases preferentially. Therefore, the effective range of separation decreases as the voltage is increased.
2. Electrophoresis buffers are generally prepared as concentrated stock solutions, as shown in **Table 3**.
   A precipitate may form when 10X TBE is stored for extended periods of time; therefore, it should be stored in brown glass bottles at room temperature and discarded if a precipitate develops.
3. The unavoidable loss of water that occurs during the heating of the gel means that, in practice, the percentage value is not precise.
4. If the gel is to be handled extensively, it may be convenient to place a sheet of hydrophilic plastic support at the bottom of the gel mould, aiding in the handling of the gel once set.
5. Oxygen at above trace levels acts as an inhibitor; therefore, many people advocate the deaeration of stock acrylamide solutions. Gelation should ideally occur within 10–30 min of the addition of the catalysts, because outside of these times, uneven polymerization may result, leading to non-homogenous gels and poor separations. Because of the nature of the gel casting, inhibition of the polymerization by oxygen is confined to a narrow layer at the top of the gel.

Table 3
Formulations of Stock Electrophoresis Buffers

Electrophoresis buffer	Concentrated stock	1X Working solution
Tris-acetate (TAE)	50X Stock 242 g Tris base 57.1 mL Glacial acetic acid 100 mL of 0.5 M EDTA (pH 8.0)	40 mM Tris (pH 7.6) 20 mM Acetate 1 mM EDTA
Tris-borate (TBE)	10X Stock 108 g Tris base 55 g Boric acid 40 mL of 0.5 M EDTA (pH 8.0)	89 mM Tris (pH 7.6) 89 mM Boric acid 2 mM EDTA

6. Loading buffers are usually made as 5X to 10X concentrates and consist of three main constituents. The first is a high-density solution such as glycerol, Ficoll, or sucrose and the second is tracking dyes, such as bromophenol blue (BPB) or xylene cyanol. When choosing the loading buffer, it must be noted that it may quench the fluorescence of ethidium bromide and can obscure the presence of DNA. Chelating agents such as EDTA are also included, which complex divalent cations and stop any enzymatic reactions.

7. During storage, acrylamide and bis-acrylamide are slowly deaminated to acrylic and bisacrylic acid, catalyzed by light and alkali. The solution should be pH 7.0 or less and stored protected from light at room temperature. Fresh solutions should be prepared every few months. TEMED and persulfate are added immediately before use to initiate the polymerization process.

   Acrylamide is a potent neurotoxin and is readily absorbed through the skin. The effects of acrylamide are cumulative; therefore, gloves and a mask should be worn when working with powdered acrylamide and methylbisacrylamide. Although polyacrylamide is considered to be non-toxic, it should be handled with care, as it may contain small quantities of unpolymerized acrylamide. To avoid the hazards associated with acrylamide, stock solutions are available commercially that only require the TEMED and the persulfate to be added. Acrylamides may contain contaminating metal ions, although they can be easily removed by stirring overnight with approx 0.2 vol of monobed resin followed by filtration.

8. Some agarose gel systems enable the casting of the gel directly in the electrophoresis tank.

9. The buffer should not occupy more than 50% of the volume of the flask. Always wear protective gloves when handling heated agarose, as the solution may become superheated and boil violently when disturbed. Some evaporation of the solution may occur and can be made up with water if desired.

10. During electrophoresis, the ethidium bromide migrates toward the cathode in the opposite direction to the DNA. Extended electrophoresis can lead to removal of the ethidium bromide from the gel, making detection of smaller fragments difficult. If this occurs, the gel can be restained by soaking for 30–40 min in a solu-

Table 4
Examples of Polyacrylamide Gel Formulations for 100 mL Gel

Constituents	Gel concentration (%T)					
	3.5	5.0	7.5	10.0	15.0	20.0
Acrylamide (g)	3.24	4.7	7.13	9.6	14.55	19.5
Bis (g)	0.26	0.3	0.37	0.4	0.45	0.5
TEMED (mL)	0.1	0.1	0.1	0.1	0.1	0.1
10X Buffer stock (mL)	10.0	10.0	10.0	10.0	10.0	10.0
Water (mL)	86.4	84.9	82.4	79.9	74.9	69.9
10% Ammonium persulfate (mL)	1.0	1.0	1.0	1.0	1.0	1.0

tion containing 0.5 µg/mL ethidium bromide. The mobility of linear DNA is reduced by the presence of ethidium bromide by about 15%.

11. The electrical resistance of the gel is almost the same as that of the buffer and so a significant proportion of the current passes through the gel, but the deeper the buffer layer, the less efficient this becomes.

12. The maximum volume of solution that can be loaded is determined by the dimensions of the well. To reduce the possibility of contaminating neighboring samples, it is not advisable to fill the wells completely. The minimum amount of DNA that can be detected by ethidium-bromide-stained gels is approx 2 ng in a 5-mm-wide band, but if there is more than 500 ng of DNA, the well may become overloaded.

13. The power requirements for electrophoresis depend on the thickness and length of the gel and the concentration of agarose and buffer used. It is recommended that for maximal resolution, voltages applied to the gels should not exceed 10 V/cm, as higher voltages may preferentially increase the migration rate of higher-molecular-weight DNA and reduce the range of separation. Overnight separations using lower voltages are frequently used.

14. Extended destaining can lead to the removal of the ethidium bromide and lowering of the detection sensitivity. Insufficient de-staining will lead to a higher background of fluorescence.

15. Ultraviolet radiation is particularly dangerous to the eyes; therefore, to minimize exposure, protective goggles or a face shield that efficiently blocks ultraviolet radiation should be worn.

16. The bottom corners of the plates is where leaks are most likely to occur. An alternative method is to seal the glass plate with a strip of filter paper impregnated with catalyzed acrylamide or use a commercial gel casting apparatus.

17. Examples of typical acrylamide gel formulations are shown in **Table 4**.

18. The pore size of the matrix is affected by the temperature at which polymerization occurs and the optimum polymerization temperature is approx 25–30°C. The concentration of catalysts used to initiate the polymerization reaction and the time taken for gelation to occur also affects the pore size.

19. It is important to wash out the wells thoroughly, as any unpolymerized acrylamide in the wells may subsequently polymerize, giving rise to irregular surfaces, which lead to distorted bands.
20. It is important that the gel is not loaded symmetrically, as the orientation of the gel can become lost during subsequent steps as it is removed from the plates for visualization and staining. When loading the samples, do not attempt the expel of any remaining sample from the pipet, as the resulting air bubbles may blow out the sample from the well. It is important not to take too long to complete the gel loading process, as the samples may diffuse from the wells.

References

1. Helling, R. B., Goodman, H. M., and Boyer, H. W. (1974) Analysis of EcoRI fragments of DNA from lambdoid bacteriophages and other viruses by agarose gel electrophoresis. *J. Virol.* **14,** 1235–1244.
2. Sharp, P. A., Sugden, B., and Saunders, J. (1973) Detection of two restriction endonuclease activities in Haemophilus parainfluenzae using analytical agarose-ethidium bromide electrophoresis. *Biochemistry* **12,** 3055–3063.
3. Thomas, M. and Davis, R. W. (1975) Studies on the cleavage of bacteriophage lambda DNA with EcoRI restriction endonuclease. *J. Mol. Biol.* **91,** 315–321.
4. Hjertén, S., Jerstedt, S., and Tiselius, A. (1965) Some aspects of the use of "continuous" and "discontinuous" buffer systems in polyacrylamide gel electrophoresis. *Anal. Biochem.* **11,** 219–223.
5. Orita, M., Suzuki, Y., Sekilya, T., and Hayashi, K. (1989) Rapid and sensitive detection of point mutations and DNA polymorphisms using the polymerase chain reaction. *Genomics* **5,** 874–879.

Further Reading

Rickwood, D. and Hanes, B. D., eds. (1988) *Gel Electrophoresis of Nucleic Acids: A Practical Approach.* IRL, Oxford, UK.

Sambrook, J., Fritsch, E. F., and Maniatis, T. *Molecular Cloning: A Laboratory Manual*, 2nd ed. Cold Spring Harbor Laboratory, Cold Spring Harbour, NY.

2

Internal Labeling of DNA Probes

Ralph Rapley and Bimal D. M. Theophilus

1. Introduction

One of the most common precursors to undertaking a protocol for mutation detection is the production of a suitably labeled DNA probe *(1)*. Labeled nucleotides (radioactive or fluorescent) can be incorporated efficiently into double-stranded DNA by a number of methods. One of the most common is by a process termed *nick translation*. Nick translation works by using DNase and DNA polymerase I enzymes. DNase cuts one strand of the DNA, exposing 5'-phosphoryl and 3'-hydroxyl (OH) termini. Polymerase I adds dNTPs, including labeled dNTPs to the exposed 3'-OH strand, and at the same time, the polymerase exonuclease activity digests from the exposed 5' end. In this way, a new complementary strand, including labeled dNTPs, is produced *(2)*. It is also possible to incorporate radioactive nucleotides into a DNA using a enzymatic primer extension technique, usually termed *random primer labeling (3)*. In this method, random hexanucleotides are annealed to denatured DNA to be used as the probe. These are used as a primer for enzymatic extension in the presence of the four deoxyribonucleotides, one of which is radiolabeled. Alternative probes may be prepared where the label occurs on one of the termini of the DNA, either the 3' or the 5' end. The protocol for this type of labeling is found in Chapter 3.

2. Materials (*see* Note 1)
2.1. Nick Translation of DNA

1. 10X Nick translation buffer: 0.5 M Tris-HCl (pH 7.5), 0.1 M MgSO$_4$, 1 mM dithiothreitol, 500 mg/mL bovine serum albumin (optional).
2. DNase I: 10 ng/mL.
3. DNA polymerase I: 0.5 U/μL.

From: *Methods in Molecular Biology, vol. 187: PCR Mutation Detection Protocols*
Edited by: B. D. M. Theophilus and R. Rapley © Humana Press Inc., Totowa, NJ

4. Unlabeled dNTP: 2 mM each of dATP, dGTP, and dTTP.
5. Radiolabeled dCTP: 10 mCi/mL [α-^{32}P]dCTP, specific activity approx 3000 Ci/mmol (*see* **Note 2**). This is stored at –20°C and should be removed from the freezer approx 20 min before setting up the reaction.
6. Stop solution: 0.5 M EDTA (pH 8.0).
7. Sephadex separation spin column (*see* **Note 3**).

2.2. Random Hexamer Labeling of DNA

1. DNA probe to be labeled in TE buffer (10 mM Tris-HCl, 1 mM EDTA, pH 8.0).
2. Hexamer mix: 0.043 M each of dCTP, dTTP, and dGTP, 0.43 M HEPES, pH 7.0, 12 U/mL random hexanucleotides (Amersham Pharmacia Biotech, UK).
3. [α-^{32}P]dATP, specific activity 6000 Ci/mM (Amersham Pharmacia Biotech). This is stored at –20°C and should be removed from the freezer approx 20 min before setting up the reaction.
4. Stop solution: 0.5 M EDTA, pH 8.0.
5. DNA polymerase (e.g., Klenow fragment) (6 U/µL).

3. Methods
3.1. Nick Translation of DNA

1. Dilute DNA to be labeled to 20–200 ng/mL with sterile distilled H$_2$O and add 1 mg to a sterile microcentrifuge tube.
2. Add the following to the tube:
   a. 10 µL 10X nick translation buffer
   b. 10 µL 20 nM unlabeled dNTPs
   c. 10 µL 30 pmol labeled [α-^{32}P]dCTP
3. Add 1 ng/mL DNase (10 mL) and 2.5 U DNA polymerase I (5 mL). Gently mix by pipetting solution up and down.
4. Add water to ensure a final volume of 100 µL.
5. Incubate for 2 h at 15°C.
6. Stop the reaction by adding 10 µL EDTA.
7. The probe is now ready for hybridization. However, it may be necessary to remove any unincorporated nucleotides, using Sephadex spin columns (*see* **Notes 3 and 4**).

3.2. Random Hexamer Labeling of DNA

1. Take 25–100 ng of DNA to be labeled and adjust the volume of TE to 11 µL.
2. Denature the DNA by boiling for 5 min and transfer immediately to an ice bucket.
3. Add 11 µL of the primer mix, 2 µL of the [α-^{32}P]dATP, and 3 U of the Klenow polymerase (0.5 µL).
4. Incubate the mix at room temperature for approx 4 h.
5. Add 5 µL of stop mix to terminate the reaction.
6. At this point, the probe may be purified from free nucleotides by use of Sephadex spin columns (*see* **Notes 3** and **4**).

7. Following recovery of the labeled DNA, it must be rendered single-stranded by boiling before it may be used in hybridization experiments.

4. Notes

1. Enzymes and buffers are now available in kit forms (Amersham [Amersham Pharmacia Biotech, UK], Promega [Promega, UK]); however, slight variations exist in concentrations of enzymes and buffer ingredients.
2. Nick translation can also be used to label DNA with nonradioactive markers, including incorporation of Cy3–dCTP and fluorescein, or rhodamine–dUTP into DNA. However, radiolabeled probes are more sensitive markers for low quantities of DNA. It is also possible to label more than one dNTP if higher specific activity is required for hybridizing low amounts of DNA. However, this increases nonspecific hybridization.
3. To remove unincorporated labeled dNTPs, the probe can be purified by passing the solution through a Sephadex spin column or push column. Unincorporated dNTPs are trapped inside the Sephadex beads, whereas DNA is too large to enter the beads and passes straight through the column.
4. Percentage incorporation and the specific activity of the probe can be calculated by measuring the radioactivity in the mixture before and after separation.

$$\text{Percentage incorporation} = \frac{\text{cpm incorporated} \times 100}{\text{Total cpm}}$$

$$\text{Specific activity (cpm/mg DNA)} = \frac{\text{cpm incorporated} \times \text{dilution} \times 100}{\text{mg input DNA}}$$

References

1. Aquino de Muro, M. (1998) Gene Probes, in *The Molecular Biomethods Handbook* (Rapley, R. and Walker, J. M., eds.) Humana, Totowa, NJ.
2. Rigby, P. W. J., Dieckmann, M., Rhodes, C., and Berg, P. (1977) Labelling deoxyribonucleic acid to a high specific activity in vitro by nick translation with DNA polymerase I. *J. Mol. Biol.* **113**, 237–251.
3. Feinberg, A. P. and Vogelstein, B. (1983) A technique for radiolabelling DNA restriction endonuclease fragments to a high specific activity. *Anal. Biochem.* **132**, 6–13.

3

End-Labeling of DNA Probes

Adrian J. Harwood

1. Introduction

End-labeling is a rapid and sensitive method for radioactively, or nonisotopically, labeling DNA fragments and is useful for visualizing small amounts of DNA. End-labeling can also be used to label fragments at one end. All of the enzymes employed are specific to either the 3' or 5' termini of DNA and will, consequently, only incorporate label once per DNA strand. If double-stranded DNA is used, both ends are labeled, but single end-labeled fragments can be produced by further restriction enzyme digestion. This works well with DNA fragments cloned into polylinkers, as one labeled end can be removed as a tiny DNA fragment, making subsequent purification easier. Such single end-labeled molecules can be used to order restriction enzyme fragments and are a prerequisite for Maxam–Gilbert DNA sequencing *(1)*. End-labeled synthetic oligonucleotides have numerous applications, including sequence specific probes *(2)*, gel retardation and Southwestern assays *(3)*, and sequencing polymerase chain reaction (PCR) products *(4)*.

There are two common methods of end-labeling: the "fill-in" reaction and the "kinase" reaction. The fill-in reaction uses the Klenow fragment of *Escherichia coli* DNA polymerase *(5)* and labels DNA fragments that have been digested with a restriction enzyme to create a 5' overhang. Klenow extends the 3' recessed end of one DNA strand by using the 5' overhang of the other strand as a template (**Fig. 1A**). This is the method of choice for double-stranded DNA fragments because of its ease. When suitable restriction enzyme sites are unavailable or when the substrate is single stranded, the kinase reaction is used. The "kinase" reaction uses T4 polynucleotide kinase (T4 kinase) to transfer labeled phosphate to the 5' end of the DNA molecule *(6)* (**Fig. 1B**). This method

From: *Methods in Molecular Biology, vol. 187: PCR Mutation Detection Protocols*
Edited by: B. D. M. Theophilus and R. Rapley © Humana Press Inc., Totowa, NJ

A The Fill-in reaction

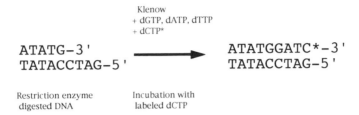

Klenow
+ dGTP, dATP, dTTP
+ dCTP*

ATATG–3'   ────────▶   ATATGGATC*–3'
TATACCTAG–5'   TATACCTAG–5'

Restriction enzyme   Incubation with
digested DNA   labeled dCTP

B The Kinase reaction

i) Removal of 5' terminal phosphate

5'pATATG..   ────────▶   p + 5'ATATG..
    TAC..   TAC..
                    Incubation with
                         CIAP

ii) Addition of labeled ^{32}P to 5' terminus

5'ATATG..   ────────▶   p + 5'*pATATG..
    TAC..   TAC..
                    Incubation with
                    ^{32}P-γ-ATP
                    and T4 kinase

Fig. 1. (**A**) The fill-in reaction; (**B**) the kinase reaction.

is ideal for labeling oligonucleotides, which are normally synthesized without a 5' phosphate. To label restriction-enzyme-digested DNA fragments, the terminal phosphate must first be removed by using a phosphatase, such as calf intestinal alkaline phosphatase (CIP). All of these reactions can be used without labeled nucleotides to modify the DNA fragments for further recombinant DNA manipulations.

2. Materials

Molecular-biology-grade reagents should be utilized whenever possible. Manipulations are performed in 1.5-mL disposable, sterile polypropylene tubes, with screw tops to prevent leakage of radioactivity. Local safety precautions must be obeyed when using radioactivity.

2.1. End-Labeling with Klenow

1. 10X Klenow buffer: 200 mM Tris-HCl, pH 7.6, 100 mM MgCl$_2$, 15 mM β-mercaptoethanol, 25 mM dithiothreitol.
2. Labeled nucleotide: ^{32}P-α-dNTP, most commonly supplied dATP or dCTP, but dGTP and dTTP are available. It is also possible to substitute nonisotopic label such as fluoroscein-11-dUTP and digoxygenin-11-dUTP.
3. Unlabeled dNTPs:
   a. dNTP mix: a mixture of 0.25 mM of each unlabeled dNTP, excluding that which corresponds to the labeled nucleotide (*see* **Note 1**).
   b. dNTP chase: 0.25 mM dNTP corresponding to the labeled nucleotide (*see* **Note 1**).
4. Klenow: the Klenow (large) fragment of DNA polymerase I at 1 U/μL. Store at –20°C.
5. TE: 10 mM Tris-HCl (pH 7.5), 1 mM EDTA. Autoclave and store at room temperature.
6. Phenol: Tris-HCl equilibrated phenol containing 0.1% hydroxyquinoline (as an antioxidant). Use ultrapure, redistilled phenol. Extract repeatedly with 0.5 M Tris-HCl (pH 8.0) until the aqueous phase is 8.0 and then extract once with 0.1 M Tris-HCl (pH 8.0). Can be stored at 4°C for at least 2 mo. Phenol is both caustic and toxic and should be handled with care.
7. Chloroform.
8. Phenol: chloroform mixture: A 1:1 mixture was made by adding an equal volume of chloroform to 0.1 M Tris-HCl, pH 8.0, equilibrated phenol. Can be stored at 4°C for at least 2 mo.
9. Ethanol and 70% ethanol (v/v in water).
10. 5 M Ammonium acetate, pH 7.5: Store at room temperature.

2.2. End-Labeling with T4 Kinase

11. 10X CIP buffer: 10 mM ZnCl$_2$, 10 mM MgCl$_2$, 100 mM Tris-HCl, pH 8.3.
12. CIP: calf intestinal alkaline phosphatase (Boehringer Mannhiem Gmbh) at 1 U/μL. Store at 4°C.
13. 10X Kinase buffer: 700 mM Tris-HCl, pH 7.6, 100 mM MgCl$_2$, 50 mM dithiothreitol.
14. ^{32}P-γ-ATP: Specific activity > 3000 Ci/mmol.
15. T4 kinase: T4 polynucleotide kinase at 1 U/μL. Store at –20°C.
16. Cold ATP: 1.0 mM ATP (freshly made from 20 mM stock).

3. Methods

3.1. End-Labeling with Klenow

1. Resuspend 1–1000 ng of DNA in 42 μL of dH$_2$O (*see* **Note 2**). Add 5 μL of 10X Klenow buffer, 1 μL of ^{32}P-α-dNTP, 1 μL of dNTP mix, and 1 μL of Klenow. Incubate at room temperature for 15 min (*see* **Note 3**).

2. Add 1 μL of dNTP chase. Incubate at room temperature for a further 15 min (*see* **Notes 1** and **4**).
3. Add 50 μL of TE followed by 100 μL of phenol:chloroform. Vortex briefly and separate by centrifugation at 12,000g in a microfuge (*see* **Note 5**).
4. Remove the aqueous (top) phase to a fresh tube and add 100 μL of chloroform. Separate the layers as in **step 3** and remove the aqueous phase to a fresh tube. Care must be taken, as the discarded reagents are contaminated with unincorporated ^{32}P-α-dNTP.
5. Add 60 μL (0.6 vol) of 5 M ammonium acetate and 200 μL (2 vol) of ethanol (*see* **Note 6**) and place on ice for 5 min. Centrifuge at 12,000g for 15 min. Carefully remove the supernatant (remember that it is radioactive) and wash the pellet in 70% ethanol.
6. Air-dry the pellet for 10 min and resuspend in the required amount of TE (10–100 μL).

The labeled DNA can be either immediately separated by gel electrophoresis and detected by autoradiography (*see* **Note 7**) or digested further with a second restriction enzyme. In either case, it is a good idea to count a 1-μL sample in a scintillation counter, between 5000 and 10,000 counts are required to detect the fragment by autoradiography. Possible causes of poor labeling and possible solutions are discussed in **Notes 8–10**.

3.2. End-Labeling with T4 Kinase

1. Dissolve 1–2 μg of restriction-enzyme-digested DNA in 44 μL of dH$_2$O. Add 5 μL of 10X CIP buffer and 0.05–1 U of CIP (*see* **Note 11**). Incubate for 30 min at 37°C (*see* **Notes 12** and **13**).
2. Heat-inactivate at 60°C for 10 min. Phenol extract and precipitate as in **Subheading 3.1., steps 3–5** (*see* **Notes 14** and **15**).
3. Resuspend the DNA in 17.5 μL of dH$_2$O. Add 2.5 μL of 10X kinase buffer, 5 μL of ^{32}P-γ-ATP, and 1 μL of T4 kinase. Incubate at 37°C for 30 min.
4. Add 1 μL of cold ATP and incubate for a further 30 min (*see* **Note 16**).
5. Phenol extract and precipitate as in **Subheading 3.1., steps 3–6** (*see* **Note 17**).

4. Notes

1. Unlabeled dNTPs are required for two reasons. First, the labeled nucleotide may not correspond to the first nucleotide to be filled within the restriction enzyme site. In the example shown in **Fig. 1A**, which is a BamHI site, the labeled nucleotide, dCTP*, corresponds to the fourth nucleotide; therefore, the other three nucleotides must be filled with cold dNTPs before the label is incorporated. For convenience, a general 7.5 mM mix of the unlabeled dNTPs can be used regardless of the actual composition of the restriction enzyme site. Second, a "chase" is required to generate molecules with flush ends, as the polymerase stalls in the limited concentrations of the labeled nucleotide. This step may be omitted in

cases where the heterogeneous sized termini are not a problem, (e.g., when labeling large DNA fragments for separation by agarose gel electrophoresis).

2. The fill-in reaction is very robust, and provided Mg^{2+} is present, it can be carried
out in almost any buffer. This means that it is possible to carry out the reaction by
simply adding the labeled dNTP, unlabeled dNTPs, and Klenow directly to the
restriction enzyme mix at the end of digestion.

3. As only a small region of DNA is labeled in this reaction, it proceeds very quickly.
Incubation at room temperature is sufficient, unless ^{35}S-labeled dNTP is used
when labeling should be carried out at 37°C. Prolonged incubation can result in
degradation of the DNA ends.

4. The labeled DNA may be used for gel electrophoresis at this point, but it must be
remembered that unincorporated ^{32}P-α-dNTP will be present in the DNA solution. This may increase the exposure of the operator and increase the risk of
contamination when carrying out gel electrophoresis.

5. An alternative purification is to pass the DNA through a Sephadex-G50 spin
column.

6. If only very small amounts of DNA are present, it may be necessary to add a
carrier such as 10 μg of tRNA or glycogen.

7. The gel should be fixed in 10% acetic acid or trichloroacetic acid (TCA) before
drying to prevent contamination of the gel dryer.

8. Klenow is rarely affected by inhibitors, but it rapidly loses its activity if it is
warmed in the absence of a substrate. It can be one of the first enzymes to be lost
from the general enzyme stock. If the activity of the enzyme is in doubt, carry out
a test reaction by labeling control DNA. Generally, DNA markers are good for
this, but check the structure of the ends before proceeding.

9. The structure of the end is important, as the enzyme can only "fill-in" those bases
present in the site. Recheck the sequence of the single-strand end produced by
restriction enzyme digestion. It may be possible to exchange the ^{32}P-α-dNTP for
another which has a higher specific activity.

10. The Klenow "fill-in" reaction only incorporates a small number of ^{32}P-labeled
nucleotides per DNA molecule. If higher levels of incorporation are required, T4
DNA polymerase may be used. T4 DNA polymerase has a 200-fold higher 3'–5'
exonuclease activity than Klenow. If the DNA fragments are incubation in the
absence of dNTPs, this enzyme will produce a region of single-stranded DNA,
which can be subsequently labeled with a higher incorporation by the addition of
^{32}P-α-dNTP and cold dNTPs to the mix (*6*).

11. One unit of CIP dephosphorylates 50 pmol of ends in 1 h (for a 5-kb fragment,
1 pmol of ends is approx 2 μg).

12. The efficiency of dephosphorylation of blunt and 5' recessed ends is improved by
incubating the reaction at 55°C.

13. The phosphatase reaction can be carried out in restriction enzyme buffer by the
addition of 0.1 vol of 500 m*M* Tris-HCl, pH 8.9, 1 m*M* EDTA, and the required
amount of enzyme.

14. It is important to remove all phosphatase in order to prevent removal of the newly incorporated labeled phosphate.

15. The T4 kinase reaction is very sensitive to inhibitors such as those found in agarose. Care should be taken to ensure that the DNA is inhibitor-free. In addition, T4 kinase will readily phosphorylate RNA molecules; therefore, the presence of RNA should be avoided, as this will severely reduce the incorporation of labeled ^{32}P into the DNA.

16. The labeling reaction is only approx 10% efficient. To get all of the molecules phosphorylated, it is necessary to chase the reaction with excess cold ATP.

17 This is a poor way to purify oligonulceotides, instead I recommend a Sephadex-G25 spin column.

References

1. Pickersky, E. (1996) Terminal labelling for Maxam-Gilbert sequencing, in *Basic DNA and RNA Protocols* (Harwood, A. J., ed.), Methods in Molecular Biology, vol. 58, Humana, Totowa, NJ.

2. Wallace, R. B., Shaffer, J., Murphy, R. F., Bonner, J., Hirose, T., and Itakura, K. (1979) Hybridisation of synthetic oligodeoxyribonucleotides to phi chi 174 DNA: the effect of single base pair mismatch. *Nucl. Acid Res.* **6**, 3543.

3. Harwood, A. J., ed. (1994) *Protocols for Gene Analysis*, in Methods in Molecular Biology, vol. 31. Humana, Totowa, NJ.

4. Harwood, J. C. and Phear, G. A. (1996) *Direct sequencing of PCR products,* in *Basic DNA and RNA Protocols*, (Harwood, A. J., ed.), Methods in Molecular Biology, vol. 58, Humana, Totowa, NJ.

5. Klenow, H., Overgaard-Hansen, K., and Patkar, S. A. (1971) Proteolytic cleavage of native DNA polymerase into two different catalytic fragments. *Eur. J. Biochem.* **22**, 371–381.

6. Challberg, M. D. and Englund, P. T. (1980) Specific labelling of 3' termini with T4 DNA polymerase. *Meth. Enzymol.* **65**, 39–43.

4

Southern Blotting of Agarose Gels by Capillary Transfer

Ralph Rapley and Ian J. Williams

1. Introduction

The detection of specific nucleic acid species following electrophoretic separation of a complex sample may be undertaken by the use of Southern blotting *(1)*. Genomic DNA fragments are separated according to size by agarose gel electrophoresis following digestion with suitable restriction enzymes (*see* Chapter 5).

To facilitate the transfer of larger DNA fragments, the immobilized DNA contained within the gel matrix is partially cleaved by depurination with HCl. Subsequent soaking of the gel in NaOH denatures the double-stranded DNA to produce single strands, which may be probed with an appropriately labeled single-stranded DNA fragment *(2)*.

Traditionally, the DNA is transferred to a nitrocellulose filter, although now the membrane is usually constructed of nylon. Nylon has an improved capacity for DNA binding and is more robust allowing reprobing to be undertaken. The simplest and least expensive method of transfer utilizes capillary action to draw liquid through the gel matrix, transferring the nucleic acid fragments onto the nylon membrane. The nylon-bound immobilized DNA fragments provide an exact representation of their original location following agarose gel electrophoresis. Alternative methods of transfer such as vacuum blotting or electroblotting may provide a more efficient method of transfer and reduce blotting time, but they are generally more expensive. Following transfer, the DNA is covalently crosslinked to the nylon membrane by exposure to ultraviolet irradiation, after which the blot may be stored or probed.

From: *Methods in Molecular Biology, vol. 187: PCR Mutation Detection Protocols*
Edited by: B. D. M. Theophilus and R. Rapley © Humana Press Inc., Totowa, NJ

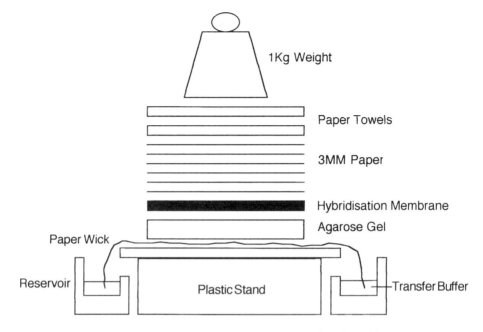

Fig. 1. A typical setup for capillary action Southern blot.

2. Materials

1. Suitable apparatus for blotting, two buffer tanks, paper.
2. Towels, Whatmann paper (*see* **Fig. 1**).
3. Nylon hybridization membrane (e.g., Hybond-N⁺).
4. Depurination buffer: 0.25 *M* HCl.
5. Denaturation buffer: 1.5 *M* NaCl, 0.5 *M* NaOH.
6. Transfer buffer: 1.5 *M* NaCl, 0.25 *M* NaOH.
7. 20X SSC: 3 *M* NaCl, 0.3 *M* trisodium citrate, pH 7.0.
8. Ultraviolet (UV) light transilluminator, 302-nm output.
9. Fixing solution: 0.4 *M* NaOH.
10. Rinsing solution: 5X SSC.

3. Methods

1. Isolate DNA with an appropriate extraction technique.
2. Digest total genomic DNA with desired restriction enzyme.
3. Separate the digested DNA fragments by agarose gel electrophoresis (*see* Chapter 1) (*see* **Notes 1** and **2**).
4. Following electrophoresis, visualize the gel on a UV transilluminator and photograph (*see* **Note 3**).
5. Trim the gel with a clean scalpel blade to remove any unused areas (e.g., gel wells and sides) (*see* **Note 4**).

6. Soak the gel in three gel volumes of depurination buffer for 30 min at room temperature with gentle agitation on an orbital shaker (*see* **Notes 5** and **6**).
7. Decant depurination buffer and rinse the gel in distilled H_2O.
8. Discard H_2O and soak in three gel volumes of denaturation buffer. Incubate with gentle agitation at room temperature for 30 min.
9. Decant denaturation buffer. Replace with three gel volumes of transfer buffer. Equilibrate the gel with gentle agitation at room temperature for 30 min.
10. Prepare Southern blotting system (**Fig. 1**). Cut a wick from 3MM paper, sufficient in width to cover the area of the gel platform and in length to reach both reservoirs (*see* **Note 7**).
11. Soak the 3MM paper wick in transfer buffer and place over the gel platform. Ensure that both ends of the wick reach the reservoirs and that all air bubbles are removed from the wick by gently smoothing with a gloved finger.
12. Remove gel from transfer buffer and place, face up, on the gel platform of the capillary transfer system. Fill both reservoirs with transfer buffer.
13. Cut a piece of Hybond-N⁺ nylon membrane to the exact size as the gel (*see* **Note 8**) and mark the side of the membrane that will be in contact with the gel (*see* **Note 9**).
14. Wet the membrane by floating it on distilled water. Rinse in transfer buffer and place the membrane on the gel, smoothing out any air bubbles between the gel and the membrane (*see* **Note 10**).
15. Cut three sheets of 3MM paper to the exact size of the gel/membrane sandwich and wet with transfer buffer. Place on top of the membrane and smooth out any air bubbles.
16. Cut a stack of absorbent paper towels to the size of the gel and place on top of the 3MM paper. Compress completed setup with a 1-kg weight to allow the transfer to proceed and leave for at least 12 h (*see* **Notes 11** and **12**).
17. After blotting, carefully disassemble the gel and membrane from the transfer system. Before separating the gel and the membrane, mark the position of the gel slots with a pencil, as this will allow orientation following autoradiography (*see* **Note 13**).
18. Carefully remove the membrane (*see* **Note 8**) and rinse the filter in 2X SSC.
19. Covalently crosslink the DNA fragments to the matrix by exposure to a 302-nm UV light transilluminator. Place the filter, DNA side down, on a piece of cling film, and expose for 2–3 min (*see* **Notes 14** and **15**). The filter can be used immediately or stored at 4°C, in cling-film until required.

4. Notes

1. DNA fragment separation may be improved by varying electrophoresis conditions. Overnight runs at low voltages will provide good resolution.
2. It is possible to determine the size of the hybridizing band following autoradiography by comparison with standard or marker DNA (e.g., λDNA/*Hind*III, 1-kb marker, 123-bp marker). This needs to be end-labeled with a radioactive or nonradioactive marker (*see* Chapter 3).

3. Ethidium bromide stain can be incorporated into the gel or the buffer tank during electrophoresis. Similarly, the gel can be stained after the run is complete.

4. Trimming away unwanted areas of the gel reduces the size of the nylon membrane required to cover the gel.

5. The depurination step partially cleaves large DNA fragments within the gel matrix. The smaller DNA fragments are transferred more efficiently during the blotting procedure. When the xylene cyanol loading dye changes color to a greenish color or the bromophenol blue turns yellow, the depurination buffer can be removed. Alternatively, this step can be achieved by exposing the gel to UV light (302 nm) for 30 s, to cleave high-molecular-weight DNA.

6. The blotting apparatus can be set up during the 30-min incubation periods in **steps 6–9**.

7. The width of the 3MM paper wick is cut to accommodate the width or length of each agarose gel. The gel platform can vary in size to accommodate different gel sizes.

8. Avoid touching the surface of the nylon membrane as any dirt or grease may affect the result. Hold the membrane at the edges and wear gloves.

9. A small portion of the corner can be cut off for orientation.

10. Air bubbles trapped between the gel and nylon membrane sandwich will affect DNA transfer.

11. A glass plate can be placed on top of the stack in order to distribute the weight evenly, allowing a more even transfer of DNA. It is necessary to cover the blotting apparatus with cling film to avoid evaporation of transfer buffer.

12. With the completed setup, ensure that only the gel is in contact with the wick. To ensure correct and only vertical transfer of DNA fragments from the gel to the nylon membrane, contact of blotting items within the stack should only be with the layer above or below. In some cases, the wick can be covered or "sectioned off" using cling film. This will also prevent evaporation from the wick and the reservoirs.

13. Alternatives to the capillary system include vacuum blotting or electroblotting. There are a number of manufacturers that produce equipment for this purpose, and although they are more expensive, they reduce the transfer process to as little as 1 h. In some cases, a more even transfer takes place.

14. For neutral nylon membranes (e.g., Amersham, Hybond-N), crosslinking is necessary; however, for positively charged membranes (e.g., Amersham Hybond-N$^+$) crosslinking may be undertaken by placing the membrane in 0.4 N NaOH for 30 min and rinsing in 5X SSC with gentle agitation for 1 min. If using nitrocellulose, it is necessary to bake the filter at 80°C for 20–60 min.

15. Efficient crosslinking of DNA to nylon filters is achieved with an optimal amount of exposure to UV light. Some manufacturers (e.g., Stratagene) produce UV crosslinkers (Stratalinker) that exposes the filter to the radiation for the optimal amount of time. It is useful if no equipment such as this is available to calibrate a UV source before use. This can be done by exposing filters with identical amounts of DNA on each filter to UV for different lengths of time.

Hybridization to the same probe will reveal the strongest signal that can be used to establish the optimal time for exposure. With standard UV transilluminator, regular recalibration is required.

References

1. Southern, E. M. (1975) Detection of specific sequences among DNA fragments separated by gel electrophoresis. *J. Mol. Biol.* **98,** 503–517.
2. Evans, M. R., Bertera, A. L., and Harris, D. W. (1994) The Southern blot: an update. *Mol. Biotechnol.* **1,** 1–12.

5

Restriction Fragment Length Polymorphism

Mohammad S. Enayat

1. Introduction

DNA sequence changes within a gene result either in polymorphism or mutation, causing different diseases. Some of these polymorphisms that occur with a
high frequency within the population can be a useful tool for gene tracking for a
given disease. Such investigations have initially been done by Southern blot
techniques, but, where possible, they have now been replaced by polymerase
chain reaction (PCR)-based methodology. The nucleotide substitutions can be
identified in two ways:

1. By use of restriction enzyme analysis or restriction fragment length polymorphisms
   (RFLPs).
2. Allele specific oligonucleotide hybridization (ASO-H) or similar techniques.

Another type of polymorphism, a polymorphic tandem dinucleotide repeat
sequence or variable number tandem repeat (VNTR) can also be used for gene
tracing in a familial disease. In these cases, a segment containing the repeats is
amplified and the fragment size differences are detected by gel electrophoresis.

Hemophilia A or Factor VIII deficiency is the most common inherited bleed-
ing disorder in humans. This X-chromosome-linked disorder affects approx 1
in every 10,000 males, and within the families of these patients, the females
are at risk of being carriers of this disorder. Factor VIII is a component of the
intrinsic coagulation pathway and the *FVIII* gene is a large gene, encompasses
186 kb at Xq28. It has 26 exons encoding a mRNA of 9 kb (*1*). Both RFLP and
VNTR analysis have been extensively used in carrier detection and antenatal
diagnosis in families with classical or familial hemophilia A. So far, 10 useful
polymorphisms have been identified within (intragenic) or flanking (extragenic)

From: *Methods in Molecular Biology, vol. 187: PCR Mutation Detection Protocols*
Edited by: B. D. M. Theophilus and R. Rapley © Humana Press Inc., Totowa, NJ

Table 1
DNA Polymorphisms Within or Flanking the Factor VIII Gene (2)

Restriction enzyme	Site	Detection		Heterozygosity in Caucasian
		PCR	Probe	
*Bcl*I	Intron 18	+	+	0.43
*Xba*I	Intron 22	+	+	0.49
*Hin*dIII	Intron 19	+	+	0.38
*Msp*I	Intron 22	–	+	0.01
*Taq*I	5'	–	+	0.40
*Bgl*I	3'	–	+	0.25
*Msp*I	3'	–	+	0.43
(CA repeat)	Intron 13	+	–	(10 alleles approx 0.80)
(CA repeat)	Intron 22	+	–	(6 alleles approx 0.55)
(G/A)	Intron 7	+	–	0.33

the *FVIII* gene (**Table 1**). Seven of these polymorphisms are diallelic RFLP and one, within the intron 7, is a nucleotide substitution (G/A) usually detected by ASO-H (3).

The most useful intragenic polymorphic sites with high heterozygosity in different ethnic populations are in intron 18 and intron 22, recognized with the *Bcl*I and *Xba*I restriction enzymes, respectively. A closely linked polymorphism recognized by the *Bgl*II restriction is also highly informative, but with a 5% theoretical chance of recombination. All three of these RFLPs have originally been identified and analyzed by Southern blotting. However, this method is time-consuming and may need the radioactive method for DNA band visualization. To this end, the *Bcl*I intragenic RFLP method has now been replaced by a fast and nonradioactive polymerase chain reaction (PCR) analysis (4,5).

1.1. Use of Southern Blotting in BclI RFLP Analysis of Hemophilia A

This method involves a series of techniques some of which are dealt with in detail elsewhere (6). These techniques include extraction of DNA from blood samples, digestion with appropriate restriction enzyme and electrophoresis, preparation, extraction, isolation, purification, and radiolabeling of the DNA probe, Southern blotting, hybridization, and, finally, autoradiography for DNA band visualization.

2. Materials

2.1. Restriction of DNA with the BclI Enzyme

1. Assay buffer (10X): The composition of this buffer varies from one manufacturer to another. For example, the composition of Amersham Pharmacia Bio-

tech (Amersham Pharmacia Biotech, Buckinghamshire, UK) reaction buffer called One-Phor-All buffer PLUS (OPA) is 10 mM Tris-HCl (pH 8.0), 100 mM KCl, 10 mM MgCl2, 10 mM β-Mercaptoethanol (β-ME), 100 μg bovine serum albumin (BSA)/mL.
2. 200 μg/mL BSA.
3. Loading buffer (LB): 30% xylene in 30% glycerol.
4. *See* **Notes 1–4** for the restriction enzyme.

2.2. Southern Blotting

1. 0.25 M HCl.
2. Standard saline citrate (SSC) (20X): 175.3 g/L NaCl and 88.2 g/L trisodium citrate. Adjust to pH 7.0 with concentrated HCl.
3. Denaturation buffer: 87.66 g/L NaCl and 20 g/L NaOH.
4. Neutralization buffer: 60.55 g/L Tris and 87.66 g/L NaCl. Adjust to pH 7.4 with concentrated HCl.
5. 0.4 M NaOH.
6. Amersham Hybond-N$^+$ (nylon) (Amersham Pharmacia Biotech, Buckinghamshire, UK) as transfer membrane
7. Whatman 3MM chromatography paper (Whatman International, Ltd., Maidstone, Kent, UK).

3. Methods

3.1. Restriction of DNA

1. Isolate and purify DNA using standard methods.
2. Pipet a desired amount (about 25 μg) of DNA into a small Eppendorf tube and dilute to 20 μL with distilled water.
3. Add 3 μL of the appropriate 10X assay buffer, 5 μL of 200 μg/mL BSA, and an appropriate number of units (usually 10 U) of the restriction enzyme diluted in dilution buffer (usually supplied with the enzymes) as desired, in a 2-μL volume (*see* **Notes 2** and **3**).
4. Mix by pipetting and incubate at desired temperature for at least 60 min, preferably 3–4 h.
5. Centrifuge contents in bench-top microcentrifuge at full speed to recover the full content of the tubes.
6. Add 6 μL of loading buffer mixture and mix thoroughly.
7. Load the samples into a suitable size submarine gel well without touching the sides (*see* **Note 4**).
8. Electrophorese the gel at 30 V for 8 h, but usually overnight.
9. Observe the gels on an ultraviolet (UV) transilluminator and make a permanent record by taking a photograph of the gel. If the DNA has been digested properly, a smearing from the well to the bottom of the gel should be present.

3.2. Southern Blotting

This method is used for the determination of the molecular sizes of the DNA fragments after digestion with restricted enzymes and gel electrophoresis. DNA fragments are transferred to a nylon membrane for reaction with a labeled probe for band visualization and molecular-weight sizing of each of the fragments.

All of the procedures are done at room temperature and the buffers used do not have to be sterile.

1. After electrophoresis, trim away unwanted areas in the gel. Mark on the corner for gel orientation and identification.
2. Soak the gel in 0.25 *M* HCl for 15 min with gentle agitation on an orbital shaker (*see* **Note 5**).
3. Wash the gel twice with denaturation buffer for 30 min.
4. Neutralize the gel by replacing fluid with neutralization buffer and soak as in **step 2** for 30 min and repeat (*see* **Note 6**).
5. While the gel is in final soak, construct a bridge for blotting. Cut a piece of the 3MM paper to the same width as the base glass plate but long enough to form a wick into the buffer compartment over the edges of the bridge.
6. After the final soak in the neutralization buffer, pour off the excess fluid and take up the gel onto a spare piece of 3MM paper.
7. Place the gel onto the bridge with the DNA side up.
8. Smooth out the gel gently with a gloved finger to remove any air bubbles between the bridge and the gel.
9. Cut piece of Hybond-N⁺ membrane to the approximate size of the gel and place on the gel. Trim to the exact size of the gel, again ensuring that no air bubbles are trapped underneath the membrane.
10. Cut two or three pieces of paper to the size of the gel and then presoak briefly in 2X SSC. Layer on top of membrane.
11. Surround the bridge/gel with Saran Wrap to prevent buffer bypass and evaporation.
12. Cut a stack of paper towels to size and place on top of the presoaked papers. Finally, compress with a glass plate and a 1- to 1.5-kg weight.
13. Add transfer buffer (approx 400 mL) 0.4 *M* NaOH and allow the DNA to transfer overnight.
14. After blotting, carefully remove the membrane and soak in 2X SSC to remove any adherent agarose.
15. Briefly blot dry the membrane, which is now ready for either storage at 4°C or immediate hybridization.

3.3. Hybridization and Autoradiography

There are many different methods of hybridization that are dealt with elsewhere *(6)*. However, after hybridization, the filter is probed with a ³²P-labeled DNA fragment from the *FVIII* gene. This genomic probe, called p114.12, is a 647-bp *StuI/SacI*-restricted *FVIII (7)*. The probed filter is exposed to an X-ray film (Hyperfilm MP, Amersham) for 4–7 d at –70°C in a cassette fitted with an intensifying screen. In this polymorphism, a restriction fragment of variable length of 879 bp and/or 1165 bp can be detected in Southern blots of genomic DNA. About 42% of females are heterozygous at this locus. **Figure 1** shows the

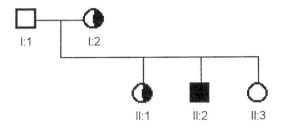

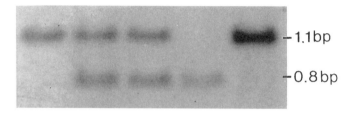

Fig. 1. A family with history of hemophilia A was investigated for *Bcl*I polymorphism and found to be fully informative. Patient (II:2) has inherited the hemophilic haplotype (0.8 bp) from her mother, who is the daughter of a hemophiliac and an obligate carrier. The patient's sister (II:1) is also a carrier and has the hemophilic haplotype, whereas his other sister (II:3) is unaffected. She has inherited the unaffected 1.1-bp haplotype from her mother.

Southern blot using restricted DNA samples with *Bcl*I and probed with the ^{32}P-labeled p114.12 probe.

3.4. Use of PCR and BclI RFLP Analysis in Hemophilia A

The same RFLP, identified by *Bcl*I Southern blotting, has now been demonstrated by PCR followed by digestion with the restriction enzyme (**6**). The PCR product of this highly polymorphic allele gives a 142-bp (–) allele and 99+43-bp (+) allele after restriction with the enzyme (*see* **Fig. 2**).

4. Notes

1. Keep the restriction enzyme cold at all times; if removed from the freezer, it should be immediately kept on ice. In the majority of cases, it can be used straight from the freezer.
2. Reaction volume here is fixed at 30 μL, as it is manageable. Generally, the smaller the volume, the better. Note that if the reaction volume is changed:
   a. Change the volume 10X assay buffer.
   b. Ensure that the enzyme added is <10% of the reaction volume.
   c. LB mixture added is one-fifth the reaction volume.

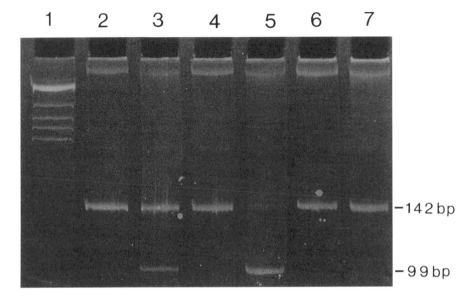

Fig. 2. *Bcl*I-restricted (lanes 3, 5, and 7) and -unrestricted (lanes 2, 4, and 6) PCR products, showing 142-bp (–) and 99-bp (+) fragments. Lane 1 is the molecular-weight marker.

3. Less enzyme can be used if the incubation period is lengthened.
4. To ensure adequate digestion before full-size gel electrophoresis, a minigel should be run. During digestion, remove 2–3 µL of reaction mix and add 6 µL LB mixture. Load into 1% minigel and run at 50–60 mA for 1 h. View under UV illumination to check digestion (*see* Chapter 1).
5. HCl acid denaturation allows large >13-kb fragments to be transferred more efficiently by breaking the DNA into smaller fragments. Do not leave in HCl for more than 30 min or smaller DNA fragments will also be broken up into <300 bp, significantly reducing the ability to DNA to bind covalently to the membrane.
6. Gels may be left in neutralization buffer for longer than 1 h with no adverse effects if kept at 4°C so as to limit diffusion. Maximum time in neutralization buffer is 4 h.

References

1. White, G. C. and Shoemaker, C. B. (1989) Factor VIII and haemophilia A. *Blood* **73,** 1–12.
2. Peake, I. R., Lillicrap, D. P., Boulyjenkov, V., Briet, E., Chan, V., et al. (1993) Report of a joint WHO/WFH meeting on the control of haemophilia: carrier detection and prenatal diagnosis. *Blood Coag. Fibrinol.* **4,** 313–344.
3. Peake, I. (1995) Molecular genetics and counselling in haemophilia. *Thromb. Haemost.* **74,** 40–44.

4. Gitschier, J., Drayna, D., Tuddenham, E. G. D., White, R. L., and Lawn, R. M. (1985) Genetic mapping and diagnosis of haemophilia A achieved through a BclI polymorphism in the factor VIII gene. *Nature* **314,** 738–740.

5 Gitschier, J., Lawn, R. M., Rotblat, F., and Goldman, E. (1985) Antenatal diagnosis and carrier detection of haemophilia A using factor VIII gene probe. *Lancet* **1,** 1093–1094.

6. Rapley, R., ed. (2000) *The Nucleic Acid Protocol Handbook.* Humana Press, Totowa, NJ.

7. Kogan, S. C., Doherty, N., and Gitschier, J. (1987) An improved method for prenatal diagnosis of genetic diseases by analysis of amplified DNA sequences: application to haemophilia A. *New Engl. J. Med.* **317,** 980–990.

6

PCR

Principles, Procedures, and Parameters

Nicola Louise Jones

1. Introduction

In 1983, the Cetus scientist Kary Mullis developed an ingenious "in vitro" nucleic acid amplification technique termed the *polymerase chain reaction* (PCR). This technique involves the use of a pair of short (usually 20 bp long) pieces of synthesized DNA called primers and a thermostable DNA polymerase to achieve near-exponential enzymatic amplification of target DNA. Because of the sensitivity of this technique, DNA of relatively poor condition may be amplified, as only short intact sequences are required. Therefore, it is not always necessary to carry out lengthy template sample preparation. For example, a simple boiling step is often enough to release DNA from blood samples *(1)*. The starting material for PCR may be DNA from a variety of sources such as blood, tissues, paraffin-embedded material, ancient archaeological samples, or forensic material. The PCR may also be used to amplify RNA, which must first be converted into cDNA by the enzyme reverse transcriptase (RT-PCR). In contrast to DNA, great care must be taken in the preparation and handling of RNA because of its instability and susceptibility to degradation.

Polymerase chain reaction proceeds in three stages:

1. Denaturation of double-stranded DNA. (This initial denaturation step is not necessary when amplifying RNA because it is a smaller target molecule.)
2. Primer annealing.
3. Extension of the annealed primers.

The target DNA is suspended in a reaction mixture consisting of distilled water, buffer (containing $MgCl_2$, which is necessary for the polymerase to work

From: *Methods in Molecular Biology, vol. 187: PCR Mutation Detection Protocols*
Edited by: B. D. M. Theophilus and R. Rapley © Humana Press Inc., Totowa, NJ

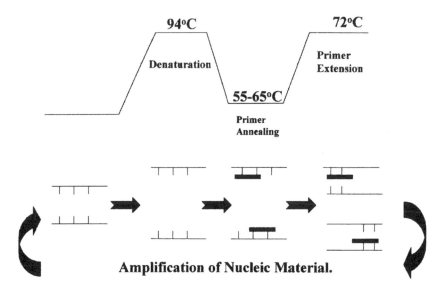

Amplification of Nucleic Material.

Fig. 1. Basic steps in the PCR cycle. Hatched lines indicate target DNA. Filled boxes represent annealed primers. *See* text for further details.

efficiently), the thermostable *Taq* polymerase, and each of the four deoxynucleotide triphosphates (dNTPs). Also present are a pair of primers whose sequences are complementary to that of the DNA flanking the target region. Numerous parameters must be taken into consideration when designing the primers (*see* **Note 1**).

The reaction mixture is first heated to denature the double-stranded DNA into single strands, then cooled to an optimum temperature to facilitate primer annealing. The primer pair consists of a forward "sense" primer that binds to its complementary sequence upstream of the region to be amplified and a reverse "antisense" primer that binds downstream, both with their 3' ends facing inward. During primer extension the DNA polymerase progressively adds dNTP's, complementary to the target, to the 3' end of each primer so that the target sequence is copied. The 5' ends of the primers defines the length of the PCR product. These three steps constitute a PCR cycle (*see* **Fig. 1**). Usually, 30 cycles are performed in a programmable thermal cycler, with each cycle theoretically doubling the quantity of target sequence. This exponential increase is not achieved practically however because of the exhaustion of the PCR components and the accumulation and reannealing of product strands.

Polymerase chain reaction is a very sensitive and specific molecular biology technique. Its versatility is demonstrated by the growing array of technical modifications that lend PCR to multiple applications. Such variations include RT-PCR, nested, multiplex, long-range, and allele-specific PCR.

1. RT-PCR: RNA may be amplified following its conversion to cDNA by the enzyme reverse transcriptase. The use of RNA as a starting material ensures that only the coding regions, or exons, are amplified during the PCR reaction. RT-PCR not only provides a useful tool for analyzing the transcriptional activity of genes but also enables the investigation of many contiguous exons in a single analysis.

2. Nested PCR: This involves the use of two sets of primers. The first set (external primers), which flanks the region of interest, allows for a first round of amplification. A small aliquot from the first round of PCR is then used as the target for a second round of amplification, primed by a second pair of primers that lie internal to the first. This technique is designed to increase the sensitivity of the PCR reaction and is particularly useful for the amplification of small quantities of target material from readily accessible cell types expressing very low levels of mRNA. For example, Factor VIII mRNA present at a quantity of 1 molecule per 500–1000 cells can be amplified in this manner from the lymphocytes of hemophilia A patients *(2)*.

3. Multiplex PCR: More than one target sequence can be amplified in a single PCR reaction by using multiple primer pairs. Multiplex PCR has been utilized in the analysis of deletions, mutations, and polymorphisms as well as in RT-PCR and quantitative assays *(3)*. A specific application of multiplex PCR is in deletion screening of the *dystrophin* gene in patients with Duchenne muscular dystrophy.

4. Long-range PCR: It is possible to amplify long fragments of DNA (>20 kb) using an enzyme mixture of *Taq* and *Pwo* DNA polymerases. High yields of PCR products from episomal and genomic DNA can be obtained from this powerful polymerase mixture. An example of its application is in the routine diagnosis of Fragile X syndrome. The mutation in Fragile X syndrome involves the expansion of CGG repeats in the *FMR*1 gene. These repeats are resistant to amplification by ordinary PCR methods possibly because of the stopping or pausing of the *Taq* polymerase enzyme during the amplification of regions containing high levels of CGG repeats *(4)*. The presence of *Pwo* polymerase, with its proofreading ability, can reduce this error frequency, whereas improved cycle and buffer conditions can overcome the length limitation for PCR. Thiel et al. *(5)* have reported a system encompassing the effective amplification of up to 20 kb of viral genomic RNA. The subject of long-range PCR is reviewed in Chapter 8.

5. Allele-specific PCR: PCR can be used to discriminate between alleles by using allele specific oligonucleotides as primers. Successful PCR amplification requires that the 3' end of the primers are complementary to the target DNA sequence; any mismatch results in the inability of *Taq* polymerase to extend the primer. This phenomenon can be exploited by designing primers that contain a base at the 3' end that matches either that of a known mutation or its wild-type counterpart. These primers will only anneal to and amplify target DNA containing that par-

ticular sequence. Allele identity may therefore by determined by the presence or absence of PCR products. Diseases that encompass only one mutation are particularly well suited to this technique. Examples include the Prothrombin 20210A mutation *(6)*, sickle cell anemia, Fc-γ receptor IIA polymorphism *(7)*, and α-1antitrypsin deficiency, where the technique is termed the *amplification refractory mutation system* (ARMS). Allele-specific PCR is reviewed in Chapter 7.

Polymerase chain reaction plays two roles in molecular biology. First, it provides enough material to allow further technical manipulations such as for the detection of mutations, conformation sensitive gel electrophoresis (e.g., in mutation screening methods such as chemical mismatch cleavage, single-strand conformational polymorphism, denaturing gradient gel electrophoresis, and DNA sequencing). Second, it can be used analytically as a direct tool in mutation detection as in allele-specific PCR and multiplex for deletion analysis. Following are given two standard methods for PCR: first, for amplification of DNA and, second, for amplification of RNA by first converting it to cDNA. The many variations on the standard method are covered in later chapters of this volume.

2. Materials
2.1. DNA PCR

1. Programmable thermal cycler.
2. Sterile distilled water (use as fresh).
3. Solution A: containing $MgCl_2$ and a source of buffer/salt (usually KCl and Tris-HCl) at a pH of 8.3. In our laboratory, solution A comprises 670 m*M* Tris-HCl, 166 m*M* ammonium sulfate, and 67 m*M* magnesium chloride. Twenty-milliliter batches are usually prepared and stored at room temperature for up to 1 mo (*see* **Note 2**).
4. 10X Reaction buffer: To 1 mL of solution A, add 34 μL of 5% bovine serum albumin (BSA) and 7 μL of 14.4 *M* β-mercaptoethanol. (*Note*: β-Mercaptoethanol is toxic; use in a fume cupboard.) Store at –20°C for up to 1 mo (*see* **Note 3**).
5. 5 m*M* dNTPs: dilute 50 m*M* stock solution of dNTPs (i.e., dATP, dCTP, dGTP, and dTTP) in distilled water. Store at -20°C for up to 1 mo.
6. 15 pmol primers 1 (forward) and 2 (reverse): approx 0.1 μg/μL for a 20-mer.
7. 5 U/μL *Taq* polymerase.
8. Mineral oil. Store at room temperature.

2.2. RT-PCR

1. Programmable thermal cycler.
2. Sterile distilled water. Use fresh and keep supply separate from that used for DNA PCR.
3. 15 pmol of primers 1 (forward) and 2 (reverse): approx 0.1 μg/μL for a 20-mer. Store at –20°C.

4. 10X Reaction buffer (*see* **Subheading 2.1.**, **items 3** and **4**).
5. 5 m*M* dNTPs (*see* **Subheading 2.1.**, **items 5**).
6. 5X Reverse transcriptase (RT) buffer containing 250 m*M* Tris-HCl, pH 8.3, 375 m*M* KCl, and 15 m*M* MgCl$_2$. This is usually supplied with the reverse transcriptase. Store at –20°C.
7. Maloney Murine Leukaemia Virus (MMLV) Reverse Transcriptase (200 U/µL). Store at –20°C.
8. 100 m*M* Dithiothreitol (DTT). Store at –20°C.
9. Rnase inhibitor (50 U/µL). Store at –20°C.
10. 5 U/µL *Taq* polymerase. Store at –20°C.
11. Mineral oil. Store at room temperature.

3. Methods

3.1. DNA PCR

Note: Always wear, and frequently change, gloves to prevent contamination by DNA from human skin cells. As likely sources of contamination include other samples and previous amplification products, separate rooms, away from sites of DNA/RNA extraction, manipulation, and recovery of PCR products, should ideally be allocated for the setup of PCR. Designate separate sets of pipets for DNA and RT-PCR and use filter tips whenever possible. Solutions and buffers must be stored in sterile containers and always prepared with fresh distilled water.

1. Into a sterile tube, prepare the following master mix according to the number of samples to be analyzed:

    10 µL of 10X reaction buffer;
    10 µL of 5 m*M* dNTPs;
    15 pmol primer 1 (approx 0.1 µg for a 20-mer);
    15 pmol primer 2 (approx 0.1 µg for a 20-mer);
    0.5 µL (2.5 U) *Taq* polymerase;
    Distilled water to a volume of 98 µL per sample.

2. Aliquot 98 µL of master mix into a PCR reaction tube. Add 2 µL genomic DNA (50 µg/mL) to give a final volume of 100 µL. PCR may be done in much smaller volumes down to 10 µL, especially for analytical PCR, by reducing the reaction components proportionately.
3. Always perform a blank control alongside the samples by replacing the genomic DNA with dH$_2$O. This will check for contamination.
4. Overlay the reaction mixture with 1 drop of mineral oil to prevent evaporation, unless using an "oil-free" thermal cycler that has a heated lid to prevent evaporation.
5. Denature the template for 5 min at 94°C.
6. PCR at the following settings for 30 cycles:

a. Denaturation: 94°C for 1 min.
b. Annealing: 55–65°C for 1 min (adjust temperature according to the calculated T_m of the primers; *see* **Note 1**).
c. Extension: 72°C for 3 min (optimum temperature for *Taq* polymerase activity) (*see* **Note 4**).

7. Analyze 10 μL of the sample on a 1–2% agarose gel containing ethidium bromide. (**Ethidium bromide is carcinogenic.**) *See* Chapter 1.

3.2. RT-PCR

Note: Follow the same precautions outlined in the note at the beginning of **Subheading 3.1.** The sterile technique is even more critical with RT-PCR than DNA PCR in order to prevent RNA degradation as well as protecting against contamination.

1. Pipet 5 μL (0.2–0.5 μg) RNA into a sterile PCR reaction tube. Always run a blank alongside the sample by replacing the RNA with distilled water.
2. Prepare a premix of 15 pmol primer 2 (reverse primer) and distilled water to a volume of 2.5 μL per sample. Add 2.5 μL of premix to the RNA. Overlay with mineral oil (*see* **Note 5**).
3. Incubate at 65°C for 10 min.
4. Prepare the following premix according to the number of samples to be analyzed:
   4 μL of 5X RT buffer;
   2 μL of 100 mM DTT;
   1 μL of MMLV Reverse Transcriptase (200 U);
   0.5 μL of RNase inhibitor (25 U);
   5 μL of 5 mM dNTPs.
5. Add 12.5 μL of this second premix to each sample. Incubate at 42°C for 1 h.
6. Prepare a PCR mix:
   5 μL of 10X reaction buffer;
   15 pmol primer 1 (forward primer);
   2.5 U *Taq* polymerase;
   Distilled water to a volume of 30 μL per sample.
7. Add 30 μL of the PCR mix to each sample.
8. Perform 30 PCR cycles at the following settings:
   a. Denaturation: 93°C for 1 min.
   b. Annealing: 55–65°C for 1 min (adjust according to the calculated T_m of the primers; *see* **Note 1**).
   c. Extension: 72°C for 5 min (*see* **Note 4**).
9. Remove 10 μL of sample for analysis on a 1–2% agarose gel containing ethidium bromide (**carcinogen**). *See* Chapter 1.

3.3. Troubleshooting

1. Inadequate design of the primers may result in the formation of a hairpin loop. This is where internal complementarity of the primer sequence enables it to bind to itself rather than the DNA template so that the PCR cannot proceed.

2. Complementarity between a primer pair allows them to bind to each other, thus forming "primer-dimers." As with internal complementarity, the primers are no longer free to bind to the DNA template, resulting in failure of the PCR. Primer-dimers can also be produced as a byproduct of a successful PCR if the primers are present in excess *(8)*.

3. Heparin, porphyrins, and high concentrations of ionic detergent (e.g., Proteinase K, phenol, and sodium dodecyl sulfate), inhibit PCR.

4. As RNase enzymes can degrade target RNA, RT-PCR requires the use of Rnase-free reagents and equipment to prevent false-negative results.

5. Polymerase chain reaction failure or poor yield may result from a low target copy number (e.g., where the RNA is from cell types expressing low levels of target material). This may be remedied by using a second set of primers, as in nested PCR (*see* **Note 6**).

6. It has been noted that mineral oil breaks down under 254-nm UV light, with breakdown products inhibiting PCR. Remedial suggestions include the addition of 0.1% of the antioxidant 8-hydroxyquinoline to the mineral oil prior to UV treatment *(8)*.

7. False negatives may also result from "stalling" of the *Taq* DNA polymerase during primer extension. This occurs when regions of target DNA form secondary structures *(9)*. Suggestions for overcoming this problem include adding glycerol or tetramethylammonium chloride (TMAC) to the PCR mix. Elimination of buffer components stabilizing these secondary structures (e.g., KCl) may also help.

8. If the annealing temperature is too high, the primers will not bind to the target DNA, and PCR failure may ensue. Conversely, if the temperature is too low, nonspecific binding will occur.

9. It has been suggested that an increase in amplification products from GC-rich sequences may be obtained following the addition of either betaine or trimethylamine *N*-oxide to the PCR reaction. Betaine is thought to reduce the formation of secondary structures caused by GC-rich regions, however, the effect of betaine on the fidelity of *Taq* polymerase is unknown *(10)*.

10. Reagents and plasticware may be treated with UV irradiation to convert contaminating DNA into a nonamplifiable form. Times and conditions for irradiation are dependant upon the energy of irradiation, length of contaminating DNA, and its thymidine content.

11. To control contamination, PCR may be performed routinely with dUTP substituted for dTTP. Treatment of PCR reactions with uracil *N*-glycosylase then prevents contamination by carry over of PCR products from a previous reaction.

12. In order to eliminate false primer binding in the initial stages of DNA PCR, which may produce nonspecific products, a technique known as "hot-start" PCR can be used. Hot-start PCR ensures the physical separation of one essential component of the PCR reaction (e.g., primers or *Taq* polymerase) prior to denaturation of the DNA template. Using a solid wax bead keeps the reagents separate. Upon heating, the wax bead melts and the components mix. An alternative approach involves the use of a variant *Taq* polymerase that is inactive at room

temperature but is activated at 94°C. Conversely, the missing component can be added manually immediately following the initial 5-min denaturation at 94°C

4. Notes

1. Primers are usually 20 bp in length, but typically between 15 and 30 bp with approx 50% GC content. It is desirable to have a G or C at the 3' end, as this binds to the target DNA with a triple hydrogen bond, thus anchoring the primer more effectively than A or T. The primers should not contain self-complementary sequences that might produce loop back or hairpin structures. The annealing temperature of the primers varies according to their sequence, but, it is usually 50–65°C. For primers up 20 bp in length, the theoretical melting temperature (T_m) (in °C) can be calculated approximately as:

$$2\times \text{(Number of A/T bases)} + 4\times \text{(Number of G/C bases)}$$

   The starting point for determining the optimal annealing temperature is approx 5–10°C below that of the primer with the lowest T_m. Ideally, the two primers should have similar T_m's. The primers usually completely complement the sequence of interest, although for specific purposes such as allele-specific PCR, primers with mismatches to the target sequence may be used. *See* Chapter 7.

2. The pH of the reaction buffer is important, as enzymes are sensitive to pH changes. The optimal pH for *Taq* polymerase activity is 8.3. The magnesium concentration of the buffer is also critical. Insufficient Mg^{2+} leads to low product yields, whereas excessive Mg^{2+} gives nonspecific products. Free magnesium, in addition to that complexed to dNTPs, is required for the polymerase to work efficiently. Therefore, because the concentration of dNTPs affects the amount of free Mg^{2+}, optimum concentrations of $MgCl_2$ should be determined for each PCR protocol.

3. The addition of denaturants and other reagents such as dimethyl sulfoxide (DMSO), formamide, glycerol, polyethylene glycol (PEG), Triton X-100, bovine serum albumin, and spermadine appear to enhance the PCR reaction. Such compounds may stabilize the polymerase, prevent loop-back sequences, and increase the specificity of primer binding.

4. Owing to the superior quality of modern *Taq* polymerases and the technological advances in thermal cyclers, it is often possible to reduce the extension time to approx 1 min/kb.

5. In the protocol described, the primer that is used subsequently as the reverse primer in the PCR reaction is also used to prime the cDNA reaction. As an alternative, the cDNA may be primed by an oligo-dT primer, which binds to the poly A tail present at the 3' end of most mammalian mRNAs, or by "random hexamers." In these cases, the whole mRNA population is converted to cDNA, and both forward and reverse PCR primers must be added subsequently to amplify the sequence of interest.

6. Nested RT-PCR: Follow **steps 1–8** in **Subheading 3.2.**, then make the following master mix:

10 µL of 10X reaction buffer;
10 µL of 5 m*M* dNTP;
15 pmol primer 3 (internal forward primer);
15 pmol primer 4 (internal reverse primer);
2.5 U *Taq* polymerase;
Distilled water to a volume of 98 µL per sample.

To 98 µL of master mix, add 2 µL of the primary PCR product (this may also be done in a 50 µL or smaller reaction). Overlay with mineral oil (except when using an oil-free thermal cycler that has a heated lid to prevent evaporation). Perform 30 cycles at the following settings:

a. Denaturation: 93°C for 1 min.
b. Annealing: 55–65°C for 1 min (adjust according to the calculated T_m of the primers; *see* **Note 1**).
c. Extension: 72°C for 5 min (*see* **Note 4**).

Link to a program of:

a. 72°C for 2 min.
b. 4°C soak.

Remove 10 µL of sample for analysis on a 1–2% agarose gel containing ethidium bromide (**carcinogenic**). *See* Chapter 1.

Acknowledgments

The author wishes to express her gratitude to Mr. Charles Williams (Head MLSO, Haematology Department, Birmingham Children's Hospital) for his help and support in the writing of this chapter.

References

1. Rapley, R. (1998) Polymerase chain reaction, in *Molecular Biomethods Handbook* (Rapley, R. and Walker, J. M., ed.), Humana, Totowa, NJ, pp. 305–325.
2. Cooper, D. N. and Krawczak, M., ed. (1993) *Human Gene Mutation*, BIOS Scientific, Oxford.
3. Henegariu, O., Heerema, N. A., Dlouhy, S. R., Vance, G. H., and Vogt, P. H. (1997) Multiplex PCR: critical parameters and step-by-step protocol. *Biotechniques* **23,** 504–511.
4. Hecimovic, S., Barisic, I., Muller, A., Petkovic, I., Baric, I., Ligutic, I., et al. (1997) Expand Long PCR for Fragile X mutation detection. *Clin. Genet.* **52,** 147–154.
5. Thiel, V., Rashtchian, A., Herold, J., Schuster, D. M., Guan, N., and Siddell, S.G. (1997) Effective amplification of 20-kb DNA by reverse transcription PCR. *Anal. Biochem.* **252,** 62–70.
6. Poort, S. R., Bertina, R. M., and Vos, H. L. (1997) Rapid detection of the prothrombin 20210A variation by allele specific PCR. *Thromb. and Haemost.* **78,** 1157–1163.

7. Flesch, B. K., Bauer, F., and Neppert, J. (1998) Rapid typing of the human Fcγ receptor IIA polymorphism by polymerase chain reaction amplification with allele-specific primers. *Transfusion* **38,** 174–176.
8. Dzik, S. (1998) The power of primers. *Transfusion* **38,** 118–121.
9. Hengen, P. N. (1997) Optimizing multiplex and LA-PCR with betaine. *Trends Biomed. Sci.* **22,** 225–226.
10. Henke, W., Herdel, K., Jung, K., Schnorr, D., and Loening, S. A. (1997) Betaine improves the PCR amplification of GC-rich DNA sequences. *Nucl. Acids Res.* **25(19),** 3957–3958.

Further Reading

Bauer, P., Rolfs, A., Regitz-Zagrosek, V., Hildebrandt, A., and Fleck, E. (1997) Use of manganese in RT-PCR eliminates PCR artifacts resulting from Dnase I digestion 2. *Biotechniques* **2,** 1128–1132.

Maudru, T. and Peden, K. (1997) Elimination of background signals in a modified polymerase chain reaction-based reverse transcriptase assay. *J. Virol. Meth.* **66,** 247–261.

Tang, Y., Procop, G. W., and Persing, D. H. (1997) Molecular diagnostics of infectious diseases. *Clin. Chem.* **43(11),** 2021–2038.

Vaneechoutte, M. and Van Eldere, J. (1997) The possibilities and limitations of nucleic acid amplification technology in diagnostic microbiology. *J. Med. Microbiol.* **46,** 188–194.

7

Allele-Specific Oligonucleotide PCR

Elaine K. Green

1. Introduction

Many single-base substitutions that lead to inherited diseases, the predisposition to genetic disorders, and cancer are increasingly being discovered. The ability to amplify specific DNA sequences by the polymerase chain reaction (PCR) *(1)* has made it possible to rapidly and accurately diagnose many inheritable diseases.

Prior to the use of PCR, point mutations were identified by using direct cloning and sequencing, Southern blotting and hybridization with labeled oligonucleotide probes centered on the site of the mutation or digestion with restriction endonucleases. These methods have been greatly enhanced by PCR, which allows amplification of DNA fragments containing the polymorphic sites from minute quantities of DNA. However, these techniques tend to be time-consuming, complex, require the use of radioactive label, and, in the case of the restriction endonuclease detection, are only applicable when the mutation alters a known cleavage site.

Polymerase chain reaction using allele-specific oligonucleotides (ASOs) is an alternative method for the detection of mutations in which only the perfectly matched oligonucleotide is able to act as a primer for amplification. The advantage of ASO PCR is that it is a rapid, simple, and nonradioactive method. ASO-PCR, otherwise known as the amplification refractory mutation system (ARMS) was first described for the detection of mutations in the α_1-antitrypsin gene *(2)*. It has since been adopted in the study of a number of genes, including prenatal diagnosis of cystic fibrosis *(3)*, polymorphisms of apolipoprotein E *(4,5)*, and point mutations in the ras oncogene *(6)*. In this technique oligonucleotide primers are designed such that they are complementary to either the nor-

From: *Methods in Molecular Biology, vol. 187: PCR Mutation Detection Protocols*
Edited by: B. D. M. Theophilus and R. Rapley © Humana Press Inc., Totowa, NJ

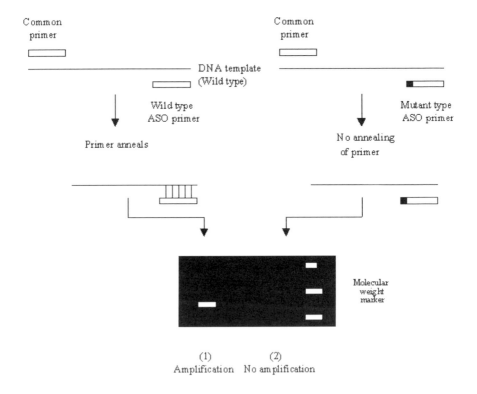

Fig. 1. Principle of allele-specific oligonucleotide PCR. Allele-specific oligonucle-otides allow amplification of a sequence of DNA to which they are perfectly homolo-gous *(1)*, but not one containing a mismatch *(2)*.

mal (wild-type) or mutant sequence, and both are used in conjunction with a common primer. Because DNA polymerase lacks a 3' exonuclease activity, it is unable to repair a single-base mismatch between the pimer and the template at the 3' end of the DNA primers. Thus, if oligonucleotide primers are designed to contain mismatches close to or at the 3' end, the primer will or will not be extended depending on which alternative single-base polymorphisms are present in the target sequence. Hence, under the appropriately stringent condi-tions, only target DNA exactly complementary to the primer will be amplified, as shown in **Fig. 1**.

2. Materials

All reagents should be of molecular biology grade and solutions made up with sterile distilled water.

1. PCR reaction buffer (10X): 100 m*M* Tris-HCl, pH 8.3, 500 m*M* KCl, 15 m*M* MgCl$_2$, 0.01% gelatin. Autoclave and store at –20°C (*see* **Note 1**).

2. Nucleotide mix: 200 mM each of dATP, dCTP, dGTP, and dTTP in sterile distilled water and store at –20°C.
3. Allele-specific oligonucleotide primers and common primer: 10 μM. Store at –20°C (*see* **Note 2**).
4. DNA *Taq* polymerase.
5. Sterile distilled water.
6. Sample DNA (*see* **Note 3**).
7. Mineral oil.

3. Methods

1. To 0.5 mL Eppendorf tube, add 5 μL 10X reaction buffer, 5 μL nucleotide mix, 5 μL ASO primer (either wild or mutant type primer) (*see* **Note 4**) and 5 μL common primer, 100 ng template DNA, and 2 U *Taq* polymerase (*see* **Note 5**), make up to a final volume of 50 μL with sterile distilled water (*see* **Note 6**) and overlay with mineral oil to prevent evaporation.
2. Place Eppendorf tubes on the thermal cycler to amplify the DNA by repeated cycles of denaturation, annealing, and extension: initial denaturation of 94°C for 5 min, followed by 30 cycles of 94°C for 30 s, 55°C for 30 s and 72°C for 1 min (*see* **Note 7**).
3. Following thermal cycling, electrophoresis 10 μL of the reaction sample through an agarose gel, with a DNA size marker and stain with ethidium bromide. A typical result is shown in **Fig. 1** (*see* **Note 8**).

4. Notes

1. The concentration of MgCl$_2$ may be altered (0.5–5 mM) to optimize the specificity and yield of the reaction.
2. The design of the ASO PCR primer is essential for specific amplification of the template. Primers are synthesized in two forms (the wild or normal type and the mutant), with the correspondingly different bases at the 3' end. However, a single mismatch is often not enough to prevent nonspecific amplification and the introduction of additional deliberate mismatches near the 3' terminal end (e.g., four bases from the 3' end) of the primers may overcome this problem. Several investigations have examined the effect of the type of the 3' terminal primer–template mismatches on the PCR amplification (*2–7*), however it appears to differ depending on the gene being studied. Where possible, select primers of random-base distribution and approx 50% GC content. The primers should not be complementary to each other or contain a sequence with significant secondary structure. The common primer should be designed to give a product of suitable size (e.g., 200 bp).
3. A concentration 100 ng of template DNA is usually sufficient to amplify.
4. Two reactions are required for the detection of a point mutation: one including the wild type and common primers, and the other with the mutant type and common primers.
5. Addition of the *Taq* polymerase following the initial denaturation step while the PCR reaction is held at 80°C prior to the cycling reactions may increase the specificity of the PCR products.

6. The volume of PCR reaction can be altered according to requirements: 10–100 µL.
7. These are standard PCR reaction conditions, which may not amplify the template specifically. By varying the conditions and constituents of the reaction (altering the magnesium [0.5–5 mM], deoxynucleotide triphospates [dNTPs] [50–200 µM], ASO primer concentration, DNA template, and *Taq* polymerase concentration *(1,8)*, and increasing the annealing temperature), this may be overcome. A good indication of the correct annealing temperature is the melting template of the oligonucleotide primers. This can calculated using the formula 64.9 + 0.41 (%C + %G) –600/n. The addition of specificity enhancers such as DMSO (10%) *(9)*, may also increase the specificity. However, it may be necessary to redesign the primers altering the 3' mismatches.
8. Amplification of a control DNA template with known point mutations will aid the establishment of the ASO PCR. By including an internal control reaction, such as β-globin amplification, the risk of false negatives will be reduced.

References

1. Saiki, R. K., Scharf, S., Faloona, F., Mullis, K. B., Horn, G. T., Erlich, H. A. et al. (1985) Enzymatic amplification of β-globin genomic sequences and restriction site analysis for diagnosis of sickle cell anaemia. *Science* **230**, 1350–1354.
2. Newton, C. R., Graham, A., Heptinsall, L. E., Powell, S. J., Kalsheker, K., Smith, J. C. et al. (1989) Analysis of point mutation in DNA. The amplification refractory mutation system (ARMS). *Nucl. Acids. Res.* **17**, 2503–2516.
3. Newton, C. R., Summers, C., Schwarz, M., Graham, A., Heptinstall, LE., Super, M., et al. (1989) Amplification refractory mutation system for prenatal diagnosis and carrier assessment in cystic fibrosis. *Lancet* **II**, 1481–1483.
4. Green, E. K., Bain, S. C., Day, P. J. R., Barnett, A. H., Charleson, F., Jones, A. F. et al. (1991) Detection of human apolipoprotein E3, E2, and E4 genotypes by an allele-specific oligonucleotide-primed polymerase chain reaction assay: development and validation. *Clin. Chem.* **37**, 1263–1268.
5. Main, B. J., Jones, P. J., MacGillivary, R. T. A., and Banfield, D. K. (1991) Apoliprotein E genotyping using the polymerase chain reaction and allele-specific oligonucleotide primers. *J. Lipid Res.* **32**, 183–187.
6. Ehlen, T. and Dubeau, L. (1989) Detection of Ras point mutations by polymerase chain reaction using mutation-specific, inosine-containing oligonucleotide primers. *Biochem. Biophys. Res. Commun.* **160**, 441–447.
7. Kwok, S., Kellog, DE., McKinney, N., Spasic, D., Goda, L., Levenson, C., et al. (1990) Effects of primer–template mismatches on the polymerase chain reaction: human immunodeficiency virus type I model studies. *Nucl. Acids Res.* **18**, 999–1005.
8. Rychlik, W., Spencer, W. J., and Rhoads, R. E. (1990) Optimisation of the annealing temperature for DNA amplification in vitro. *Nucl. Acids Res.* **18**, 6904–6912.
9. Winship, P. R. (1989) An improved method for directly sequencing PCR amplified material using dimethyl sulphoxide. *Nucl. Acids Res.* **17**, 1266.

8

Long-Range PCR

Peter A. Davies and George Gray

1. Introduction

The heat-stable DNA polymerase utilized in the polymerase chain reaction (PCR) has been widely used to amplify DNA fragments since its conception by Kerry Mullis in 1985. However, it soon became apparent that there was a constraint on the maximum size of amplified fragments. For genomic DNA, this was 3–4 kb (*1*), whereas for phage lambda DNA amplifications of up to 15 kb have been possible (*2*).

The basis for this constraint is the rate of errors of misincorporation of *Taq* DNA polymerase, which have been shown to be 2×10^{-4} to 2×10^{-5} (*3*) mutations per nucleotide per cycle. Incorporation of a mismatched base causes strand extension to stall and the *Taq* DNA polymerase to dissociate from the template strand. The longer the PCR product, the more likely misincorporation is to occur.

Barnes (*4*) demonstrated that inclusion of a proofreading enzyme with the *Taq* polymerase could greatly increase the length of amplified DNA strands. The proofreader removes misincorporated nucleotides and allows the *Taq* polymerase to continue. Although capable of high-fidelity DNA synthesis, proofreading enzymes alone produce a low yield of amplified template. Combining *Taq* polymerase with the proofreader produces an increased yield of a high-fidelity long-range PCR product.

Using this mixture of enzymes, it is possible to obtain amplifications of 40 kb from phage DNA clones and up to 22 kb from genomic DNA (*5*).

Applications of long-range PCR (L-PCR) include looking for structural rearrangements of mitochondrial DNA (mtDNA) in patients with neuromuscular diseases (*6*), rapid alignment of clones in gene libraries, amplification and mapping of chromosomal translocation break points, and successful amplifi-

From: *Methods in Molecular Biology, vol. 187: PCR Mutation Detection Protocols*
Edited by: B. D. M. Theophilus and R. Rapley © Humana Press Inc., Totowa, NJ

Table 1
Examples of Commercially Available L-PCR Kits

Product name	Manufacturer	DNA polymerase	Proofreading enzyme
Expand™ Long Template PCR System	Boehringer Mannheim	*Taq*	Pwo
Gene Ampl XL PCR Kit	Perkin-Elmer	r Tth	Vent
Taq Plus™	Stratagene	*Taq*	Pfu

Note: L-PCR reaction mixes are used with the Expand™ Long Templates Kit as used for mtDNA amplication in our department.

cation of long stretches of trinucleotide repeat expansions *(7)*. L-PCR has also been used to amplify RNA templates by first converting the RNA into a DNA intermediate by use of a reverse-transcriptase (RT) enzyme (long RT-PCR). The enzyme is a genetically engineered version of the Moloney Murine Leukaemia Virus reverse transcriptase. Extension of the cDNA occurs in the 3' → 5' direction. A point mutation within the RNase H sequence prevents 5' → 3' exonuclease digestion of the growing cDNA strand. Long RT-PCR applications include looking for deletions in exons and preliminary enrichment of sequences prior to subsequent PCR and mutation detection.

2. Materials

1. Template DNA: The quality of template DNA is of utmost importance. Many methods are now available for DNA extraction. Care should be taken to select a method that will provide the following:
   a. Template DNA of a sufficiently high molecular weight for the intended size of the L-PCR product.
   b. Template of a sufficiently high purity.
2. Optimal primer design: This is essential to achieve successful amplification (*see* **Note 1**).
3. dNTP and magnesium concentrations: To ensure efficient incorporation, the concentration of dNTP's is usually higher in L-PCR (300–500 µmol/L) than in normal PCR (200 µmol/L), as is the magnesium concentration (*see* **Note 2**).
4. Polymerase mix: Since Barnes described the "narrow window of success" for achieving amplification of long templates *(4)*, many commercial kits have been developed with optimal mixes of *Taq* DNA polymerase and proofreader enzyme (*see* **Table 1**). The amount of enzyme mix used can affect specificity. An excess of enzyme can promote extension of oligonucleotides bound to nontarget sequences.
5. Reaction buffer: Different buffers are supplied in kits, each recommended for amplification of fragments within a specific size range.

Table 2
Reaction Mix 1

Reactant	Volume (μL)	Final concentration
Primer 1 (20 pmol/μL)	0.5	0.4 pmol/μL
Primer 2 (20 pmol/μL)	0.5	0.4 pmol/μL
10X L-PCR reaction buffer	2.5	1X L-PCR reaction buffer
Sterile distilled/ double-distilled water	As required to 18 μL final volume	

Final concentrations refer to the concentrations in the reaction mix after adding Reaction Mix 2.

6. Thermocycler: A machine with an option of increasing the extension time with each successive cycle is recommended, such as the Perkin-Elmer 9600. Rapid ramping between temperatures is also desirable.
7. PCR reaction tubes: Thin-walled tubes allow rapid heat transfer and so minimize the time reactions are held at the relatively high denaturing temperature, prolonging enzyme life.

3. Methods
3.1. Template DNA Preparation

Having selected an appropriate DNA extraction method, the following points should be observed:

1. To reduce shearing and maintain high-molecular-weight DNA, tube transfers should be carried out using sterile plastic pastettes, rather than pipet tips.
2. To prevent shearing, any mixing should be performed by gentle repeated inversion by hand rather than by a vortex mixer.
3. If phenol is used in the DNA extraction protocol, it should be buffered at pH 8.0, as nicking of template DNA may occur at acid pH. Auto-oxidation of phenol can also lead to degradation of the DNA.

3.2. Long-Range PCR Procedure

1. Thaw all the reaction components at room temperature. Leave the enzyme mix at –20°C until needed.
2. Pipet 4 μL of appropriately diluted template DNA into PCR tubes (100 ng per reaction).
3. Prepare a mix containing diluted primers in an appropriate 10X reaction buffer and water (*see* **Table 2**). Mix thoroughly by gentle inversion and pipet 18 μL into PCR tubes. Overlay with 1 drop of mineral oil (*see* **Note 3**). Prepare a separate mix of dNTPs and combined *Taq* and proofreader (*see* **Table 3**). Pipet into the PCR tubes immediately following a "hot-start" procedure. After the initial long denaturing step of the cycle, the temperature of the PCR heating block should be held at 80°C

Table 3
Reaction Mix 2

Reaction reactant	Volume (µL)	Final concentration
dNTPs–Mix 5 mmol/L of each nucleotide	2.5	500 µmol/L
DNA polymerase plus proofreader	0.2–0.7	0.028–0.098 U/µL
Sterile pharmacy water	As required to 3 µL final volume	

Final concentration refers to the concentration in the mix after adding Reaction Mix 2.

while 3 µL of the DNA polymerase mix is added to each tube. After addition of the enzyme, the thermal cycler should proceed to the amplification part of the program.

3.3. Thermocycling Conditions

The optimal annealing temperature will vary with the primers used and is dependent on their T_m (*see* Chapter 6). The following cycle conditions give an indication of the time intervals for which temperatures should be maintained:

Initial denaturation		92°C	2 min
Hold "hot-start"		80°C	10 min
Cycle 1	Denature	92°C	50 s
	Anneal and extend	68°C	10 min
Cycles 2–30	Denature	92°C	10 s
	Anneal and extend	68°C	10 min plus 30 s per cycle (*see* **Note 4**)

3.4. Analysis of PCR Products

The concentration of agarose used, the voltage applied, and the time of electrophoresis will vary according to the size of the product to be resolved (*see* Chapter 1). For mtDNA, the following condition can be used:

Mix 5 µL of PCR products with 5 µL of loading dye (15% Ficoll [w/v], in sterile double-distilled water). Load onto a 0.8% agarose gel (in 1X TBE) and electrophorese at 90 V for 2 h.

To maximize the yield of product produced in each successive cycle, the combined annealing and extension time is increased by 30 s per cycle.

4. Notes

1. Primer sequences should be chosen such that the melting temperature (T_m) of the primer is between 65°C and 70°C. The length of L-PCR primers is typically 38 bases. The high T_m maximizes the specificity of primer and template binding and

minimizes internal mispriming over long template reads. The T_m values for each primer in the pair should not vary by more than 1°C or 2°C.

2. Nucleotides and magnesium ions form a complex that forms the substrate for the DNA polymerase. A relatively high magnesium ion concentration (2.5 mM) in the reaction buffer ensures that there is a sufficient supply of substrate for incorporation.

3. Use of a mineral oil overlay is recommended even with oil-free PCR machines.

4. The combined annealing and extension time is increased by 30 s per cycle to maximize the yield of product produced in each successive cycle.

References

1. Ehrlich, H. A., Gelfand, D., and Skinsky, J. J. (1991) Recent advances in the polymerase chain reaction. *Science* **252,** 1643–1651.

2. Kainz, P., Schmiedtechner, A., and Strack, H. B. (1992) In vitro amplification of DNA fragments greater than 10 kb. *Anal. Biochem.* **202,** 46–49.

3. Lundberg, K. S., Shoemaker, D. D., Adam, M. W. W., Short, J. M., Sorge, J. A., and Mathur, E. J. (1991) High fidelity amplification using a thermostable DNA polymerase isolated from Pyrococcus furiosus. *Gene* **108,** 1–6.

4. Barnes, W. M. (1994) PCR amplification of up to 35 kb DNA with high fidelity and high yield from lambda bacteriophage templates. *Proc. Natl. Acad. Sci. USA* **91,** 2216–2220.

5. Cheng, S., Fockler, C., Barnes, W.M., and Higuchi, R. (1994) Effective amplification of long targets from cloned inserts and human genome DNA. *Proc. Natl. Acad. Sci. USA* **99,** 5695–5699.

6. Li, Y. Y., Hengstenberg, C., and Maisch, B. (1995) Whole mitochondrial genome amplification reveals basal level multiple deletions in mtDNA of patients with dilated cardiomyopathy. *Biochem. Biophys. Res. Comm.* **210,** 211–218.

7. Campuzano, V., Montermini, L., Molto, M. D., Pianese, L., Cossee, M., Cavalcanti, F., et al. (1996) Friedreichs ataxia—an autosomal recessive disorder caused by an intronic GAA triplet repeat expansion. *Science* **271,** 1423–1427.

9

Analysis of Nucleotide Sequence Variations by Solid-Phase Minisequencing

Anu Wartiovaara and Ann-Christine Syvänen

1. Introduction

The Sanger dideoxy-nucleotide sequencing method has been simplified by a number of methodological improvements, such as the use of the polymerase chain reaction (PCR) technique for generating DNA templates in sufficient quantities for sequencing, the use of affinity-capture techniques for convenient and efficient purification of the PCR fragments for sequencing, the development of laboratory robots for carrying out the sequencing reactions, and the development of instruments for automatic on-line analysis of fluorescent products of the sequencing reactions. Despite these technical improvements, the requirement for gel electrophoretic separation remains an obstacle when sequence analysis of large numbers of samples are needed, as in DNA diagnosis, or in the analysis of sequence variation for genetic, evolutionary, or epidemiological studies.

We have developed a method for analysis of DNA fragments differing from each other in one or a few nucleotide positions (1), denoted as solid-phase minisequencing, in which gel electrophoretic separation is avoided. Analogous to the methods for solid-phase sequencing of PCR products, the solid-phase minisequencing method is based on PCR amplification using one biotinylated and one unbiotinylated primer, followed by affinity capture of the biotinylated PCR product on an avidin- or streptavidin-coated solid support. The nucleotide at the variable site is detected in the immobilized DNA fragment by a primer extension reaction: A detection step primer that anneals immediately adjacent to the nucleotide to be analyzed is extended by a DNA polymerase with a single labeled nucleotide complementary to the nucleotide at the variable site (**Fig. 1**).

From: *Methods in Molecular Biology, vol. 187: PCR Mutation Detection Protocols*
Edited by: B. D. M. Theophilus and R. Rapley © Humana Press Inc., Totowa, NJ

The amount of the incorporated label is measured, and it serves as a specific indicator of the nucleotide present at the variable site.

We have used the solid-phase minisequencing method for detecting numerous mutations causing human genetic disorders *(2)*, for analyzing allelic variation in genetic linkage studies, and for identification of individuals *(3)*. The protocol presented here is generally applicable for detecting any variable nucleotide. The method suits well for analyzing large numbers of samples because it comprises simple manipulations in a microtiter plate or test-tube format and the result of the assay is obtained as an objective numeric value, which is easy to interpret. Furthermore, the solid-phase minisequencing method allows quantitative detection of a sequence variant present as a minority of less than 1% in a sample *(4)*. We have utilized the possibility of the sensitive quantitative analysis for detecting point mutations in malignant cells present as a minority in a cell population *(4)* and for analyzing heteroplasmic mutations of mitochondrial DNA *(5,6)*. The high sensitivity is an advantage of the minisequencing method, compared to dideoxy-nucleotide sequencing, in which a sequence variant must be present as 10–20% of a mixed sample to be detectable. On the other hand, a limitation of the solid-phase minisequencing method is that it is restricted to analyzing variable nucleotides only at positions predefined by the detection step primers used. The minisequencing reaction principle is utilized in a variety of other assay fomats than the one described here, for reviews, *see* **refs.** *2* and *12*.

2. Materials
2.1. Equipment and Materials

1. One of the PCR primers should be biotinylated at its 5' end during the oligonucleotide synthesis, and the other primer is not biotinylated, resulting in a PCR product with one biotinylated strand (*see* **Note 1**).
2. Detection step primer: an oligonucleotide complementary to the biotinylated strand, designed to hybridize with its 3' end with the nucleotide adjacent to the variant nucleotide to be analyzed (*see* **Fig. 1** and **Note 2**).
3. Facilities for PCR.
4. Microtiter plates with streptavidin-coated wells (e.g., Combiplate 8, Labsystems, Finland) (*see* **Note 3**).
5. Shaker at 37°C.
6. Water bath or incubator at 50°C.
7. Liquid scintillation counter.
8. Multichannel pipet and microtiter plate washer (optional).

2.2. Reagents

All of the reagents should be of standard molecular biology grade. Use sterile distilled or deionized water.

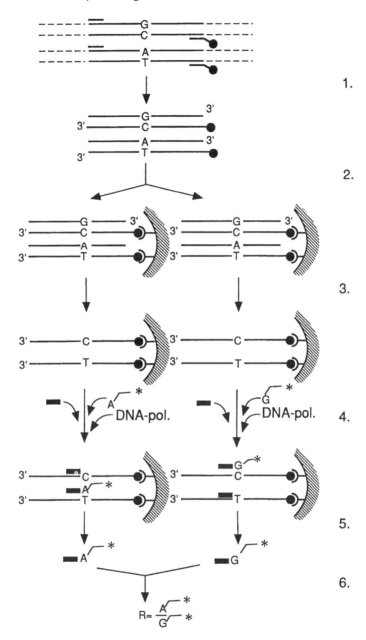

Fig. 1. Steps of the solid-phase minisequencing method. (1) PCR with one biotinylated (black ball) and one unbiotinylated primer; (2) affinity capture of the biotinylated PCR product in streptavidin-coated microtiter wells; (3) washing and denaturation; (4) the minisequencing primer extension reaction; (5) measurement of the incorporated label; (6) calculation of the result.

1. PBS/Tween solution: 20 mM sodium phosphate buffer, pH 7.5, 0.1% (v/v) Tween-20. Store at 4°C; 50 mL is enough for several full-plate analyses.
2. TENT solution: 40 mM Tris-HCl, pH 8.8, 1 mM EDTA, 50 mM NaCl, 0.1% (v/v) Tween-20. Store at 4°C. Prepare 1–2 L at a time, which is enough for several full-plate analyses.
3. 50 mM NaOH (make fresh every 4 wk); store at room temperature (about 20°C). Prepare 50 mL.
4. Thermostable DNA polymerase: *Thermus aquaticus* (*Taq*) DNA polymerase (Promega, 5 U/µL) (*see* **Note 4**).
5. 10X concentrated *Taq* DNA polymerase buffer: 500 mM Tris-HCl, pH 8.8, 150 mM $(NH_4)_2SO_4$, 15 mM $MgCl_2$, 0.1% (v/v) Triton X-100, 0.01% (w/v) gelatin; store at –20°C.
6. [³H]-labeled deoxynucleotides (dNTPs): dATP to detect a T at the variant site, dCTP to detect a G, and so forth (Amersham; [³H] dATP, TRK 625; dCTP, TRK 576; dGTP, TRK 627; dTTP, TRK 633); store at –20°C (*see* **Note 5**).
7. Scintillation reagent (e.g., Hi-Safe II, Wallac).

3. Method

3.1. PCR for Solid-Phase Minisequencing Analysis

The PCR is done according to routine protocols, except that the amount of the biotin-labeled primer used should be reduced so as not to exceed the biotin-binding capacity of the microtiter well (*see* **Note 3**). For a 50-µL PCR reaction, we use 10 pmol of biotin-labeled primer and 50 pmol of the unbiotinylated primer. The PCR should be optimized (i.e., the annealing temperature and template amount) to be efficient and specific; to be able to use [³H] dNTPs, which are low-energy emitters, for the minisequencing analysis, one tenth of the PCR product should produce a single visible band after agarose gel electrophoresis, stained with ethidium bromide. After optimization, there is no need for purification of the PCR product.

3.2. Solid-Phase Minisequencing Analysis

1. Affinity capture: Transfer 10-µL aliquots of the PCR product and 40 µL of the PBS/Tween solution to two streptavidin-coated microtiter wells (*see* **Note 6**). Include a control reaction (i.e., a well with no PCR product). Seal the wells with a sticker and incubate the plate at 37°C for 1.5 h with gentle shaking.
2. Discard the liquid from the wells and tap the wells dry against a tissue paper.
3. Wash the wells three times at room temperature as follows: pipet 200 µL of TENT solution to each well, discard the washing solution, and empty the wells thoroughly between the washings (*see* **Note 7**).
4. Denature the captured PCR product by adding 100 µL of 50 mM NaOH to each well, incubate at room temperature for 3 min. Discard the NaOH and wash the wells as in **step 3**.

5. For each DNA fragment to be analysed, prepare two 50-μL mixtures of nucle-otide-specific minisequencing solution—one for detection of the normal and one for the mutant nucleotide (*see* **Note 8**). Mix 5 μL of 10X *Taq* DNA poly-merase buffer, 10 pmol of detection step primer, 0.2 μCi (usually equal to 0.2 μL) of one [^{3}H] dNTP, 0.1 U of *Taq* DNA polymerase, and dH$_2$O to a total volume of 50 μL. It is obviously convenient to prepare master mixes for the desired number of analyses with a certain nucleotide.

6. Pipet 50 μL of one nucleotide-specific mixture per well, incubate the plate at 50°C for 10 min in a water bath or 20 min in an oven (*see* **Note 9**).

7. Discard the contents of the wells and wash them as in **step 3**.

8. Release the detection step primer from the template by adding 60 μL of 50 m*M* NaOH and incubating for 3 min at room temperature.

9. Transfer the eluted primer to the scintillation vials, add scintillation reagent, and measure the radioactivity (i.e., the amount of incorporated label) in a liquid scin-tillation counter (*see* **Note 10**).

10. The result is obtained as counts per minute (cpm) values. The cpm value of each reaction expresses the amount of the incorporated [^{3}H] dNTP. Calculate the ratio (*R*) between the mutant and normal nucleotide cpms. In a sample of a subject homozygous for the mutant nucleotide, the *R* will be >10; in a homozygote for the normal nucleotide, *R*<0.1; and in the case of a heterozygote, *R* varies between 0.5 and 2.0, depending on the specific activities of the [^{3}H] dNTPs (*see* **Note 11**).

4. Notes

1. The efficiency of the 5'-biotinylation of an oligonucleotide on a DNA synthesizer is most often 80–90%. The biotin-labeled oligonucleotides can be purified from the unbiotinylated ones either by high-performance liquid chromatography *(7)*, polyacrylamide gel electrophoresis *(8)*, or ion-exchange columns manufactured for this purpose (Perkin-Elmer/ABI). If the biotin-labeled primer is not purified, the biotinylation should be confirmed after the PCR by affinity capture of the biotinylated PCR product, followed by detection of possible unbound products by agarose gel electrophoresis; *see* Chapter 1.

2. The detection step primer for our standard protocol is a 20-mer. It is advisable to use a nested primer as a detection step primer, to ensure that possible unspecific PCR products remain undetected. The primer should be at least five nucleotides nested in relation to the unbiotinylated PCR primer.

3. The binding capacity of a streptavidin-coated microtiter well is 2–5 pmol of biotinylated oligonucleotide. If higher binding capacity is desired, avidin-coated polystyrene beads (Fluoricon, 0.99 μm [IDEXX Corp., Portland ME]; biotin-binding capacity over 2 nmol of oligonucleotide/mg beads) or streptavidin-coated magnetic polystyrene beads (Dynabeads M-280, streptavidin; biotin-binding capacity 300 pmol/mg) can be used *(9)*. The biotin-binding capacity of a microtiter well allows reliable detection of up to 2% of a sequence variant present in the sample *(6)*, whereas a detection sensitivity of less than 0.1% is obtained with the bead-based format *(4)*.

4. It is advantageous to use a thermostable DNA polymerase for the single-nucle-otide primer extension reaction, because a high temperature, favorable for the simultaneous primer annealing reaction, can be used.

5. Although the specific activities of the [^{3}H] dNTPs are low, their half-lives are long (13 yr) and the necessary precautions for working with [^{3}H] should be taken. Also dNTPs or dideoxy-nucleotides labeled with other isotopes ([^{35}S] or [^{32}P]) or with haptens can be used (*1,10*).

6. Each nucleotide to be detected at the variant site is analyzed in a separate well. Thus, at least two wells are needed per PCR product.

7. The washings can be performed utilizing an automated microtiter plate washer or by manually pipetting the washing solution to the wells, discarding the liquid, and tapping the plate against a tissue paper. Thorough emptying of the wells is important to avoid unspecific nucleotide incorporation.

8. The minisequencing reaction mixture can be stored at room temperature for 1–2 h. It is convenient to prepare it during the incubation of **step 1**.

9. The conditions for hybridizing the detection step primer are not stringent and the temperature of 50°C can be applied to the analysis of most PCR products irre-spective of the sequence of the detection step primer. If the primer, however, is considerably shorter than a 20-mer or its GC content is low (melting temperature close to 50°C), lower temperatures for the primer annealing may be required.

10. Streptavidin-coated microtiter plates made of scintillating polystyrene are avail-able (ScintiStrips, Wallac, Finland). When these plates are used, the final wash-ing, denaturation, and transfer of the eluted detection primer can be omitted, but a scintillation counter for microtiter plates is needed (*11*).

11. The ratio between the cpm values for the two nucleotides reflects the ratio between the two sequences in the original sample. Therefore, the solid-phase minisequencing method can be used for quantitative PCR analyses (*4–6*). The *R* value is affected by the specific activities of the [^{3}H] dNTPs used, and if either the mutant or the normal sequence allows the detection step primer to be extended by more than one [^{3}H] dNTP, this will obviously also affect the *R* value. Both of these factors can easily be corrected for when calculating the ratio between the two sequences. Another possibility is to construct a standard curve by mixing the two sequences in known ratios and plotting the obtained cpm val-ues as a function of the ratios to obtain a linear standard curve (*5,6*). The test results can then be interpreted from the standard curve without the need to take the specific activities of the number of [^{3}H] dNTPs incorporated into account.

References

1. Syvänen, A.-C., Aalto-Setälä, K., Harju, L., Kontula, K., and Söderlund, H. (1990) A primer-guided nucleotide incorporation assay in the genotyping of apolipoprotein E. *Genomics* **8,** 684–692.
2. Syvänen, A.-C. (1999) From gels to chips: "minisequencing" primer extension for analysis of point mutations and single nucleotide polymorphisms. *Hum. Mutat.* **13,** 1–10.

3. Syvänen, A.-C., Sajantila, A., and Lukka M. (1993) Identification of individuals by analysis of biallelic DNA markers, using PCR and solid-phase minisequencing. *Am. J. Hum. Genet.* **52,** 46–59.

4. Syvänen, A.-C., Söderlund, H., Laaksonen, E., Bengtström, M., Turunen, M., and Palotie, A. (1992) N-ras gene mutations in acute myeloid leukemia: accurate detection by solid-phase minisequencing. *Int. J. Cancer* **50,** 713–718.

5. Suomalainen, A., Kollmann, P., Octave, J.-N., Söderlund, H., and Syvänen, A.-C. (1993) Quantification of mitochondrial DNA carrying the $tRNA_{8344}^{Lys}$ point mutation in myoclonus epilepsy and ragged-red-fiber disease. *Eur. J. Hum. Genet.* **1,** 88–95.

6. Suomalainen, A., Majander, A., Pihko, H., Peltonen, L., and Syvänen, A.-C. (1993) Quantification of $tRNA_{3243}^{Leu}$ point mutation of mitochondrial DNA in MELAS patients and its effects on mitochondrial transcription. *Hum. Mol. Genet.* **2,** 525–534.

7. Bengtström M., Jungell-Nortamo, A., and Syvänen, A.-C. (1990) Biotinylation of oligonucleotides using a water soluble biotin ester. *Nucleosides Nucleotides* **9,** 123–127.

8. Wu, R., Wu, N.-H., Hanna, Z., Georges, F., and Narang, S. (1984) *Oligonucleotide Synthesis: A Practical Approach.* (Gait, M. J., ed.), IRL, Oxford, p. 135.

9. Syvänen, A.-C. and Söderlund, H. (1993) Quantification of polymerase chain reaction products by affinity-based collection. *Meth. Enzymol.* **218,** 474–490.

10. Harju, L., Weber, T., Alexandrova, L., Lukin, M., Ranki, M., and Jalanko, A. (1993) Colorimetric solid-phase minisequencing assay illustrated by detection of alpha-1-antitrypsin Z mutation. *Clin. Chem.* **39,** 2282–2287.

11. Ihalainen, J., Siitari, H., Laine, S., Syvänen, A.-C., and Palotie, A. (1994) Towards automatic detection of point mutations: use of scintillating microplates in solid-phase minisequencing. *BioTechniques* **16,** 938–943.

12. Syvänen, A.-C. (2001) Accessing genetic variation: genotyping single nucleotide polymorphisms. *Nat. Rev. Genet.* **2,** 930–942.

10

Cycle Sequencing of PCR Products

G. K. Surdhar

1. Introduction

The development of the polymerase chain reaction (PCR) has allowed the rapid isolation of DNA sequences utilizing the hybridization of two oligonucleotide primers and subsequent amplification of the intervening sequences by *Taq* polymerase. This technique has facilitated the identification and typing of single-nucleotide substitutions in the analysis of DNA sequence polymorphisms and the screening of large numbers of samples to either detect known mutations or search for unknown mutations at a defined locus *(1–4)*.

Since the introduction of PCR, a variety of methods for sequencing PCR-generated fragments have been described, based on the Sanger chain-terminating dideoxynucleotide sequencing method *(5)*. Cycle sequencing *(6–10)* is a kind of PCR sequencing approach. Like standard PCR, it utilizes a thermostable DNA polymerase and a temperature cycling format of denaturation, annealing, and DNA synthesis. Cycle sequencing, also called the linear amplification process, in contrast with a traditional PCR reaction where the increase is exponential, employs a single primer so that the amount of product DNA increases linearly with the number of cycles in contrast with a traditional PCR reaction where the increase is exponential.

Initially, cycle sequencing utilized ^{32}P-labeled primers and a nonthermostable polymerase, which needed to be added after every denaturation cycle *(6)*. Further progress was made with the introduction of thermostable *Taq* polymerase *(7,8)* and the replacement of labeled primers with internal labeling using α-labeled ^{35}S or [32]P dATPs and mixtures of nucleotides similar to those used originally by Sanger *(9,10)*. The latest manual method available is the termination cycle sequencing (Amersham Pharmacia Biotech), which incorporates the isotopic label into the sequencing reaction products by the use of four

From: *Methods in Molecular Biology, vol. 187: PCR Mutation Detection Protocols*
Edited by: B. D. M. Theophilus and R. Rapley © Humana Press Inc., Totowa, NJ

[α-^{33}P]dideoxynucleotide (ddNTP) terminators (G, A, T, and C). This method is the most efficient in terms of radioactivity usage, as the labeled nucleotides label only the properly terminated DNA chains. As a result, prematurely terminated chains are not labeled, therefore, "stop" artifacts that result in bands across all four lanes and most background bands are eliminated. All results obtained from the above-mentioned methods are visualized by running the samples on a polyacrylamide gel followed by exposure to X-ray film for 24 h or more depending on the age of the radioactivity and amount of template used.

Fluorescent or automated sequencing methodology is commonly used now in order to avoid the use of radioactive materials. The other advantage of this method is that approx 500 bp can be read per reaction on an automated DNA sequencer with 99.3% accuracy compared to 150–200 bp on an average manual sequencing gel *(11,12)*. The automated sequencing market is divided between 4-dye and 1-dye technology. With the 4-dye technology, there is one tube per reaction with the four dyes corresponding to the four bases; therefore, each sample can be run on a gel in one lane. This is in contrast to the 1-dye four-lane approach, where there are four tubes per reaction, each tube corresponding to one of the four bases and is, therefore, run on a gel in four lanes. The former single-lane approach has been the option for both large-scale and large-sample-number application groups. The 4-dye single-lane instrumentation allows for higher sample throughput and consistent results among the large numbers of samples that are processed on a single run. The other advantage of the 4-dye single-lane approach compared to the 1-dye four-lane approach is that variations in electrophoretic mobility across the four lanes with the single-dye technology may result in artifacts of sequence miscalling between adjacent lanes. The Big Dye™ Terminators from Perkin-Elmer and the DYEnamic dyes from Amersham Pharmacia Biotech are examples of the 4-dye and single-lane approach respectively. The Big Dyes have been an improvement on previous systems, as they produce more even peak heights, which facilitates the identification of heterozygosity. Prior to this system, the recommended method for recognizing heterozygotes was to use the dye primer technology, which gave even peak heights. This method of sequencing requires the primer to be fluorescently labeled with a dye, so four separate reactions have to be carried out. It is possible to label each primer four times, each time using a different dye, and then pool the reactions together for sequencing, but this turns out to be very expensive. There have been vast improvements in sequencing technology such that the 4-dye terminator kits available on the market now result in very even peak heights, which allows for an easier detection of heterozygotes.

This chapter describes the two recent technologies for cycle sequencing: manual, radioactive termination cycle sequencing, and automated sequencing using the Big Dye™ Terminators (Perkin-Elmer Applied Biosystems). Both of these methods produce sequences of high quality.

2. Materials

2.1. Manual Cycle Sequencing

2.1.1. Purification of PCR Products

1. Geneclean Kit II (Anachem).

2.1.2. Termination Cycle Sequencing

1. Termination cylce sequencing kit (Amersham). The following components are provided in the kit:
    a. 10X Reaction buffer concentrate: 260 mM Tris-HCl,pH 9.5, 65 mM MgCl$_2$.
    b. dGTP termination master mix: 7.5 µM dATP, dCTP, dGTP, dTTP.
    c. dITP termination master mix: 7.5 µM dATP, dCTP, dTTP, 37.5 µM dITP.
    d. Four Redivue™ ^{33}P-labeled terminators which consist of the following, which are all 0.3 µM [α-^{33}P] ddNTP (1500 Ci/mmol, 450 µCi/mL), 11.25 µCi:

    ddGTP
    ddATP
    ddTTP
    ddCTP

2.1.3. Treatment of PCR Products After Cycle Sequencing

1. Stop solution: 95% formamide, 20 mM EDTA, 0.05% bromophenol blue, 0.05% xylene cyanol (store at 4°C).

2.1.4. 6% Polyacrylamide Sequencing Gels

1. 20X Glycerol tolerant gel buffer (1 L): 216 g Tris base, 72 g taurine, 4 g Na$_2$EDTA·2H$_2$O (stable at room temperature).
2. 30 mL 30% (w/v) acrylamide stock solution (19:1 acrylamide/bis-acrylamide) (store at 4°C).
3. 63 g Urea.
4. TEMED (*N,N,N',N'*-tetramethylethylenediamine).
5. 25% Ammonium persulfate (APS) (made fresh).
6. Loading buffer: stop solution as in **Subheading 2.1.3.**

2.1.5. Autoradiography

1. X-ray Hyperfilm 18 cm × 43 cm (Sigma) and cassettes.

2.2. Automated Fluorescent Cycle Sequencing

2.2.1. Purification of PCR Products Before Cycle Sequencing

1. Microcon-100 Microconcentrators (Amicon).

2.2.2. Cycle Sequencing

1. Terminator Ready Reaction Mix, Applied Biosystems (ABI) contains the following:
    A-Dye terminator labeled with dichloro [R6G];
    C-Dye terminator labeled with dichloro [ROX];

G-Dye terminator labeled with dichloro [R110];
T-Dye terminator labeled with dichloro [TAMRA];
Deoxynucleoside triphosphates (dATP, dCTP, dITP, dUTP);
AmpliTaq DNA polymerase, FS, with thermally stable pyrophosphatase;
$MgCl_2$;
Tris-HCl buffer, pH 9.0.

2.2.3. Purifying Extension Products

1. 3 *M* Sodium acetate, pH 5.2.
2. Absolute ethanol.
3. 70% Ethanol.
4. Distilled water.

2.2.4. Preparing and Loading Samples

1. Template suppression reagent (ABI) (store at 4°C).

3. Methods

3.1. Manual Cycle Sequencing

3.1.1. Purification of PCR Products Before Cycle Sequencing

Purify the PCR products prior to sequencing using a Geneclean Kit II (Anachem) according to manufacturer's instructions (*see* **Note 1**).

3.1.2. Termination Cycle Sequencing

The amount of genecleaned PCR products should be roughly estimated on an agarose gel before sequencing. The approximate range of template required is about 2–50 ng/μL. A general outline of the protocol is as follows (details are available in the manual provided by the kit [Amersham Pharmacia Biotech]):

1. Prepare four termination mixes (one each for ddA, ddC, ddG and ddT):

	×1
Termination mix dGTP (or dITP) (*see* **Note 2**)	2 μL
α^{33}P ddNTP	0.5 μL
	2.5 μL

   This can be multiplied by the number of samples to make a master mix and 2.5 μL aliquoted into each tube.
2. Dispense 2.5 μL each of ddA, ddC, ddG, and ddT termination mix into respectively labeled tubes for each template to be sequenced.
3. Prepare a reaction mix:

Reaction buffer	2 μL
DNA (2–50 ng/μL)	μL*
Primer (0.1 μg/μL) (*see* **Note 3**)	1 μL

*The concentration of the PCR product should be as stated above.

Water	μL⁺
Thermo Sequenase (4 U/μL)	2 μL
	20 μL

Add 4.5 μL of this reaction mix to each of the termination mixes. Mix contents of the tube with a pipet and then add one drop of mineral oil.
4. Incubate the reactions in a thermal cycler using the following conditions: 50 cycles at 95°C, 30 s; 50 cycles at 50°C, 30 s; and 50 cycles at 72°C, 60 s (*see* **Note 4**).

3.1.3. Treatment of PCR Products After Cycle Sequencing

Add 4 μL stop solution to each tube and store at –20°C until ready to run on a gel.

3.1.4. 6% Polyacrlyamide Sequencing Gels

1. To make 150 mL of acrylamide solution, add 63 g urea, 30 mL of 30% acrylamide stock solution, and 12 mL of 20X glycerol tolerant buffer and make up to 150 mL with distilled water.
2. Just before pouring the gel, add 50 μL of 25% APS and 50 μL TEMED to 50 mL of the acrylamide solution to initiate polymerization. Pour this into a sandwich of two 45 inches long gel plates that have been sealed at the bottom with 10 mL 6% acrylamide, 50 μL 25% APS and 50 μL TEMED.
3. Leave gel to set for approx 1 h or overnight.
4. Set up the gel apparatus, using 0.8X glycerol tolerant buffer (*see* **Note 5**).
5. Load 4 μL of each sample into each well.
6. Run the gel at 50°C, approx 50 W until the sample dyes have migrated the required distance. As a general guide, bromophenol blue migrates with a DNA fragment of approx 26 bp and xylene cyanol with a fragment of approx 106 bp in a 6% gel.

3.1.5. Autoradiography

Following electrophesis, allow the glass plates to cool to room temperature. Carefully separate the glass plates, leaving the gel adhered to one of the plates. Transfer the gel to a piece of Whatmann 3MM paper cut out to the same size of the gel by slowly placing it over the gel. Slowly peel off the paper with the attached gel.

Cover the gel with cling film and dry in a slab gel dryer at 80°C for 30–120 min until the gel is dry. Remove the cling film and autoradiograph the gel against a high-speed X-ray film in an appropriate cassette.

Exposure of the film at room temperature varies from about 1 d to 6 d depending on the amount of starting DNA template used and the age of the radioactivity.

3.2. Automated Cycle Sequencing

3.2.1. Purification of PCR Products Before Cycle Sequencing

There are various methods used for this procedure. The Microcon-100 spin columns are recommended by ABI to give good sequencing results, as they

⁺Adjust the amount of water to make up to a total volume of 20 μL.

remove salt which can interfere with the sequencing reaction. There are alternative methods available (*see* **Note 1**).

3.2.2. Cycle Sequencing

1. For each reaction, mix the following reagents in a labeled tube:
   (This is a half-reaction; one-eighth and one-quarter reactions can also be carried out, but this requires a propietary buffer from ABI that maintains a reasonable reaction volume.)

Terminator-ready reaction mix	4 µL
PCR product (30–90 ng)	x µL
Primer (0.02 µg/µL) (*see* **Note 3**)	1 µL
dH$_2$O	y µL
	10 µL

   Set up and keep the reaction on ice until ready to transfer to the thermal cycler.
2. Perform the sequencing reactions in a thermal cycler using the following conditions: 96°C for 10 s; 25 cycles of 96°C for 10 s, 50°C for 5 s, 60°C for 4 min (*see* **Note 4**).
   Make up to 20 µL with water following the reaction.

3.2.3. Purifying Extension Products

This method utilizes ethanol and sodium acetate to purify the extension products (*see* **Note 6**).

1. After completion of the sequencing reactions, transfer the 20 µL extension products to 0.5 µL tubes.
2. Add the following to the tubes:
   2 µL of 3 *M* NaOAc, pH 5.2;
   50 µL of 95% EtOH at room temperature (*see* **Note 7**).
3. Cap the tubes and vortex briefly.
4. Incubate the tubes at room temperature for 15 min.
5. Place the capped tubes in a microcentrifuge, and spin the tubes for 20 min at maximum speed. This should be at least 1400g but less than 3000g.
6. Without disturbing the pellet, carefully aspirate the supernatant with a pipettor and discard it.
7. Rinse the pellet by adding 200 µL of 70% EtOH. Cap the tubes and vortex briefly. Centrifuge for 5 min at maximum speed.
8. Without disturbing the pellet, carefully aspirate the supernatant with a pipettor and discard.
9. Dry the pellet in a vacuum centrifuge for 10–15 min, or remove the caps and place the tubes in a heat block at 90°C for 1 min.

3.2.4. Preparing and Loading Samples

Resuspend each pellet in 20 µL of Template Suppression Reagent (ABI) and heat denature at 95°C on a heating block for 2 min, 30 s; immediately place on ice. Load sample on to automated analyzer according to manufacturer's instructions.

4. Notes

1. Geneclean Kit II (Anachem) or any other silica-based purification methods, purification with Exonuclease and Shrimp Alkaline Phosphatase, and spin columns that concentrate and desalt the PCR products, such as the Microcon-100 spin columns (Millipore), may be used. For automated fluorescent sequencing, the latter method of purification is more appropriate as silica from the Geneclean can interfere with the sequencing reaction, such that a shorter read is obtained. When there are multiple bands present because of nonspecific amplification, then Geneclean is used to purify the sample, in which case it is very important to not get any silica carryover into the sequencing reaction. The enzyme method of PCR purification gives good sequencing, as it removes the excess dNTPs and primers: 8 μL of a 50-μL PCR product is digested with 1 μL of Exonuclease I (10 U/μL) and 1 μL of shrimp alkaline phosphatase (2 U/μL) (Amersham supplies both enzymes together as a presequencing kit), incubated at 37°C for 15 min and then inactivated at 80°C for 15 min; 5 μL is used for the sequencing reaction.

2. Compressions occur when the DNA (usually G-C rich) synthesized by the DNA polymerase does not remain fully denatured during electrophoresis. dITP can be used instead of dGTP to reduce compressions in the sequence.

3. Primers should be about 18–25 nucleotides long. It is also important to check the sequence of the primer for possible self-complementarity.

4. The number of cycles required depends on the amount of template used for sequencing, as well as the purity and sensitivity of autoradiographic detection. In general, when the amount of template is low, more cycles should be used. Most reactions can be carried out at the 50°C default annealing temperature. Polymerization is optimal at 70–75°C, except when using dITP, which requires a maximum temperature between 55°C and 60°C. Big Dye Terminators™ utilize dITP; therefore, the polymerization is carried out at 60°C.

5. The 0.8X glycerol tolerant buffer is recommended, as it results in faster gel migration.

6. When using Big Dye™ Terminators, there are two other methods recommended for purifying extension products; using either 60% ± 5% isopropanol or 60% ± 3% ethanol. These methods are described further in the ABI manual. All three methods have been tested simultaneously and the described ethanol/sodium acetate method was found to be the most efficient for purifying the extension products.

7. When exposed to air, absolute ethanol absorbs moisture and becomes more dilute. Also, slight variations in concentration occur when ethanol is diluted to 95%. These variations can result in increased residual dyes. Use of a commercially prepared molecular-biology-grade 95% ethanol is therefore recommended to eliminate variations. Alternatively, 100% molecular-grade ethanol can be diluted to 95% with molecular-grade isopropanol.

References

1. Orita, M., Suzuki, Y., Sekeiya, T., and Hayashi, K. (1989) Rapid and sensitive detection of point mutations and DNA polymorphisms using the polymerase chain reaction. *Genomics* **5,** 874–879.
2. Kwok, P. Y., Carlson, C., Yager, T. D., Ankener, W., and Nickerson, D. A. (1994) Comparative analysis of human DNA variations by fluorescence based sequencing of PCR products. *Genomics* **23,** 138–144.
3. Martin-Gallardo, A., McCombie, W. R., Gocayne, J. D., Fitzgerald, M. G., Wallace, S., Lee, B. M. B., et al. (1992) Automated DNA sequencing and analysis of 106 kilobases from human chromosome 19q 13.3. *Nature Genet.* **1,** 34–39.
4. NIH/CEPH Collaborative Mapping Group (1992) A comprehensive genetic linkage map of the human genome. *Science* **258,** 67–86.
5. Sanger, F., Nicklen, S., and Coulson, A. R. (1977) DNA sequencing with chain terminating inhibitors. *Proc. Natl. Acad. Sci. USA* **74,** 5463–5467.
6. McMahon, G., Davis, E., and Wogan, G. N. (1987) Characterization of c-Ki-ras oncogene alleles by direct sequencing of enzymatically amplified DNA from carcinogen induced tumours. *Proc. Natl. Acad. Sci. USA* **84(14),** 4974–4978.
7. Carothers, A. M., Urlab, G., Mucha, J., Grunberger, D., and Chasin, L. A. (1989) Point mutation analysis in a mammalian gene: rapid preparation of total RNA, PCR amplification of cDNA, and Taq sequencing by a novel method. *BioTechniques* **7(5),** 494–499.
8. Murray, V. (1989) Improved double-stranded DNA sequencing using the linear polymerase chain reaction. *Nucl. Acids Res.* **17(21),** 8889.
9. Levedakou, E. N., Landegren, U., and Hood, L. E. (1989) A strategy to study gene polymorphism by direct sequence analysis of cosmid clones and amplified genomic DNA. *BioTechniques* **7(5),** 438–442.
10. Lee, J. S. (1991) Alternative dideoxy sequencing of double-stranded DNA by cyclic reactions using Taq polymerase. *DNA Cell Biol.* **10(1),** 67–73.
11. McBride, L. J., Koepf, S. M., Gibbs, R. A., Nyugen, P., Salser, W., Mayrand, P. E., et al. (1989) Automated DNA sequencing methods involving PCR. *Clin. Chem.* **35,** 2196–2201.
12. Tracy, T. E. and Mulcahy, L. S. (1991) A simple method for direct automated sequencing of PCR products. *BioTechniques* **11(1),** 68–75.

11

Fluorescent *In Situ* Hybridization

Sara A. Dyer and Elaine K. Green

1. Introduction

Single-stranded DNA will recognize a complementary strand with high specificity under suitably controlled conditions. *In situ* hybridization (ISH) exploits this phenomenon by hybridizing an appropriately labeled single-stranded DNA "probe" to target sequences *in situ* in either dissociated cell preparations or tissue sections.

Early ISH protocols used radioactivity to label probes, whereas current techniques use fluorescent labels giving rise to the term "fluorescent *in situ* hybridization" (FISH). FISH probes may be directly labeled by the incorporation of a fluorochrome-conjugated molecule (e.g., fluoroscein d-UTP or Texas red d-UTP) or indirectly labeled by the incorporation of a reporter molecule (e.g., biotin d-UTP or digoxygenin d-UTP). If probes are indirectly labeled, they must be detected posthybridization using a reporter-binding fluorescent molecule. A schematic representation of the FISH principle is shown in **Fig. 1**.

Fluorescent *in situ* hybridization probes fall into three main categories depending on the region of the genome to which they hybridize: repetitive-sequence probes, single-copy probes, and whole-chromosome libraries. Repetitive-sequence probes are specific for repetitive DNA regions found within the human genome. Centromeric probes are a family of repetitive-sequence probes hybridizing to mainly α-satellite DNA found near the centromeres of all human chromosomes. Subtelomeric probes are also repetitive-sequence probes. Single-copy probes are specific for unique sequences within the genome. Whole-chromosome libraries are made up of a mixture of DNA sequences comprising the entire length of a specific chromosome; these probes are also known as chromosome paints.

From: *Methods in Molecular Biology, vol. 187: PCR Mutation Detection Protocols*
Edited by: B. D. M. Theophilus and R. Rapley © Humana Press Inc., Totowa, NJ

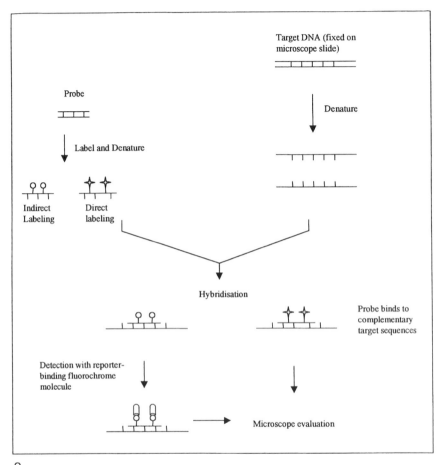

Fig. 1. Schematic representation of the FISH principle.

Many FISH probes are commercially available prelabeled with a choice of direct or indirect labels. A list of suppliers used by the authors is given in **Table 1**. Alternatively, probes may be grown in the laboratory from appropriate DNA stocks and labeled with a desired fluorochrome or reporter molecule.

Target preparations for FISH may take one of many forms, including cell suspensions, touch preparations, or paraffin-sectioned material and the technique is applicable to both dividing and nondividing cells (i.e., metaphases or interphases, respectively).

Table 1
Suggested Probe Suppliers

Supplier	Address
Appligene Oncor	Qbiogene
	Salamander Quay West
	Park Lane
	Harefield
	Middlesex UB9 6NZ
	United Kingdom
Vysis Ltd.	Vysis SA
	UK Customer Department
	81470 Maurens Scopont
	France

Over the last decade, FISH has proved to be an invaluable technique in both diagnostic and research laboratories. The appropriate choice of probe and target can lead to the characterization of chromosomal and gene rearrangements and the ascertainment of chromosome copy number. Single-copy unique-sequence probes have proved useful in the detection of microdeletions such as those found on chromosomes 7, 15, and 22 in Williams, Prader Willi, and Di George syndromes, respectively. Single-copy probes have also been widely used in recent years in the identification of gene rearrangements such as *BCR/ABL* and *PML/RARA* fusion in blood and bone marrow samples from patients with leukemia. Chromosome libraries and subtelomeric probes have proved to be valuable in the characterization of complex and sometimes subtle chromosomal rearrangements in both constitutional and acquired genetic abnormalities. Centromeric probes have been widely used to determine the chromosomal origin of small marker chromosomes that are unidentifiable by conventional Giemsa-banding analysis. These repetitive-sequence probes are also of great value in determining the chromosome copy number in nondividing cells. In recent years, "multi-probe" devices have been developed that allow simultaneous detection of all human telomeres or centromeres on a single microscope slide.

Over the past 10 years the basic FISH principle has been used to develop new molecular cytogenetic techniques, including multicolor FISH (M-FISH) *(1)* and comparative genomic hybridization (CGH) *(2)*. M-FISH allows each chromosome to be visualized in a different color, enabling the rapid identification of chromosome rearrangements, whereas CGH allows a whole genome to be scanned for genetic imbalances in a single hybridization. Both M-FISH and CGH rely on sophisticated image-analysis equipment, which is becoming increasingly important in the growing field of molecular cytogenetics.

Practically, FISH involves probe labeling, denaturation of probe and target DNA, hybridization of probe to target DNA, stringency washing to remove excess probe, detection (of indirectly labeled probes), and counterstaining. Basic hybridization methods are given in the following sections, together with protocols for growing and labeling probes in the laboratory. All methods are designed as general guidelines only, and in the case of commercially purchased probes, manufacturer's guidelines should always be followed. Particularly hazardous reagents have been highlighted but appropriate safety precautions should be followed for all laboratory work.

2. Materials
2.1. Growing Probes in the Laboratory

1. Plasmid Maxi kit (Qiagen, cat. no. 12162) containing Qiagen 100 tips, and buffers: P1 (resuspension buffer), P2 (lysis buffer), P3 (neutralization buffer), QBT (equilibration buffer), QC (wash buffer) and QF (elution buffer).
2. LB broth (Luria–Bertani medium): Dissolve 10 g bacto-tryptone, 5 g bacto-yeast extract, and 10 g NaCl in 950 mL of distilled water, adjust the pH to 7.0 with 5 N NaOH, and make up to 1 L with distilled water. Sterilize by autoclaving.
3. Kanamycin (50 mg/mL) store at $-20°C$.
4. Shaking incubator.
5. Sorvall centrifuge (GSA and SS-34 rotors).
6. 250 mL and 50 mL centrifuge tubes.
7. Isopropanol.
8. 70% Ethanol; store at $-20°C$.

2.2. Probe Labeling

1. Nick translation kit (e.g., Roche Diagnostics, cat. no. 976776, containing dNTPs, buffer, DNA polymerase/DNase I enzyme mix).
2. Labeled dUTP: digoxygenin-11-dUTP, biotin-11-dUTP, fluoroscein-11-dUTP, or Texas Red-11-dUTP at a concentration of 0.4 mM (*see* **Note 1**).
3. 0.2 M EDTA, pH 8.0.
4. Herring sperm DNA (Gibco-BRL) (5 μg/mL).
5. Human COT-1 DNA (Gibco-BRL) (1 mg/mL).
6. 100% Ethanol.
7. 3 M Sodium acetate, pH 5.2.
8. Glycogen: 20 mg/mL (Gibco-BRL).
9. 20X SSC: Dissolve 87.66 g NaCl and 44.11 g Na$_3$ citrate in 500 mL distilled H$_2$O (dH$_2$O). Adjust pH to 7.0. Autoclave and store at room temperature.
10. For use with repetitive-sequence probes: Hybridization buffer 1 (HB1), containing 50% formamide. Dissolve 1 g dextran sulfate in 5 mL deionized formamide (*see* **Note 2**) by heating to 70°C. Add 1 mL 20X SSC and adjust pH to 7.0. Make up final volume to 10 mL with dH$_2$O. Store at $-20°C$.

Table 2
Suggested Stringency Washing Conditions

	Repetitive sequence	Single copy	Chromosome library
SSC concentration[a]	0.5X SSC	2X SSC	1X SSC
Temperature[b]	75°C	75°C	75°C
Time	5 min	5 min	2×5 min

[a]To prepare various SSC concentrations use 20X SSC (*see* **Subheading 2.2.**) and dilute as appropriate (e.g., to make 2X SSC, use 1 vol of 20X SSC to 9 vol H_2O).
[b]Incubate wash chamber containing SSC in a 75°C water bath for at least 30 min prior to washing. Check temperature inside the chamber before commencing washing.

11. For use with single-copy probes or whole-chromosome libraries: Hybridization buffer 2 (HB2), containing 71% formamide. Dissolve 1 g dextran sulfate in 5 mL deionized formamide (*see* **Note 2**) by heating to 70°C. Add 1 mL of 20X SSC and adjust pH to 7.0. Make up final volume to 7 mL with H_20. Store at –20°C.
12. Water bath, 15°C.
13. Microcentrifuge.

2.3. Hybridization

1. Labeled probe(s).
2. Methanol (100%, 95%, 75%).
3. For use with repetitive-sequence probes: HB1 (*see* **Subheading 2.2.**).
4. For use with single-copy probes and whole-chromosome libraries: HB2 (*see* **Subheading 2.2.**).
5. Denaturation solution: 70% deionized formamide/2X SSC. To 35 mL deionized formamide (*see* **Note 2**), add 5 mL of 20X SSC (*see* **Subheading 2.2.**) and 10 mL dH₂O, adjust pH to 7.0. *Note*: Denaturation solution is not required when using repetitive-sequence probes.
6. Diamond pen to mark target slides.
7. Sealable plastic hybridization chamber lined with moist tissue.
8. Hot plate, 50°C.
9. Water bath, 75°C.
10. Incubator, 37°C, 42°C.

2.4. Stringency Washing

1. SSC (*see* **Table 2** for concentration).
2. PN buffer: Titrate 4 L of 0.1 *M* sodium phosphate dibasic (Na_2HPO_4) solution with 0.1 *M* sodium phosphate monobasic (NaH_2PO_4) solution to give a pH of 8.0. Add 4.5 mL nonidet-P-40 (detergent). Mix well and store at room temperature for up to 2 wk.
3. Water bath, 75°C.

2.5. Detection

1. PNM buffer: To 250 mL PN buffer (*see* **Subheading 2.4.**), add 12.5 g nonfat dried milk (or other blocking agent [e.g., bovine serum albumin] and 0.05 g sodium azide; *see* **Note 3**), mix well, and incubate at 37°C for 60 min. Cool and store at 4°C for up to 2 mo.
2. Fluoroscein avidin (Vector Laboratories, Inc.) (2 mg/mL).
3. Biotinylated anti-avidin D, affinity purified (Roche Diagnostics) (0.5 mg/mL).
4. Mouse monoclonal anti-digoxin (Sigma Biosciences) (100 µg/mL).
5. Sheep anti-mouse Ig-digoxygenin, F(ab)$_2$-fragment (Roche Diagnostics) (200 µg/mL).
6. Sheep anti-digoxygenin–rhodamine, FAB fragments (Roche Diagnostics) (200 µg/mL).
7. PN buffer (*see* **Subheading 2.4.**).

2.6. Counterstaining

1. Propidium iodide (PI) (Vector Laboratories, Inc.) in an antifade solution (e.g., Vectashield, Vector Laboratories, Inc.) to give a final concentration of 1 µg/mL (*see* **Note 4**).
2. 6-Diamidine-2-phenylindole (DAPI) (Vector Laboratories, Inc.) in an antifade solution (e.g., Vectashield, Vector Laboratories, Inc.) to give a final concentration of 0.5 µg/mL.

3. Methods
3.1. Growing Probes in the Laboratory

Plasmid, cosmid, and P1 clone (PAC) DNA suitable for labeling as FISH probes can be isolated using a Qiagen Plasmid kit. In the following methodology, the isolation of PACs will be described (*see* **Note 5**), for plasmid and cosmid preparations, *see* the Qiagen Plasmid purification handbook.

1. Grow a single colony, picked from a freshly streaked LB (kanamycin, 25 µg/mL) plate, in a starter culture of 5 mL LB medium containing kanamycin (25 µg/mL) and incubate in a shaking incubator for approx 8 h at 37°C with vigorous shaking at 300 rpm.
2. Inoculate a 500-mL LB kanamycin (25 µg/mL) culture with 1 mL of the starter culture and grow as in **step 1** for 12–16 h.
3. Transfer the culture to clean centrifuge tubes and pellet the bacterial cells at 7500 rpm for 15 min.
4. Resuspend the bacterial pellet in 10 mL of P1 buffer containing RNase A (100 µg/mL) until there are no visible clumps.
5. Add 20 mL of P2 buffer and mix gently by inverting the tube 10 times and leave at room temperature for 5 min.
6. Add 20 mL of chilled P3 buffer, mix immediately by inverting four to five times and incubate on ice for 20 min.

7. Centrifuge at 9500 rpm for 30 min at 4°C.
8. In the meantime, prepare the Qiagen tip by adding 10 mL of equilibration buffer QBT and allowing to drip through by gravity flow.
9. If the supernatant is clear, pipet onto the Qiagen tip and allow to enter the resin by gravity flow. If the supernatant is not clear, centrifuge again before applying to the Qiagen tip.
10. Wash the Qiagen tip with 2 × 10 mL buffer QC.
11. To elute the DNA, apply 5 mL buffer QF prewarmed to 50°C (*see* **Note 6**).
12. Precipitate the DNA with 0.7 vol isopropanol at room temperature, centrifuge immediately at 4°C for 30 min and carefully remove the supernatant (*see* **Note 7**).
13. Wash the pellet by adding 2 mL cold 70% ethanol and inverting. Centrifuge at 11,000 rpm for 10 min.
14. Remove the supernatant, air-dry the pellet, and redissolve in sterile water.

3.2. Probe Labeling

Nick translation is the method of choice for labeling FISH probes. The technique involves the use of two enzymes; a DNase that randomly nicks probe DNA and a DNA polymerase that repairs the nicked DNA and, at the same time, incorporates labeled dNTP molecules into the probe. The following protocol uses a nick translation kit and describes a basic method suitable for indirect or direct labeling of FISH probes.

1. To label 1 µg of DNA, add the following to a 0.5-mL Eppendorf tube on ice:
   1 µg DNA;
   10 µL dNTP mix (1 vol labeled dUTP[0.4 mmol/L], 2 vol dTTP [0.1 m*M*], 3 vol each dATP, dCTP, dGTP [0.1 m*M*]);
   2 µL 10X buffer;
   2 µL enzyme mix (*see* **Note 8**);
   Sterile water to give final volume of 20 µL.
2. Mix well and incubate at 15°C for 90 min.
3. Add 2 µL EDTA (0.2 *M*) to stop enzyme activity.
4. For nick translation of single-copy probes or whole-chromosome libraries, add 25 µL Human COT-1 DNA (1 mg/mL) (*see* **Note 9**).
5. Add 10 µL Herring sperm DNA (5 µg/mL) (*see* **Note 10**)
6. To precipitate DNA, add the following:
   0.1 × total volume sodium acetate (3 *M*, pH 5.2);
   1 µL glycogen (20 mg/mL);
   2 × total volume 100% ethanol.
7. Mix well and incubate at −20°C for 30 min to aid DNA precipitation.
8. Spin at 13,000 rpm for 30 min to pellet the DNA.
9. Discard aqueous phase and air-dry DNA pellet at room temperature for 15 min or until supernatant is no longer visible.
10. Resuspend dried pellet in 20 µL HB1 (for centromeric probes) or HB2 (for single-copy probes and whole-chromosome libraries) to give a final concentration of 50 ng/µL.

11. Incubate at 37°C for 30 min to aid resuspension of DNA. Store probe at −20°C while not in use.

3.3. Hybridization

The methodology for hybridization depends on the nature of the probe. Three hybridization protocols are described for use with laboratory-grown repetitive-sequence, single-copy, and whole-chromosome library probes. The protocols are designed for use on fixed cell preparations.

Target slides should be checked prior to hybridization under a phase-contrast microscope. The cells should not be overlapping and there should be minimal visible cytoplasm. Chromosomes, if present, should be of an appropriate length and adequately spread. They should appear a dark gray color; light gray chromosomes or those with a "glassy" appearance are likely to result in poor hybridization. Optimum results are achieved on unbaked slides that are less than 1 mo old.

3.3.1. Hybridization Protocol 1:
For Use with Repetitive Sequence Probes

Repetitive-sequence probes can be applied with equal success to metaphase or interphase preparations. By using differentially labeled probes, it is possible to apply more than one probe to a single target slide. The following protocol involves codenaturation of target and probe DNA.

1. Prewarm the hybridization chamber in a 75°C water bath.
2. Mark the hybridization area on the underside of each target slide using a diamond pen.
3. Dehydrate the target slide(s) by incubating through an alcohol series; 2 min in each of 75%, 95%, and 100% methanol at room temperature; dry briefly on a 50°C hot plate.
4. For each slide, aliquot 0.5 μL of each labeled probe into an Eppendorf tube and make volume up to 10 μL with HB1 (e.g., if simultaneously applying two differentially labeled probes to two slides, use 1 μL of each probe and 18 μL of HB1); mix well.
5. Apply 10 μL of probe mix to the marked hybridization area of each dehydrated slide, cover the hybridization area with a 22 × 22-mm coverslip (*see* **Note 11**), and seal the edges with rubber sealant.
6. Place the slide (s) in the prewarmed hybridization chamber, seal the lid, and incubate at 75°C for exactly 10 min in order to codenature the probe and target DNA.
7. Incubate the hybridization chamber containing the slide(s) at 37°C overnight.

3.3.2. Hybridization Protocol 2: For Use with Single-Copy Probes

Single-copy probes can be applied to metaphase or interphase preparations. The use of differentially labeled probes allows the simultaneous application of more than one probe to a single target slide.

1. At least 30 min prior to hybridization, prewarm a lidded plastic Coplin jar containing denaturation solution in a 75°C water bath. The water bath should be sited in a Class 1 fume hood to contain formamide fumes.
2. Mark the hybridization area on the underside of each target slide using a diamond pen.
3. For each slide, aliquot 1 μL of each labeled single-copy probe into an Eppendorf tube and make up volume to 10 μL with HB2 (e.g., if simultaneously applying two differentially labeled probes to two slides, use 1 μL of each probe and 18 μL of HB2); mix well.
4. Denature by incubating the probe mix at 75°C for exactly 5 min, snap-chill on ice for 30 s and prehybridize by incubating at 37°C for 30 min (*see* **Note 12**).
5. Shortly before the end of prehybridization, check that the temperature of the denaturation solution is 75°C. Denature the target slide(s) by incubating in the denaturation solution for exactly 2 min (*see* **Note 13**). Dehydrate the target slide(s) by incubating through an ice-cold alcohol series; 2 min in each of 75%, 95%, and 100% methanol and dry briefly on a 50°C hot plate.
6. Apply 10 μL of the denatured, prehybridized probe mix to the marked hybridization area of each denatured slide. Cover the hybridization area with a 22 × 22-mm coverslip and seal the edges with rubber sealant (*see* **Note 11**). Place slide(s) in a hybridization chamber and hybridize overnight at 37°C.

3.3.3. Hybridization Protocol 3:
For Use with Whole-Chromosome Libraries

Whole-chromosome libraries are usually used for the characterization of metaphase chromosomes rather than interphases cells, where the signals appear as indistinct, often overlapping domains. Dual-color painting refers to the simultaneous application of two differentially labeled chromosome libraries to target chromosome spreads.

1. At least 30 min prior to hybridization, prewarm a lidded plastic Coplin jar containing denaturation solution in a 75°C water bath. The water bath should be sited in a Class 1 fume hood to contain formamide fumes.
2. Mark the hybridization area on the underside of each target slide using a diamond pen.
3. If using a single library, dilute 3 μL of labeled library with 7 μL HB2 (per slide); mix well. For dual-color painting, use 3 μL of each of two differentially labeled chromosome libraries and 4 μL HB2 (per slide). Denature the probe(s) at 75°C for exactly 5 min. Snap-chill on ice for 30 s and prehybridize at 37°C for at least 2 h (*see* **Note 12**).
4. Shortly before the end of prehybridization, check that the temperature of the denaturation solution is 75°C. Denature the target slide(s) in denaturation solution for exactly 2 min. Dehydrate the target slide(s) by incubating through an ice-cold alcohol series; 2 min in each of 75%, 95%, and 100% methanol and dry briefly on a 50°C hot plate.
5. Apply 10 μL denatured probe mix to the marked hybridization area of each denatured slide. Cover the hybridization area with a 22 × 22 mm coverslip (*see*

Note 11) and seal the edges with rubber sealant. Place the slide(s) in a hybridization chamber and hybridize overnight at 42°C.

3.4. Stringency Washing

Stringency washing removes unbound probe from the target slide. The amount of probe left on the slide depends on the stringency of the wash; high-stringency washes have a low salt concentration and/or high temperature, whereas low-stringency washes have a high salt concentration and/or low temperature.

1. After hybridization, gently remove coverslips and rubber sealant using fine forceps and wash the slide(s) in the appropriate SSC concentration at 75°C (*see* **Note 13**) for the required time (*see* **Table 2**).
2. Rinse briefly in PN buffer and then allow the slide to stand in PN buffer for at least 10 min. If using indirectly labeled probes, proceed to the detection stage (*see* **Subheading 3.5.**). If using directly labeled probes, proceed to the counterstaining stage (*see* **Subheading 3.6.**).

3.5. Detection

This stage is only necessary for slides hybridized with indirectly labeled probes; if directly labeled probes have been used, slides may be counterstained immediately (*see* **Subheading 3.6.**). Protocols are given for the detection of both biotin- and digoxygenin-labeled probes and involve "amplifying" detection reagents to give high-intensity signals. Biotin is detected with a green fluorescent label (fluoroscein) and digoxygenin with a red label (rhodamine).

1. Prepare detection reagents I, II, and III in light-protected plastic Coplin jars as shown in **Table 3**. The detection reagents can be reused over a 6 wk period and should be stored at 4°C while not in use.
2. After washing, incubate the hybridized slide(s) in detection reagent I at room temperature for 20 min. Rinse briefly in PN buffer and leave standing at room temperature in fresh PN buffer for 10 min. Do not allow the slides to dry out at any stage.
3. Repeat **step 2** using detection reagents II and III, consecutively. Leave slide(s) standing in PN buffer prior to the counterstaining stage.

3.6. Counterstaining

For optimum contrast, probes labeled or detected with a green label are best viewed on a red (PI) background, whereas red probes are best viewed on a blue (DAPI) background. If both red and green probes have been applied, a DAPI counterstain should be used.

1. Remove slide(s) from PN buffer (i.e., the final stage of either stringency washing or detection for directly or indirectly labeled probes, respectively) and drain off

Table 3
Preparation of Detection Reagents

	Detection of biotin only	Detection of digoxygenin only	Simultaneous detection of biotin and digoxygenin
Detection Reagent I	50 mL PNM buffer, 125 μL fluoroscein avidin[a] (2 mg/mL)	50 mL PNM buffer, 100 μL mouse monoclonal anti-digoxin (100 mg/mL)	50 mL PNM buffer, 125 μL fluoroscein avidin[a] (2 mg/mL), 100 μL mouse monoclonal anti-digoxin (100 mg/mL)
Detection Reagent II	50 mL PNM buffer, 500 μL biotinylated anti-avidin D (0.5 mg/mL)	50 mL PNM buffer, 500 μL sheep anti-mouse Ig-digoxygenin (200 mg/mL) 500 μL sheep anti-mouse Ig-digoxygenin (200 mg/mL)	50 mL PNM buffer, 500 μL biotinylated anti-avidin D (0.5 mg/mL),
Detection Reagent III	50 mL PNM buffer, 125 μL fluoroscein avidin[a] (2 mg/mL)	50 mL PNM buffer, 500 μL sheep anti-digoxygenin–rhodamine[a] (200 mg/mL)	50 mL PNM buffer, 125 μL fluoroscein avidin[a] (2 mg/mL), 500 μL sheep anti-digoxygenin–rhodamine[a] (200 mg/mL)

[a]These chemicals are light sensitive, handle in reduced-light conditions at all times.

excess buffer by touching the end of each slide to a paper towel. Do not allow the slide(s) to dry out completely.

2. Apply 10 µL of the appropriate counterstain to the marked hybridization area of each target slide.

3. Cover the hybridization area with a 22 × 32-mm coverslip (avoiding air bubbles) and store at 4°C in the dark until ready for microscopic evaluation.

3.7. Microscope Evaluation

A fluorescent microscope is required for analysis of FISH preparations, and good signal intensities are achieved using a 100-W mercury bulb. The microscope needs to be fitted with excitation and barrier filters appropriate for the fluorescent labels used. At this stage, slides are prone to fading if exposed to light and should be kept in reduced-light conditions.

During microscopic evaluation, probe signals should appear as distinct signals with minimal background fluorescence. Single-copy probes should be seen as small bright signals and there should be virtually no background nonspecific hybridization. Repetitive-sequence probes tend to show some degree of cross-hybridization across the genome, but specific signals should still be bright and easily recognizable. It should be noted that certain repetitive-sequence centromeric probes hybridize to more than one centromeric pair because of extensive sequence homology (e.g., the chromosome 1 centromeric probe hybridizes to the centromeric regions of chromosomes 1, 5, and 19 whereas centromeric probes specific for chromosomes 13 and 14 also hybridize to the centromeres of chromosomes 21 and 22, respectively). Whole-chromosome libraries should hybridize specifically along the length of a chromosome pair although it is not uncommon to see low levels of cross-hybridization across the rest of the genome. Occasionally, FISH preparations show no signal or excessive cross-hybridization; there is usually a simple reason for this and some troubleshooting tips are given in **Table 4**.

4. Notes

1. If using direct labels (i.e., fluoroscein-11-dUTP or Texas Red-11-dUTP), handle in reduced-light conditions at all times.

2. Formamide is a known teratogen and should be handled with extreme care.

3. Sodium azide is a known mutagen and should be handled with extreme caution.

4. Propidium iodide is a potential carcinogen and should be treated with care.

5. Yield for PAC DNA can be as little as 10 µg from a 500-mL culture.

6. After precipitation, if no DNA appears to have been eluted, apply a higher-pH buffer, pH 9.0, to the Qiagen tip and precipitate as before.

7. Isopropanol pellets have a glassy appearance and tend to be easily dislodged. If this occurs, use a standard ethanol precipitation in an Eppendorf and add the following:

Table 4
Troubleshooting FISH

Problem	Possible causes and solutions
No signal or dim signal at microscope evaluation	1. Probe concentration may have been too weak—increase the amount of probe in the probe mix. 2. The target slide was not sufficiently denatured—remake the denaturation solution using fresh reagents and ensure temperature is 75°C. If using the codenaturation method, ensure that the temperature is 75°C. 3. The wash was too stringent—increase the SSC concentration. 4. Incorrect filters were used for slide evaluation—check that filters and fluorescences are compatible. 5. Detection solutions were incorrectly prepared (indirectly labeled probes only)—remake detection solutions I, II, and III. 6. Probe labeling may have failed—repeat using fresh nick translation reagents.
High background to probe signal	1. Too much probe was used—decrease the amount of probe in the probe mixture. 2. The wash was not stringent enough—decrease the SSC concentration.
Chromosomes have poor morphology	The slide may have been overdenatured—remake the denaturation solution, ensure that the temperature is 75°C and that slide is denatured for 2 min only. If codenaturing, ensure that the temperature of hybridization chamber is 75°C. If the slides still have poor morphology, decrease the denaturation time or denaturation solution temperature.

      $0.1 \times$ total volume 3 M sodium acetacte, pH 5.2;
      $2 \times$ total volume cold 100% ethanol.

Mix well and incubate at $-20°C$ for 30 min and centrifuge for 30 min at 13,000 rpm. Remove the supernatant and wash the pellet with 200 μL cold 70% ethanol; recentrifuge for 10 min. Remove the supernatant and air-dry the pellet before resuspending in distilled sterile water.

8. The enzyme mixture is heat sensitive; keep at $-20°C$ until just before use and replace immediately after use.
9. Human COT-1 DNA blocks repetitive sequences within single-copy probes or whole chromosome libraries.
10. Herring sperm DNA acts as carrier DNA and aids the precipitation of probe DNA. It also blocks the nonspecific attachment of probe to the slide surface.

11. Take care to avoid trapping air bubbles, as the probe will not hybridize to target DNA under the bubbles.
12. This step allows COT-1 DNA included in the probe mixture to hybridize to repetitive sequences within the probe DNA, thus blocking these region from hybridizing to target DNA. This process is known as competitive *in situ* suppression (CISS) *(3)*.
13. A maximum of four slides should be processed at any one time to avoid a significant decrease in temperature.

References

1. Speicher, M. R., Ballard, S. G., and Ward, D. C. (1996) Karyotyping human chromosomes by combinatorial multi-fluor FISH. *Nature Genet.* **12,** 368–375.
2. Kallioniemi, A., Kallioniemi O-P., Sudar, D., Rutovitz, D. Gray, J. W., Waldman F., et al. (1992) Comparative genomic hybridization for molecular cytogenetic analysis of solid tumours. *Science* **258,** 818–821.
3. Lichter, P., Cremer, T., Borden, J., Manuelidis, L., and Ward, D. C. (1988) Delineation of individual human chromosomes in metaphase and interphase cells by in situ suppression hybridization using recombinant DNA libraries. *Hum. Genet.* **80,** 224–234.

Further Reading

Barch, M. J., Knutsen, T., and Spurbeck, J., eds. (1997) *The AGT Cytogenetics Laboratory Manual,*. Lippincott–Raven, PA.

Rooney, D. E. and Czepulkowski, B. H., eds. (1992) *Human Cytogenetics, Volume I*, IRL, Oxford.

12

The Protein Truncation Test

Carol A. Hardy

1. Introduction

The protein truncation test (PTT) *(1)*, occasionally referred to as the in vitro synthesized-protein (IVSP) assay *(2)*, is a method for screening the coding region of a gene for mutations that result in the premature termination of mRNA translation. The techniques involved in performing PTT are relatively straightforward and begin with the isolation of genomic DNA or RNA. The polymerase chain reaction (PCR) is used to amplify a DNA template, usually of 1–3 kb in size, that is tested in an in vitro transcription and translation assay. Truncated proteins are identified by sodium dodecyl sulfate polyacrylamide gel electrophoresis (SDS-PAGE) and autoradiography or fluorography. A single, large exon of a gene can be amplified directly from genomic DNA in several overlapping fragments. The complete coding sequence of a large gene, with many small exons, can be amplified in several overlapping fragments by reverse-transcription PCR (RT-PCR) starting from RNA. Amplifying the gene in several segments that overlap each other by 300–500 bp increases the sensitivity of PTT. A truncating mutation located toward the 3' end of one segment will also occur near the 5' end of the next overlapping segment, thus increasing the likelihood of identifying a truncated protein. A specially modified forward PCR primer is required for PTT. In addition to the in-frame gene-specific primer sequence, an extension at the 5' end includes the T7 promoter sequence for RNA transcription by T7 RNA polymerase, the eukaryotic consensus sequence for the initiation of protein translation, and the ATG protein translation start site. The yield and specificity of the PCR amplification is checked by agarose gel electrophoresis. Then, an aliquot of the PCR product is used as the template for in vitro transcription and translation, using RNA polymerase and rabbit reticulocyte lysate or wheat-germ extract. A radiolabeled amino acid is

From: *Methods in Molecular Biology, vol. 187: PCR Mutation Detection Protocols*
Edited by: B. D. M. Theophilus and R. Rapley © Humana Press Inc., Totowa, NJ

usually incorporated into the translated protein, to allow detection by autorad-iography, although nonradioactive PTT methods have also been described *(3,4)*. The translated proteins are separated by SDS-PAGE, and autoradiogra-phy is performed. The presence of a protein that is smaller than the full-length protein identifies a translation terminating mutation (*see* **Figs. 1** and **2**). The size of the truncated protein indicates the position of the premature stop codon, and DNA sequencing of genomic DNA is performed to confirm the presence of a mutation. The mutations identified are principally frameshift and nonsense mutations, but may also include mutations that disrupt the normal splicing of exons. There are two main advantages of PTT compared to most other mutation detection methods. Several kilobase segments of a gene can be rapidly screened in a single reaction, and PTT only identifies those mutations that have a clear pathological effect on protein function (i.e., those that result in a truncated protein and are likely to result in loss of function). Missense mutations and neutral polymorphisms are not identified. If it is important to identify missense mutations, another mutation detection method, such as single-strand conformation polymorphism (SSCP) (*see* Chapter 16), would need to be used in conjunction with PTT.

The protein truncation test was first described in 1993 as a technique for identifying translation terminating mutations in the dystrophin gene *(1)*, which cause the X-linked disorder Duchenne muscular dystrophy (DMD). The technique was subsequently applied to identification of truncating mutations in the *APC* gene *(2)* that cause familial adenomatous polyposis. Since then, PTT has been used successfully in the identification of translation terminating mutations in many other genes that result in human hereditary disorders, some examples are shown in **Table 1**. The dystrophin gene illustrates some of the problems encountered in screening large genes for mutations and shows how these problems are largely overcome by the use of PTT. The dystrophin gene is exceptionally large, with 79 small exons contained within a region of 2400 kb. The majority of dystrophin mutations in DMD patients are easily identifiable deletions of one or more exons *(1)*. However, approximately one-third of dystrophin mutations are nonsense mutations or small frameshift mutations that result in premature truncation of the dystrophin gene, less than 2% of DMD cases are the result of a pathogenic missense mutation *(1,5)*. Using conven-tional methods, such as SSCP, to screen for these mutations is exceedingly time-consuming and is usually less than 100% sensitive. Moreover, neutral polymorphisms and missense mutations of uncertain pathogenicity are identi-fied by these techniques. Direct sequencing of the gene is both time-consuming and expensive. PTT was developed to screen the dystrophin gene specifically for translation-terminating mutations, following amplification of large seg-ments of the gene by nested RT-PCR *(1)*. The whole of the dystrophin coding

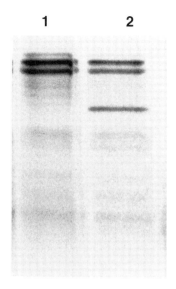

Fig. 1. PTT for exon 11 of the *BRCA1* gene, amplified from genomic DNA. From left to right: lane 1, a normal control with only the full-length translated protein; lane 2, a breast cancer patient with a 5 bp deletion of nucleotides 1623–1627, which results in a premature stop at codon 503. (PTT performed by Ms. Kim Hampson, DNA Laboratory, Regional Genetics Service, Birmingham Women's Hospital, Birmingham.)

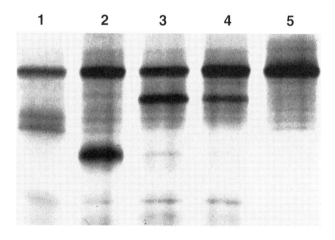

Fig. 2. PTT for the *PTCH* gene in patients affected with naevoid basal cell carcinoma syndrome (NBCCS or Gorlin syndrome). RNA was isolated from lymphoblastoid cell lines, and nested RT-PCR was used to amplify a 1.6-kb fragment of the gene, which includes exons 10–16. Truncated proteins were identified in NBCCS patients with the following mutations (from left to right): lane 1, 2439insC; lane 2, 2101del19bp; lanes 3 and 4, siblings with 2683insC mutation; lane 5, normal control.

Table 1
Some Examples of Human Hereditary Disorders Frequently Caused by Translating Terminating Mutations that Can Be Identified by PTT

Disorder	Gene	Forward primer 5' extension	Ref.
Duchenne muscular dystrophy	DMD	ggatccTAATACGACTCACTATAGGaacagaCCACCATG	1
Duchenne muscular dystrophy	DMD	ggatccTAATACGACTCACTATAGGaacagaCCACCATG	5
FAP	APC	ggatccTAATACGACTCACTATAGGaacagaCCACCATG	7,8
Breast/ovarian cancer	BRCA1	gcTAATACGACTCACTATAGGaacagaCCACCATGG	9
Breast/ovarian cancer	BRCA1	ggatccTAATACGACTCACTATAGGgagaCCACCATGG	10
Breast/ovarian cancer	BRCA1/BRCA2	TAATACGACTCACTATAGGgagaCCACCATG	11
Breast cancer	BRCA1/BRCA2	ggatccTAATACGACTCACTATAGGacagaCCACCATG	13
HNPCC	hMLH1	ggatccTAATACGACTCACTATAGGgagaCCACCATGG	14
HNPCC	hMSH2	ggatccTAATACGACTCACTATAGGgagaCCACCATGG	15
Cystic fibrosis	CFTR	ggatccTAATACGACTCACTATAGGaacagaCCACCATG	16
Neurofibromatosis 1	NF1	ggatccTAATACGACTCACTATAGGgagaCCACCATG	17
Tuberous sclerosis	TSC2	ggatccTAATACGACTCACTATAGGaacagaCCACCATG	18
Polycystic kidney disease	PKD1	ggatccTAATACGACTCACTATAGGaacacaCCACCATG	19
Gastric cancer	E-cadherin	aagcttATTAACCCTCACTAAAGGGA[a]gagaCCACCATGG	4

Note: The forward primer 5' extensions used in each case are shown. Restriction sites and spacer sequences are shown in lowercase letters; the T7 promoter ([a] is the T3 promoter) is shown in uppercase letters and underlined, the eukaryotic translation initiation site is shown in uppercase letters.

sequence can by amplified and screened by PTT in just 10 overlapping fragments and has been successfully used to identify translation-terminating mutations in affected males and carrier females *(5)*. More recently, multiplex PTT
for the dystrophin gene has been described *(6)*. The complete coding sequence
can be analyzed in just five overlapping RT-PCR products, which are simultaneously tested for truncating mutations in a single in vitro transcription–translation reaction, translated proteins from one reaction are separated in a single
lane of an SDS-PAGE gel *(6)*.

The advantages PTT has over conventional mutation screening methods
make it an ideal choice for identifying translation-terminating mutations that
occur in several important tumor suppressor genes. Familial adenomatous polyposis (FAP) is caused by germline mutations of the *APC* gene, and somatic
mutation of the *APC* gene occurs frequently in sporadic colorectal carcinomas
(CRC). In both FAP and CRC, approx 95% of the mutations identified in the
APC gene result in premature termination of protein translation *(7)*. The *APC*
gene is large with 15 exons and a coding sequence of 8.5 kb. Exon 15 of the
APC gene is 6.5 kb in length and approx 70% of the mutations found in FAP
patients occur in the 5' end of exon 15. Approximately 65% of the somatic
mutations found in sporadic CRC are located within a small region of exon 15
known as the mutation cluster region. By direct amplification of a 2-kb segment of exon 15 of the *APC* gene from genomic DNA and a single PTT, it is
theoretically possible to identify 50% of germline mutations in FAP patients
and 75% of somatic mutations in sporadic CRC *(7)*. PTT analysis of the complete coding region of the *APC* gene can be accomplished by amplification of
the whole gene in five or six overlapping segments *(2,8)*. Exon 15 is amplified
directly from genomic DNA in four overlapping segments *(2,8)*. Exons 1–14
are amplified in one *(2)* or two overlapping segments by RT-PCR amplification of RNA *(8)*. Using this approach, translation-terminating mutation were
identified in 82% of FAP patients tested *(2)*.

A similar approach has been described for the identification of translation-
terminating mutations that occur in the *BRCA1* and *BRCA2* genes, which cause
the majority of cases of hereditary breast and ovarian cancer where a single
gene is involved *(9–11)*. *BRCA1* and *BRCA2* are similar in structure, both
are large genes with many exons, and in each case, the majority of mutations
result in the premature termination of translation. The *BRCA1* gene encodes a
7.5-kb transcript, spread over 100-kb of DNA. It has 22 coding exons, and
exon 11 alone contains 60% of the coding sequence *(10,12)*. Approximately
86% of the mutations identified in *BRCA1* are truncating mutations and approx
50% of these are located within exon 11 *(10)*. Similarly, *BRCA2* has 26 coding
exons and encodes a 10.5-kb transcript; exon 11 is exceptionally large and
contains 50% of the entire coding sequence *(12)*. Virtually all of the mutations

so far identified in *BRCA2* are protein truncating *(11)*. Thus, like the *APC* gene, translation-terminating mutations in *BRCA1* and *BRCA2* are most easily identified by PTT. In both cases, exon 11 is amplified directly from genomic DNA in one *(11)* or three *(9,10)* overlapping segments for *BRCA1* and in two to five overlapping segments for *BRCA2* *(11–13)*. Exon 10 of the *BRCA2* gene can also be amplified directly from genomic DNA as a single 1.2-kb segment *(13)*. PTT for the other, smaller exons is performed following nested RT-PCR *(9,11,12)*.

In summary, PTT is a simple, rapid, and cost-effective method of screening large genes, with a high frequency of mutations that result in premature termination of protein translation. The following sections describe how to perform mutation analysis by PTT, and consideration is given to some of the problems most often encountered when using this technique.

2. Materials

2.1. PCR Amplification of Template for In Vitro Transcription and Translation

2.1.1. Reverse Transcription of RNA

1. Total cellular RNA prepared from the appropriate tissue (*see* **Note 1**).
2. Oligo-d(N)$_6$ primer, 0.5 μg/μL (Promega Ltd.). Store at –20°C (*see* **Note 2**).
3. Diethyl pyrocarbonate (DEPC) treated distilled H$_2$O. Add 100 μL of DEPC to 100 mL of distilled H$_2$O contained in a glass bottle. Mix vigorously for 10 min and then leave at room temperature overnight. Inactivate the DEPC by autoclaving for 30 min at 120°C. Store at 4°C. DEPC should be handled with care, as it is a powerful acylating agent, always handle in a fume hood and never add to Tris buffers or solutions containing ammonia.
4. 5X Reverse transcription buffer (Invitrogen Life Technologies Ltd.). Store at –20°C.
5. 0.1 *M* Dithiothreitol (DTT) (Invitrogen Life Technologies Ltd.). Store at –20°C.
6. 10 m*M* dNTP mix in sterile DEPC-treated H$_2$O, prepared from 100 m*M* stocks of dATP, dCTP, dGTP; and dTTP (Amersham Pharmacia Biotech Ltd.). Store at –20°C.
7. RNasin ribonuclease inhibitor (Promega Ltd.). Store at –20°C.
8. M-MLV reverse transcriptase, 200 U/μL (Invitrogen Life Technologies Ltd.). Store at –20°C.

2.1.2. Nested RT-PCR

1. 10X *Taq* Extender™ reaction buffer (*see* **Note 3**), 200 m*M* Tris-HCl (pH 8.8), 100 m*M* KCl, 100 m*M* (NH$_4$)$_2$SO$_4$, 20 m*M* MgSO$_4$, 1% Triton X-100, 1 mg/mL nuclease-free bovine serum albumin (Stratagene Ltd., Cambridge, UK). Store at –20°C.
2. 2 m*M* dNTP mix in sterile H$_2$O, prepared from 100 m*M* stocks of dATP, dCTP, dGTP, and dTTP (Amersham Pharmacia Biotech Ltd). Store at –20°C.

3. Nested PCR primers pairs. The forward primer of the inner pair must have the 5' extension, which includes the T7 promoter sequence, the eukaryotic protein translation initiation sequence, followed by the in-frame ATG start codon and the gene-specific primer sequence (*see* **Table 1** and **Note 4**). Some examples of 5' extension sequences are shown in **Table 1**. Primers stock solutions are 200 pmol/μL in TE buffer (10 mM Tris-HCL, 1 mM Na$_2$EDTA, pH 8.0), store at −20°C. Working solutions are 20 pmol/μL in sterile dH$_2$O. Store at −20°C.
4. Sterile dH$_2$O.
5. *Taq* Extender™ PCR additive, 5 U/μL (*see* **Note 3**) (Stratagene Ltd., Cambridge, UK). Store at −20°C.
6. *Taq* DNA polymerase (*see* **Note 5**), 5 U/μL (Invitrogen Life Technologies Ltd.). Store at −20°C.
7. 50 mM MgCl$_2$ (Invitrogen Life Technologies Ltd.). Store at −20°C.
8. Mineral oil (Sigma Chemical Company). Store at room temperature.

2.1.3. PCR Amplification of Genomic DNA

1. Genomic DNA. (*see* **Note 6**).
2. 10X PCR buffer minus magnesium (*see* **Note 5**), 200 mM Tris-HCl (pH 8.4), 500 mM KCl (Invitrogen Life Technologies Ltd.). Store at −20°C.
3. T7 modified forward primer and reverse primer pairs. Stock solutions are 200 pmol/μL in TE buffer; store at −20°C. Working solutions are 20 pmol/μL, in sterile dH$_2$O. Store at −20°C.
4. 2 mM dNTP mix (*see* **Subheading 2.1.2.**).
5. 50 mM MgCl$_2$ (Invitrogen Life Technologies Ltd.). Store at −20°C.
6. *Taq* DNA polymerase, 5 U/μL (Invitrogen Life Technologies Ltd.). Store at −20°C.
7. Mineral oil (Sigma Chemical Company). Store at room temperature.

2.2. Agarose Gel Electrophoresis

1. 1X TBE buffer: 89 mM Tris-borate, 89 mM boric acid, 2 mM EDTA. Prepare a 10X stock solution by dissolving 108 g Trizma base, 55 g boric acid, and 9.3 g Na$_2$EDTA dissolved in a final volume of 1000 mL distilled H$_2$O without adjustment of pH. Store at room temperature.
2. Ethidium bromide solution, 5 mg/mL in dH$_2$O. Store at 4°C in a dark glass bottle. Great care should be taken when handling solutions and gels containing ethidium bromide, as it is a powerful mutagen. Gloves should be worn at all times and a mask should be worn when weighing out the solid. Solutions of ethidium bromide should be disposed of in compliance with the local safety regulations.
3. Agarose gel, 1% (w/v) agarose in 1X TBE buffer, with 0.5 μg/mL ethidium bromide.
4. 1-kb Ladder (Invitrogen Life Technologies Ltd.), supplied at a concentration of 1.0 μg/μL. Store at −20°C. Prepare a working solution of 100 ng/μL in 1X gel loading buffer Store at −20°C. Load approx 100 ng of ladder per millimeter lane width.

5. 6X DNA loading buffer, 0.25% bromophenol blue, 0.25% xylene cyanol, 30% glycerol in dH_2O. Prepare 2X DNA loading buffer by dilution in dH_2O. Store at –20°C.

2.3. In Vitro Transcription and Translation

1. TnT® T7 Quick Coupled Transcription/Translation system (Promega Ltd.) (*see* **Note 7**). Stable for 1 yr when stored at –70°C. Avoid unnecessary thawing and refreezing, as this will severely reduce the activity of the kit. The kit contains 200-μL aliquots of the TnT® T7 Quick master mix, which should be thawed and refrozen no more than twice. After thawing and using part of a 200-μL aliquot, divide the remaining master mix into smaller aliquots. For example, aliquots of 50-μL of master mix is sufficient for five 12.5 μL reactions.
2. T7 luciferase control DNA, 0.5 mg/mL. Supplied withTnT® T7 Quick Coupled Transcription/Translation system (Promega Ltd.). Store at –20°C.
3. Nuclease-free H_2O: Supplied with TnT® T7 Quick Coupled Transcription/Translation system (Promega Ltd.). Store at –20°C.
4. Redivue™ L- [^{35}S]methionine (*see* **Notes 7** and **8**), specific activity 37 TBq/mmol (1000 Ci/mmol) (product code AG1594, Amersham Life Science). Store at –20°C. Observe local regulations for handling and disposal of radioactive isotopes.

2.4. SDS-PAGE and Autoradiography

1. 0.5 *M* Tris-HCl buffer (pH 6.8), autoclave and store at room temperature.
2. 10% (w/v) Sodium dodecyl sulfate (SDS). Store at room temperature. Wear a mask and gloves when weighing out the solid SDS; if possible, use a fume cupboard. SDS is extremely irritating to the respiratory system by inhalation, and by direct contact with eyes and skin.
3. SDS-PAGE sample loading buffer. Mix 2 mL glycerol, 2 mL 10% SDS (w/v), 0.25 mg bromophenol blue, and 2.5 mL of 0.5 *M* Tris buffer (pH 6.8); add dH_2O to a final volume of 9.5 mL. Just before use, add 0.5 mL β-Mercaptoethanol. Stored at room temperature, the loading buffer is stable for about 1 wk, and then discard and prepare a fresh batch. β-Mercaptoethanol is harmful by inhalation and contact with skin, as well as having a very unpleasant smell. Therefore, only open the stock of β-Mercaptoethanol in a fume cupboard and always wear disposable latex gloves. If possible, only handle the SDS-PAGE sample loading buffer in a fume cupboard or use only in a well-ventilated laboratory.
4. 1.5 *M* Tris buffer (pH 8.8); autoclave and store at room temperature.
5. 30% Acrylamide solution (29:1 ratio of acrylamide to bis-acrylamide), store at 4°C. Both acrylamide and bis-acrylamide are neurotoxins that are readily absorbed through the skin and by inhalation of the dust. A mask, safety spectacles, and gloves must be worn when weighing them. Therefore, it is more convenient to purchase preprepared 30% acrylamide solution, which is available from a number of suppliers of chemicals and molecular-biology reagents. The solution is still extremely harmful, and great care should be taken when using it. Always wear protective clothing, disposable latex gloves, and safety spectacles.

6. 10% (w/v) Ammonium persulfate solution in dH_2O. Store at 4°C and discard after 1 wk.
7. N,N,N',N'-tetramethylethylenediamine (TEMED). Store at 4°C.
8. Polyacrylamide gel apparatus (e.g., the Mini-Protean II electrophoresis cell [*see* **Note 9**] [Bio-Rad Laboratories Ltd.]).
9. 1X Tris-glycine gel running buffer: 25 mM Tris, 250 mM glycine, 0.1% SDS. Prepare a 5X stock solution by dissolving 15.1 g Tris base and 94 g glycine dissolved in 900 mL of dH_2O, add 50 mL 10% SDS solution, adjust final volume to 1000 mL with dH_2O. Store at room temperature.
10. Rainbow™ colored protein molecular-weight markers (*see* **Note 10**), low-molecular-weight range 2.35–46 kDa or high-molecular-weight range 14.3–220 kDa (Amersham Life Sciences). Stable for at least 3 mo stored at –20°C.
11. Fix solution, 10% (v/v) acetic acid and 20% (v/v) methanol in dH_2O, freshly prepared before use.
12. 3MM Chromatography paper (Whatman).
13. Vacuum gel drier.
14. X-ray film (e.g., Hyperfilm MP [Amersham Life Science], Kodak T-mat or Kodak X-OMAT AR [Kodak]) and cassettes.

3. Methods

3.1. PCR Amplification of Template for In Vitro Transcription and Translation

3.1.1. Reverse Transcription of RNA (9)

Particular care must be taken when preparing and working with RNA to prevent contamination with RNases. The workbench should be clean. Clean disposable latex gloves should be worn at all times. New, autoclaved disposable plastic microtubes and pipet tips should be used. Some laboratories prefer to have a set of automatic pipets kept exclusively for RNA work only in order to minimize contamination problems.

1. Place approx 1–3 μg of total cellular RNA and 500 ng of oligo-d(N)$_6$ primer (*see* **Note 2**) in a 0.5-mL microtube. Make the final volume to 32 μL with DEPC-treated H_2O. Heat to 65°C for 10 min and then place on ice.
2. Add 12 μL of 5X reverse transcriptase buffer, 6 μL of 0.1 M DTT, 6 μL of 10 mM dNTP mix, 1 μL RNasin (30–40 U/μL), and 3 μL M-MLV reverse transcriptase (200 U/μL). Incubate at 42°C for 1 h. Inactivate the reverse transcriptase by heating to 95°C for 5 min.
3. Perform RT-PCR immediately or store at –20°C until required. Reverse-transcribed RNA should be stable for several months when stored at –20°C.

3.1.2. Nested RT-PCR

The gene under investigation may not be highly expressed in tissues that are easily sampled, such as blood lymphocytes, cultured skin fibroblasts, or lymphoblastoid cell lines. In some cases, the mRNA may be present only as illegitimate transcripts. This problem can usually be overcome by performing nested RT-PCR. In the first round of PCR, gene-specific primers are used to amplify a segment of the gene using the reverse-transcribed RNA as the template. A second round of PCR is then performed using an aliquot of the first PCR as template. The primers for the second round of PCR are nested within the first primer pair; the forward primer has the 5' modification required for PTT. If low-level transcription of the gene is not a problem, then one round of PCR should be adequate, using the 5' modified T7 primer and a suitable reverse primer. Obtaining good, reproducible results for PTT depends to a large extent on obtaining a single PCR product of high yield. Therefore, it is important to ensure that the PCR conditions are optimized (*see* **Note 11**).

1. Place 1–5 µL of the reverse-transcribed RNA in a 0.5 mL microtube. Store the remainder of the reverse-transcribed RNA at –20°C.
2. Add 2.5 µL of 10X *Taq* extender reaction buffer, 2.5 µL of 2 m*M* dNTPs, 20 pmol forward primer, 20 pmol reverse primer, and sterile H_2O to bring the total volume to 24 µL. When setting up a number of reactions, it is more convenient to prepare a master mix containing the appropriate volumes of the PCR components. The master mix is added to the tubes containing the reverse-transcribed RNA, thus reducing the number of pipetting steps. Mix well and overlay the reaction with 1 or 2 drops of mineral oil to prevent evaporation.
3. Perform "hot-start" PCR (*see* **Note 12**). Place the tubes in the thermal cycler and heat to 94°C for 4 min. Pause the thermal cycler at 94°C and add 1 U *Taq* DNA polymerase and 1 U *Taq* extender additive. Add the two enzymes simultaneously as a mixture diluted to 1 U/µL in 1X reaction buffer.
4. Resume thermal cycling and perform 30–35 rounds of amplification consisting of denaturing at 94°C for 1 min, annealing at 50–65°C for 1 min, and extension at 72°C for 2 min (*see* **Note 13**). End the program with a final extension step of 72°C for 10 min.
5. When the thermal cycling program has ended, check 2–5 µL from each sample by agarose gel electrophoresis (*see* **Subheading 3.2.**). It may be possible to see the first-round amplification product. If a PCR product is visible, check if it is of the expected size by comparison with the DNA size ladder.
6. Transfer 1–5 µL of the first-round PCR into a fresh 0.5 mL microtube. Store the remainder of the first-round PCR product at –20°C. Add 2.5 µL 10X *Taq* extender reaction buffer, 2.5 µL of 2 m*M* dNTPs, 20 pmol of 5' modified forward primer, 20 pmol reverse primer, and sterile H_2O to bring the total volume to 24 µL. Mix well and overlay the reaction with 1 or 2 drops of mineral oil.
7. Perform "hot-start" PCR. Add 1 U *Taq* DNA polymerase and 1 U *Taq* extender additive. Perform 30–35 rounds of amplification consisting of denaturing at 94°C for 1 min, annealing at 50–65°C for 1 min, extension at 72°C for 2 min (*see* **Note 13**). End the program with a final extension step of 72°C for 10 min.

8. Check the specificity and yield of the amplification by agarose gel electrophoresis (*see* **Subheading 3.2.**).

3.1.3. PCR Amplification of Genomic DNA

1. Place 100–500 ng of genomic DNA in a 0.5-mL microtube. Add 2.5 µL of 10X PCR buffer, 2.5 µL of 2 m*M* dNTPs, 0.75 µL of 50 m*M* MgCl$_2$ (*see* **Note 11**), 20 pmol 5' modified forward primer, 20 pmol reverse primer, and sterile H$_2$O to bring the total volume to 24 µL. Mix well and overlay the reaction with 1 or 2 drops of mineral oil.
2. Perform "hot-start" PCR. Add 1 U *Taq* DNA polymerase. Perform 30–35 rounds of amplification consisting of denaturing at 94°C for 1 min, annealing at 50–65°C for 1 min, extension at 72°C for 2 min (*see* **Note 13**). End the program with a final extension step of 72°C for 10 min.
3. Check the specificity and yield of the amplification by agarose gel electrophoresis.

3.2. Agarose Gel Electrophoresis of PCR Products

1. Mix 2–5 µL of each PCR sample with an equal volume of 2X DNA loading buffer. Load the samples onto the 1% agarose gel and load the appropriate volume of the 1-kb ladder into one lane. Separate by electrophoresis at 75 Volts for 1–2 h.
2. View the gel on an ultraviolet (UV) transilluminator. A single product of the expected size should be seen. Estimate the yield of the PCR product. This can most easily done by comparison with a similar-sized DNA band of known concentration. For example, the 1.6-kb fragment of the 1-kb ladder contains 10% of the mass loaded on the gel. It is useful to take a Polaroid photograph of the agarose gel as a record of the PCR product concentration and the specificity of the amplification.
3. A single PCR product of high yield can be used directly, without purification, in the in vitro transcription–translation reaction. The mineral oil can be removed from the PCR sample if preferred, although this is not essential when using the TnT® T7Quick kit. This can be easily done by addition of 1 or 2 drops of chloroform, vortexing, and spinning for 2 min in a microcentrifuge. Retain the upper aqueous layer and transfer to a fresh tube. Proceed directly with the in vitro transcription–translation or store the PCR samples at –20°C until required.

3.3. In Vitro Transcription and Translation

When performing the in vitro transcription and translation reaction, it is important to maintain an Rnase-free environment. Use new, sterile plastic microtubes and pipet tips; always wear clean, disposable latex gloves.

1. Remove a tube of TnT® T7 Quick master mix from the -70°C freezer. Thaw rapidly by hand warming; then place on ice. The TnT® T7 Quick master mix should always be kept on ice while the reactions are being set up and should be returned to the –70°C freezer afterward to minimize the loss of activity.

2. Place approx 250–500 ng of PCR product in a 0.5-mL microtube, if necessary, make the volume up to 2 μL with nuclease-free H_2O. Add 0.5 μL [^{35}S]methionine and 10 μL TnT® T7 Quick master mix to give a final volume of 12.5 μL (*see* **Note 14**). Mix well by gently pipetting; avoid mixing too vigorously and causing the mixture to foam, as this will reduce the activity of the lysate. When testing a number of PCR templates, it is more convenient to prepare a master mix containing the appropriate volumes of [^{35}S]methionine and TnT® T7 Quick kit. The number of pipetting steps is reduced by dispensing the master mix directly to tubes containing the PCR products. To monitor the efficiency of the in vitro transcription–translation, it is useful to include a reaction containing 250 ng of the T7 Luciferase control DNA that is supplied with the TnT® T7 Quick kit. The control DNA translation product is an intensely labeled 61-kDa protein. This provides a useful way of monitoring the activity of the kit, particularly after prolonged periods of storage, and provides an additional size marker for SDS-PAGE. Include a control reaction without any PCR product to monitor the background incorporation of labeled amino acid. If possible, include a positive control (i.e., a PCR product with a characterized translation-terminating mutation).

3. Incubate at 30°C for 90 min (*see* **Note 15**). Store at –20°C until required or proceed with the SDS-PAGE immediately.

3.4. SDS-PAGE and Autoradiography

1. When the in vitro transcription–translation reaction is complete, transfer 5 μL to a fresh tube containing 45 μL SDS-PAGE sample loading buffer. Heat at 100°C for 2 min to denature the protein; then centrifuge briefly. Store the remainder of the reaction at –20°C.

2. Ensure that the gel electrophoresis apparatus is clean and dry. Assemble the glass plates and spacers in between the side clamps. Ensure that the spacers and the glass plates are flush at the bottom end. Position the clamped plates into the gel pouring stand.

3. Prepare two 12% resolving gels (*see* **Note 16**), mixing 3.3 mL H_2O, 4.0 mL of 30% acrylamide solution, 2.5 mL of 1.5 M Tris buffer (pH 8.8), and 100 μL of 10% SDS in a 15-mL disposable plastic tube. Add 4 μL TEMED and 100 μL of 10% ammonium persulfate. Mix by inversion and pour approx 3 mL into each of the gel molds. Tap the glass plate to dislodge any air bubbles as they form. To ensure that the top of the gel sets with a straight edge, overlay the gel with 0.1% SDS solution. Leave the gel to polymerise for about 1 h.

4. When the gel is fully polymerized, pour off the 0.1% SDS solution and rinse the top of the gel with distilled H_2O. Drain away as much water as possible and soak up the remainder using a piece of Whatman 3MM paper.

5. Prepare two 5% stacking gels by mixing 2.77 mL H_2O, 0.83 mL of 30% acrylamide solution, 1.26 mL of 0.5 M Tris buffer (pH 6.8), and 50 μL of 10% SDS in a 15-mL disposable plastic tube. Add 5 μL TEMED and 50 μL of 10% ammonium persulfate. Mix by inversion and pour on top of the resolving gel, fill the gel mold

up to the level of the small glass plate. Insert the comb so that there is a depth of 1 cm between the bottom of each well and the top of the resolving gel. Leave the gel to polymerize for at least 30 min.

6. When the stacking gel is polymerized, carefully remove the comb. Remove the clamped glass plates from the pouring stand and clip in to the central cooling core. Place the central core within the electrophoresis chamber. Place approx 100 mL of 1X Tris-glycine running buffer in the upper buffer chamber and approx 200 mL in the lower buffer chamber. Flush out the wells with 1X running buffer.

7. Gently load 10–15 µL of each sample into the bottom of each well using a fine gel loading tip. Load only 5 µL of the Luciferase control reaction. Mix 3.5 µL of the Rainbow colored protein molecular-weight marker with an equal volume of SDS-PAGE sample loading buffer. Load the marker directly into one of the lanes of the gel, there is no need to heat denature the size marker.

8. Perform electrophoresis at a constant voltage of 200 V for a period of 45–60 min, or until the bromophenol dye has reached the bottom of the gel.

9. When electrophoresis is complete, dismantle the apparatus and carefully separate the glass plates containing the gel. Remove the stacking gel (the gel will be radioactive and should be disposed of in the appropriate way) and cut off a small piece of one corner, for orientation purposes.

10. Place the gel in a plastic box containing approx 200–300 mL of fix solution. Leave for 20–30 min with gentle shaking.

11. Pour off the fix solution, the solution will be radioactive and should be disposed of appropriately. Place the gel on a piece of 3MM paper that is just slightly larger than the gel. Cover the gel and 3MM paper completely with Saran Wrap and dry on a vacuum gel drier at 60°C for approx 1 h. Do not release the vacuum until the gel is completely dried; premature release of the vacuum will cause the gel to crack. The 1-mm thickness of the gel should be reduced to a negligible thickness when the gel is completely dry. Remove the gel from the drier and discard the Saran Wrap. Tape the gel inside an X-ray film cassette; and perform autoradiography overnight at room temperature (*see* **Note 17**).

12. Develop the X-ray film and align the film over the gel. Mark the position of the Rainbow colored protein molecular-weight markers on to the film with a permanent marker pen. The size of any truncated proteins can be estimated from a graph of log10 of the mass of the size standards plotted against electrophoretic mobility. A shorter or longer exposure time may be required, depending on the level of [^{35}S]methionine incorporation.

3.5. Interpretation

1. The presence of a translation-terminating mutation within the PCR-amplified fragment is indicated by the presence of a protein of lower molecular weight than the full-length protein (*see* **Figs. 1** and **2**). The size of the truncated protein, estimated by comparison with protein molecular-weight markers or with the truncated protein of a characterized mutation, indicates the position of the premature

stop codon. The position of a nonsense mutation can be determined quite accurately; whereas a frameshift mutation may be located some distance upstream from the site of the premature stop codon. Another mutation detection method, such as SSCP or heteroduplex analysis, can be used to locate the mutation more precisely and confirm the presence of the mutation in genomic DNA. Sequencing of genomic DNA should be used to definitively characterize the mutation.

2. Other labeled polypeptides, in addition to the full-length protein and any truncated protein, are often observed on the autoradiograph (*see* **Figs. 1** and **2**). These nonspecific proteins are usually weakly labeled and easily distinguishable from true truncated proteins. These extra bands often arise from protein translation initiated from internal methionine codons and do not usually affect the overall result. Occasionally, these internally initiated proteins may be quite intensely labeled and may be wrongly identified as truncated proteins. The addition of magnesium chloride to a final concentration of 1.5 mM in the in vitro transcription–translation reaction is reported to reduce nonspecific bands that result from initiation at internal methionine codons *(6)*. Translation of mRNA transcripts from the reticulocyte lysate is another cause of nonspecific protein bands appearing in all of the lanes of the gel, including the "no DNA" control. In the case of the PTT for the *BRCA1* gene, the presence of nonspecific translated proteins has been shown to be both sequence dependent and to vary with the type of in vitro translation system being used *(10)*. One of the PCR fragments tested was found to be more efficiently translated, with fewer nonspecific protein bands, when wheat-germ extract was used compared with a rabbit reticulocyte-based translation system *(10)*. The inclusion of a *myc* reporter tag (*see* **Note 8**) in the translated protein provides a means of eliminating these nonspecific protein bands *(3)*. Immunopreciptitaion can be used to isolate radiolabeled proteins containing the *myc* reporter tag. Hence, only proteins initiated from the methionine start site are identified *(3)*. Alternatively, nonradioactive PTT can be performed (*see* **Note 8**), with the *myc* reporter tag providing a means of identifying the translated proteins using Western blot methodology *(3)*.

3. An additional RT-PCR product, observed by agarose gel electrophoresis, suggests the presence of a splice site mutation. However, the presence of smaller, nonspecific PCR products in samples from both affected individuals and normal controls is frequently observed when performing nested RT-PCR from low-level mRNA transcripts and may lead to a false-positive result by PTT *(11,20,21)*. The presence of additional RT-PCR products has been correlated with the quality of the RNA preparation and is often simply a nonreproducible PCR artifact that can sometimes be eliminated by adjustment of the conditions for cDNA preparation and optimization of the PCR conditions *(11,20)*. The presence of alternative mRNA transcripts can also lead to a false-positive result and has been described in PTT for the *hMLH1* gene in hereditary nonpolyposis colorectal cancer (HNPCC) *(14)*. Alternative transcripts resulted in the identification of truncated proteins in many samples from both affected individuals and normal controls. The alternative transcripts were found to be more common in RNA samples

obtained from lymphocyte blood samples compared with RNA samples from lymphoblastoid cell lines. These alternative transcripts were often found to have a deletion of one or more exons. However, sequencing of genomic DNA failed to identify any splice site defects involving these exons, indicating that the unusual transcripts were unrelated to the HNPCC *(14)*. Thus, PCR artifacts and alternative transcripts can make the identification of translation-terminating mutations and splice site mutations complicated. However, translation-terminating mutations and splice site mutations should give a consistently reproducible result and can be confirmed by sequencing of genomic DNA.

4. For samples that show only the full-length translation product and no truncated protein, a number of possibilities must be considered. The most obvious explanation is that no translation-terminating mutation is present. However, a number of other explanations must be considered, as there are certain situations when PTT will fail to identify a truncating mutation. Truncating mutations that occur at the extreme 5' and 3' ends of the gene are not easily identified by PTT *(22)*. A truncating mutation close to the 5' end may result in a truncated protein that is too small to be seen by SDS-PAGE. A mutation close to the 3' end of the gene may result in a truncated protein that is not significantly different in size from the full-length protein, which cannot be resolved by SDS-PAGE *(22)*. As it is not possible to design overlapping PCR segments for these regions, another mutation detection method, such as SSCP or direct sequencing, should be used to screen the extreme 5' and 3' ends of the segment in question *(22)*. Mutations that alter primer binding sites may not be detected because of failure to amplify the mutated allele; although this should not present a problem when the gene is amplified in several overlapping fragments. Deletion, insertion, or duplication of several kilobases or a translocation with a break point within the gene will result in failure to amplify the mutant allele *(22)*. Consequently, these abnormalities cannot be detected by PTT. A mutation in the promotor or 3' untranslated region that reduces the level of mRNA will not be detected by PTT. Small in-frame deletions or insertions, which cause only a small change to the size of the mutant protein, may not be resolved by SDS-PAGE *(22)*. A mutation that causes instability of the mRNA may result in failure to identify a mutation by PTT when RT-PCR is used. The instability of mutant mRNA transcripts with in-frame stop codon mutations is a recognized phenomenon that is often cited as a cause of false-negative results for PTT. For example, PTT analysis of the *PAX6* gene, following RT-PCR, was found to be less sensitive than expected *(21)*. PTT failed to identify truncated proteins in a number of samples with characterized mutations, because of low levels (or, in some cases, the complete absence) of the mutant mRNA *(21)*. Similarly, Whittock et al. *(6)* postulated that the instability of mutant mRNA resulted in the failure to identify a characterized truncating mutation in an obligate female DMD carrier. In contrast, Lui et al. identified truncating mutations in the *hMSH2* gene by PTT, even though the mutant mRNA transcripts were shown to be less abundant than the normal transcript because of nonsense mediated mRNA instability *(15)*. Samples that show no evidence of a translation-terminating mutation by

PTT should be investigated by another mutation detection method, such as SSCP or denaturing gradient gel electrophoresis (DGGE), to exclude the possibility of a false-negative result and to investigate the pathogenic role of missense mutations.

4. Notes

1. Total cellular RNA can be isolated using a variety of well-established methods and a number of kits can be purchased for this purpose. The RNeasy total RNA kit (Qiagen Ltd.) is reliable and very easy to use. Using the spin columns supplied with the kit, RNA can be rapidly isolated from a wide variety of tissues with minimum preparation and without the use of organic solvents. The TRIzol reagent (Invitrogen Life Technologies Ltd.) offers another reliable method for isolating RNA. Whatever method is chosen to prepare RNA, the integrity and yield should be assessed by agarose gel electrophoresis (*see* Chapter 1).

2. Oligo-d(T) primer *(23)*, or a gene-specific primer complimentary to the 3' end of the sense strand can also be used to prime the first-strand synthesis of cDNA *(5,6)*.

3. *Taq* Extender™ PCR additive improves the efficiency of standard *Taq* DNA polymerase by increasing the number of extension reactions that go to completion by improved proof reading activity. It increases the yield of PCR product and can improve the amplification of long and difficult PCR templates. Optimized 10X *Taq* Extender™ reaction buffer is used instead of 10X *Taq* DNA polymerase reaction buffer.

4. Sequence-specific primers should be chosen to amplify a segment of between 1 and 2 kb. It is possible to amplify larger segments, although this is technically more difficult. The length of the gene-specific primer sequences should be about 18–25 bases. A number of computer programs (e.g., PRIMER) are available that can be a useful aid in designing gene-specific primer pairs. For large genes, PTT is more sensitive when the coding sequence is amplified in a number of overlapping segments. In most published methods the overlap is 300–500 bp, although overlaps of 750 bp have been reported to further improve the sensitivity of PTT *(11)*. Overlapping the PCR segments will maximize the chances of detecting all truncating mutations. A mutation occurring near the end of segment 1 is likely to result in a truncated peptide that is only fractionally shorter than the full-length product. Hence, the two proteins may not be easily resolved by SDS-PAGE. However, if segment 2 overlaps with segment 1 by 300–500 bp, then a mutation near the end of segment 1 will also occur near the start of segment 2. The smaller truncated protein should be easily identified by SDS-PAGE. The forward primer must have an extension at the 5' end, containing the sequence motifs that are essential for in vitro transcription and translation. Essentially, most PTT primers are of a similar design with minor variations. Some examples of the 5' modifications that are used in PTT are shown in **Table 1**. The extreme 5' end of the forward primer may include a restriction enzyme recognition site, often BamH1, which can be useful for cloning the PCR product into a plasmid vector *(22)*. The addition of a restriction enzyme site is not essential for the success of the PTT and may be replaced by any 2 or 3 bases. These are followed directly by the

bacteriophage T7 promoter sequence required for RNA synthesis. Alternatively, the T3 or SP6 promoter may be used, in which cases an in vitro transcription–translation system designed for use with these promoters must be used (*see* **Note 7**). The promoter sequence is followed by a small spacer sequence of 2–5 bp. The spacer is followed by the eukaryotic translation initiation sequence, the in-frame ATG translation start site, and the gene-specific sequence. It is most important to ensure that the ATG codon and the gene-specific sequence are in frame; otherwise, an incorrect translation product will be obtained and the results of the PTT will be meaningless. The forward primer sequence should be positioned in a region that contains at least one or, preferably, more, codons for the labeled amino acid *(22)* (*see* **Note 8**). The reverse primer may be modified at the 5' end so that a stop codon is incorporated at the 3' end of the PCR product. This is said to improve the efficiency of the translation reaction by preventing the ribosomes from stalling at the end of the PCR segment (TnT® T7 Quick kit user manual), although, in practice, this modification does not seem to be essential for the success of PTT.

5. *Taq* DNA polymerase is available from a variety of suppliers of molecular-biology reagents. *Taq* DNA polymerase supplied by Invitrogen Life Technologies Ltd. gives consistently good results. It is supplied with 10X *Taq* DNA polymerase buffer, a separate vial of 50 mM MgCl$_2$, and 1% W1 reagent. Added to PCRs at a final concentration of 0.05%, the W1 reagent acts to stabilize the *Taq* DNA polymerase and improve the yield of the PCR product.

6. A number of methods have been described for the isolation of genomic DNA and a variety of commercially available kits can be obtained for this purpose. Any method that gives good quality, high-molecular-weight DNA should be adequate for PCR amplification for PTT. The phenol/chloroform method, Nucleon II kit (Scotlabs Ltd.), and the Puregene DNA isolation kit (Gentra Systems) are all reliable methods of extracting DNA suitable for PCR amplification prior to PTT.

7. The TnT® T7 Quick coupled transcription–translation system is recommended because it is very straightforward to use. The TnT® T7 Quick master mix contains all the components required for in vitro transcription–translation; hence, few pipetting steps are involved. One limitation of the TnT® T7 Quick kit is that it can only be used in conjunction with T7 promoters, and [35S]methionine is the only labeled amino acid that can be used. If it is necessary to use a different radiolabeled amino acid the TnT® T7 coupled reticulocyte lysate or wheat-germ extract transcription–translation systems may be used (Promega Ltd.). These are supplied with three separate amino acid mixtures—one for use with [35S]methionine, one for [35S]cysteine, and one for [3H]leucine. These systems are also available with different RNA polymerases to allow the use of T7, T3, or SP6 promoters.

8. The choice of radiolabeled amino acid for detection of the translation products should be determined from the amino acid composition of the gene product under investigation. Ideally, the labeled amino acid should occur frequently within the protein, particularly near the start of each translated segment. Truncating mutations near the start of each segment will not be identified if they do not contain

any labeled amino acid, and this should be taken into consideration when designing PCR primers. Practically, the choice of labeled amino acid is limited by the configuration of the in vitro transcription–translation kits and the availability of labeled amino acids. Most published PTT methods use [^{35}S]methionine, [^{35}S] cysteine and [^{3}H] leucine are the two other alternatives. Nonradioactive methods have been described that involve the incorporation of biotinylated lysine, with the additional steps of electroblotting of the SDS-PAGE gel and detection of the proteins by chemiluminescence *(4)*. Commercially available kits for nonradioactive PTT are the Transcend™ nonradioactive translation detection system (Promega Ltd.), and the protein truncation test, nonradioactive kit (Boehringer Mannheim). An alternative nonradioactive method for detecting the products of the in vitro transcription–translation reaction involves the incorporation of a *myc* reporter tag into the translated protein *(3)*. This is accomplished by the inclusion of 36 basepairs of the human c-*myc* sequence in the forward primer sequence. The in-frame c-*myc* sequence is positioned directly between the ATG start site and the in-frame gene-specific primer sequence. The presence of the *myc* epitope in the translated protein allows nonradioactive detection of the translated proteins using an anti-*myc* monoclonal antibody and enhanced chemiluminescence technology *(3)*. PCR amplification of the template is performed in the usual way, using a T7 modified forward primer that also contains the *myc* tag reporter sequence. Nonradioactive in vitro transcription–translation is performed using the TnT® T7 coupled wheat-germ extract system. In vitro transcription–translation systems that contain rabbit reticulocyte lysate cannot be used because the antibodies used for the detection of the *myc* tag crossreact with the proteins in the reticulocyte lysate. The translated proteins are separated by SDS-PAGE and elec-troblotted onto a hybridization membrane. Detection of the proteins is accomplished by hybridization with an anti-*myc* monoclonal antibody, followed by a rabbit anti-mouse antibody, and, finally, with a horseradish peroxidase conjugated swine anti-rabbit antibody. The translation products are visualized by enhanced chemiluminescence. The main advantage of *myc* tag PTT is that only proteins that contain the *myc* epitope are detected. This greatly reduces the number of protein bands that are seen and simplifies the identification of truncated proteins. The advantages of these nonradioactive PTT methods are that there is no handling of radioactive isotopes, chemiluminescence detection methods are sensitive and often quicker than autoradiography, and the translated proteins can be stored indefinitely.

9. The Mini-Protean II electrophoresis cell is a convenient format for the separation of translated proteins in PTT. Two gels 7 cm in length and 8 cm in width, each with 10 or 15 lanes, can be run simultaneously in under 1 h. Any similar gel system can used (e.g., the Hoefer "Mighty Small" electrophoresis cell [SE245] is of a similar design to the Mini-Protean II model). Some laboratories prefer to use larger-format gels for the separation of PTT translation products. The Protean II electrophoresis cell (Bio-Rad Laboratories Ltd.) is an alternative for those who wish to run longer gels. Up to four gels of 16 × 16 cm, or 16 × 20 cm can be run

simultaneously to separate between 60 and 100 samples. Using the Protean II system, electrophoresis is performed over a period of 16 h at 30 mA.

10. The Rainbow™ colored protein molecular-weight markers provide a less expensive alternative to ^{14}C-radiolabeled protein molecular-weight standards. The Kaleidoscope prestained SDS-PAGE standards are a similar product (Bio-Rad Laboratories Ltd.). The main disadvantage of using the nonradioactive molecular-weight standards is that the position of the protein standards must be marked on to the X-ray film, after the film has been developed, by aligning the film with the gel. Hence, the molecular weight of the truncated proteins calculated in this way may be slightly inaccurate. The Rainbow™ colored protein molecular weight markers are also available labeled with a ^{14}C label (Amersham Life Sciences).

11. In optimizing the RT-PCR, it may be necessary to adjust the amounts of cDNA and primers added to the first round of amplification. To establish the optimal $MgCl_2$ concentration for specific amplification, it is necessary to perform a $MgCl_2$ titration. The efficiency of amplification by *Taq* DNA polymerase and the composition of 10X reaction buffer from different suppliers may vary. In establishing optimal amplification conditions, it may be useful to try different types of *Taq* DNA polymerases and buffers to see which gives the best result. Amplification of templates with a strong secondary structure can sometimes be improved by the addition of dimethyl sulfoxide to a final concentration of 5–10%.

12. Performing "hot-start" PCR reduces mispriming and thus increases the specificity of the amplification.

13. The melting temperature ^{TM} of the PCR primers can be calculated using a variety of formulas (e.g., 4[G+C] + 2[A+T]), the PCR annealing temperature is usually taken as 5°C below the T_m. In practice, the optimal annealing temperature is best determined empirically. The duration of the 72°C extension step is determined by the length of the fragment to be amplified. Allowing 1 min for each kilobase to be amplified should be sufficient time for the extension reactions to be completed. The number of cycles can be decreased if the amplification efficiency is good. The use of thin-walled PCR tubes allows the duration of the 94°C denaturing step and the annealing step to be reduced considerably, thus minimizing the loss of activity of the *Taq* DNA polymerase and reducing the time in which misannealing of primers can occur.

14. The volumes given here for performing the in vitro transcription–translation are actually one-quarter of those recommended in the TnT® Quick kit protocol. Between 0.25 and 1 μL [^{35}S]methionine can be added to the reaction. For templates that translate well, the minimum volume of [^{35}S]methionine can be used. For templates that contain few methionine residues or translate less efficiently, the volume may be increased. If it is necessary to use more than 2 μL of PCR product, the volumes of reaction components can be scaled up.

15. Incubation at 22°C for 90 min has been reported to improve the efficiency of transcription and translation in PTT analysis of the *BRCA1*, *BRCA2*, *NF1*, and *APC* genes using the Promega TnT/T7 coupled reticulocyte lysate system *(24)*.

16. A 12% SDS-PAGE gel is suitable for separating proteins in the range of approx 14–100 kDa. The volumes of reagents given in this method to prepare a 12% SDS-PAGE gel are adapted from Sambrook et al. *(25)*, which also gives the volumes of reagents required to prepare gels of 6%, 8%, 10%, and 15% concentration. Separating the translation products on several gels of different percentages (e.g., 8%, 10%, and 14% gels) *(21)* can increase the likelihood of detecting the larger and smaller truncated proteins. The use of 5–18% gradient SDS-PAGE has been reported to give better resolution for a broad size range of truncated proteins *(11)*.

17. The signal strength may be increased by treatment of the gel with Amplify™ fluorographic enhancement reagent (Amersham Life Science) for 20–30 min after the gel has been fixed. The gel is then rinsed in distilled water and dried under vacuum. Fluorography is performed with intensifying screens overnight at –70°C. Some loss in resolution may be observed following fluorographic enhancement. Fluorographic enhancement is unnecessary if a strong signal can be obtained by autoradiography.

Acknowledgments

I would like to thank all my colleagues for many helpful discussions on PTT, in particular Dr. David Bourn for his advice and comments during preparation of the manuscript and Kim Hampson for providing the *BRCA1* autorad and information used in the preparation of **Fig. 1**. I would also like to thank Professor Georgia Chenevix-Trench and Dr Sue Shanley, of the Queensland Institute of Medical Research, Brisbane, Australia, for the primer sequences used for PTT of the *PTCH* gene illustrated in **Fig. 2**.

References

1. Roest, P. A. M., Roberts, R. G., Sugino, S., van Ommen, G-J. B., and den Dunnen, J. T. (1993) Protein truncation test (PTT) for rapid detection of translation-terminating mutations. *Hum. Mol. Genet.* **2,** 1719–1721.

2. Powell, S. M., Petersen, G. M., Krush, A. J., Booker, S., Jen, J., Giardiello, F. M., et al. (1993) Molecular diagnosis of familial adenomatous polyposis. *N. Engl. J. Med.* **329,** 1982–1987.

3. Rowan, A. J. and Bodmer, W. F. (1997) Introduction of a *myc* reporter tag to improve the quality of mutation detection using the protein truncation test. *Hum. Mutat.* **9,** 172–176.

4. Becker, K.-F., Reich, U., Handschuh, G., Dalke, C., and Höfler, H. (1996) Non-radioactive protein truncation test (nrPTT) for rapid detection of gene mutations. *Trends Genet.* **12,** 250.

5. Gardner, R. J., Bobrow, M., and Roberts, R. G. (1995) The identification of point mutations in Duchenne muscular dystrophy by using reverse-transcription PCR and the protein truncation test. *Am. J. Hum. Genet.* **57,** 311–320.

6. Whittock, N. V., Roberts, R. G., Mathew, C. G., and Abbs, S. J. (1997) Dystrophin point mutation screening using a multiplexed protein truncation test. *Genet. Test.* **1,** 115–123.

7. van der Luijt, R., Khan, P. M., Vasen, H., Van Leeuwen, C., Tops, C., Roest, P., et al. (1994) Rapid detection of translation-terminating mutations at the adenomatous polyposis coli (APC) gene by direct protein truncation test. *Genomics* **20,** 1–4.

8. Prosser, J., Condie, A., Wright, M., Horn, J. M., Fantes, J. A., Wyllie, A. H., et al. (1994) APC mutation analysis by chemical cleavage of mismatch and a protein truncation assay in familial adenomatous polyposis. *Br. J. Cancer* **70,** 841–846.

9. Hogervorst, F. B. L., Cornelis, R. S., Bout, M., van Vliet, M., Oosterwijk, J. C., Olmer, R., et al. (1995) Rapid detection of BRCA1 mutations by the protein truncation test. *Nature Genetics* **10,** 208–212.

10. Plummer, S. J., Anton-Culver, H., Webster, L., Noble, B., Liao, S., Kennedy, A., et al. (1995) Detection of BRCA1 mutations by the protein truncation test. *Hum. Mol. Genet.* **4,** 1989–1991.

11. Garvin, A. M. (1998) A complete protein truncation test for BRCA1 and BRCA2. *Eur. J. Hum. Genet.* **6,** 226–234.

12. Garvin, A. M., Attenhofer-Haner, M., and Scott, R.J. (1997) BRCA1 and BRCA2 mutation analysis in 86 early onset breast/ovarian cancer patients. *J. Med. Genet.* **34,** 990–995.

13. Friedman, L. S., Gayther, S. A., Kurosaki, T., Gordon, D., Noble, B., Casey, G., et al. (1997) Mutation analysis of BRCA1 and BRCA2 in a male breast cancer population. *Am. J. Hum. Genet.* **60,** 313–319.

14. Kohonen-Corish, M., Ross, V. L., Doe, W. F., Kool, D. A., Edkins, E., Faragher, I., et al. (1996) RNA-based mutation screening in hereditary nonpolyposis colorectal cancer. *Am. J. Hum. Genet.* **59,** 818–824.

15. Lui, B., Parsons, R. E., Hamilton, S. R., Petersen, G. M., Lynch, H. T., Watson, P., et al. (1994) *hMSH2* mutations in hereditary nonpolyposis colorectal cancer kindreds. *Cancer Res.* **54,** 4590–4594.

16. Romey, M.-C., Tuffery, S., Desgeorges, M., Bienvenu, T., Demaille, J., and Clautres, M. (1996) Transcript analysis of CFTR frameshift mutations in lymphocytes using the reverse transcription-polymerase chain reaction and the protein truncation test. *Hum. Genet.* **98,** 328–332.

17. Heim, R. A., Kam-Morgan, L. N. W., Binnie, C. G., Corns, D. D., Cayouette, M. C, Farber, R. A., et al. (1995) Distribution of 13 truncating mutations in the neurofibromatosis 1 gene. *Hum. Mol. Genet.* **4,** 975–981.

18. van Bakel, I., Sepp, T., Ward, S., Yates, J. R. W., and Green, A. J. (1997) Mutations in the TSC2 gene: analysis of the complete coding sequence using the protein truncation test (PTT). *Hum. Mol. Genet.* **6,** 1409–1414.

19. Peral, B., Gamble, V., Strong, C., Ong, A. C. M., Sloane-Stanley, J., Zerres, K., et al. (1997) Identification of mutations in the duplicated region of the polycystic kidney disease 1 gene (PKD1) by a novel approach. *Am. J. Hum. Genet.* **60,** 1399–1410.

20. Fitzgerald, M. G., Bean, J. M., Hegde, S. R., Unsal, H., MacDonald, D. J., Harkin, D. P., et al. (1997) Heterozygous *ATM* mutations do not contribute to early onset of breast cancer. *Nature Genet.* **15,** 307–310.
21. Axton, R., Hanson, I., Danes, S., Sellar, G., van Heyningen, V., and Prosser, J. (1997) The incidence of PAX6 mutation in patients with simple aniridia: an evaluation of mutation detection in 12 cases. *J. Med. Genet.* **34,** 279–286.
22. Hogervorst, F. B. L. (1998) Protein truncation test (PTT). *Promega Notes* **62,** 7–10.
23. Lo Ten Foe, J. R., Rooimans, M. A., Bosnoyan-Collins, L., Alon, N., Wijker, M., Parker, L., et al. (1996) Expression cloning of a cDNA for the major fanconi anaemia gene FAA. *Nature Genet.* **14,** 320–323
24. Claes, K., Macháckova, E., Callens, T., Van der Cruyssen, G., and Messiaen, L. (1998) Improved conditions for PTT analysis of the *BRCA1, BRCA2, NF1* and *APC* genes. Elsevier Trends Journals Technical Tips Online, T01457, 22/7/98. Available from: URL: http://www.elsevier.com/homepage/sab/tto/menu.htm
25. Sambrook, J., Fritsch, E. F., and Maniatis, T. (1989) *Molecular Cloning: A Laboratory Manual*, 2nd ed., Cold Spring Harbor Laboratory, Cold Spring Harbor, NY.

13

Mutation Detection in Factor VIII cDNA from Lymphocytes of Hemophilia A Patients by Solid Phase Fluorescent Chemical Cleavage of Mismatch

Naushin H. Waseem, Richard Bagnall, Peter M. Green, and Francesco Giannelli

1. Introduction

Defects of the factor VIII gene causes (*f8*) hemophilia A, an hemorrhagic X-linked disorder. The factor VIII gene is 186 kb long with 26 exons, varying from 69 bp (exon 5) to 3106 bp (exon 14) (*1*). The factor VIII mRNA is 9028 bases in length with a coding region of 7053 nucleotides (*2*).

Hemophilia A mutations in factor VIII genes are quite heterogeneous and various methods have been used for the detection of these mutations. The most widely used are:

1. Denaturing gradient gel electrophoresis (DGGE): This is based on the ability of mutations to reduce the melting temperature of DNA domains resulting in altered mobility on a formamide and urea gradient polyacrylamide gel. This method can detect any sequence change but requires 41 PCR to amplify the 26 exons of factor VIII gene and the putative promoter region. The maximum length of the suitable PCR product is 600–700 bp and the product must have appropriate melting domain structure. Some splicing signal region were not screened by this procedure (*3*).

2. Single strand conformation polymorphism (SSCP): This method is based on the fact that single stranded DNAs differing in a single nucleotide will acquire different conformations that have different mobilities on a polyacrylamide gel. Various exons of factor VIII are amplified using polymerase chain reaction (PCR) and subjected to electrophoresis. The exon showing an altered electrophoretic mobility is then sequenced (*4*). The PCR products must be between 100–300 bp for efficient screening.

From: *Methods in Molecular Biology, vol. 187: PCR Mutation Detection Protocols*
Edited by: B. D. M. Theophilus and R. Rapley © Humana Press Inc., Totowa, NJ

3. Denaturing high performance liquid chromatography (DHPLC): This new technique for mutation detection has recently been used to scan factor VIII mutations *(5)*. The exons from a normal DNA is hybridized to the corresponding exon from a hemophilia A patient and the resulting heteroduplexes are analyzed on a partially denaturing HPLC column. This method requires 33 amplifications and the detection rate is 70% *(5)*.

   In the above procedures, the PCR products containing the mutation is identified but the position of the mutation is not determined.

4. Chemical cleavage of mismatch: In order to develop a rapid and fully effective procedure for the detection of hemophilia A mutations the method of chemical cleavage of mismatch (CCM) is combined with analysis of the traces of factor VIII mRNA present in peripheral blood lymphocytes. The method of CCM has the advantage that it detects any sequence change even in long DNA segments (1.5–1.8 kb) and indicates the position of mismatch within the segment. Hence, entire mRNA can be screened in few overlapping segments. The mRNA analysis offers a chance of detecting mutations occurring in any region of the gene, including internal region of the long introns by virtue of their effect on the structure of the mRNA, thus allowing a gain not only in the speed but also in the completeness of mutation detection. A further advantage of mRNA analysis is that it provides direct evidence of the effect gene mutation may have on the structure of the gene transcript. In the basic procedure, the factor VIII message and appropriate segments of a patient gene are specifically amplified and compared with similar products amplified from control RNA. The patient and control PCR products are then hybridized to form a hetroduplex and treated with hydroxylamine and/or osmium tetroxide, which modify C or T residues, respectively. The DNA is then cleaved with piperidine at the modified base and analyzed on a denaturing polyacrylamide gel. From the size of the cleavage fragments, the position of the mutation is estimated and the relevant exons sequenced *(6,7)*. To aid the visualization of DNA on the acrylamide gel the DNA is either radioactively or fluorescently labelled. This chapter describes the methods used in the detection of mutations by solid phase fluorescent chemical cleavage of mismatch (SPFCCM).

The factor VIII message, except the large exon 14, is reverse transcribed with AMV reverse transcriptase and amplified with Tfl DNA polymerase into four overlapping segments (**Fig. 1**). Exon 14 is amplified from genomic DNA as two additional segments so that the entire coding region is represented in six overlapping segments. The promoter region (segment P) and the polyadenylation signal region (segment T) are also amplified from the genomic DNA. The promoter region overlaps with segment 1 (**Fig. 1**). The promoter region and polyadenylation signal regions are sequenced directly whereas the rest of the segments are processed through SPFCCM. Three different fluorescent labeled dUTPs are used to label the segments: segment 1 and 2 with green, segment 3 and 4 with blue and segment 5 and 6 with yellow fluorescence. Similar segments are amplified from either control RNA or cloned factor VIII

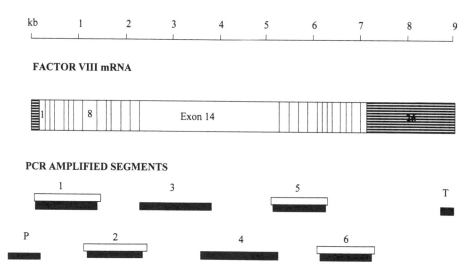

Fig. 1. Schematic diagram of factor VIII cDNA showing the positions segment 1–8. Open bars represent primary PCR whereas filled bars are nested PCR. Segment 3, 4, promoter region (P) and polyadenylation signal region (T) are amplified from DNA.

cDNA using biotinylated primers and fluorescent dUTPs. The patient and the control products are then hybridized to form heteroduplexes in two multiplex reactions. The heteroduplexes are then captured on streptavidin coated magnetic beads and treated with hydroxylamine and osmium tetroxide followed by treatment with piperidine (**Fig. 2**). The products are analyzed on an ABI PRISM 377 DNA sequencer.

2. Materials

2.1. Lymphocyte Isolation

1. Histopaque-1077 is supplied by Sigma (cat. no. 1077-1).
2. Phosphate Buffered Saline (PBS): 10 mM potassium phosphate buffer, 138 mM NaCl, 2.7 mM KCl, pH 7.4.

2.2. RNA Isolation

1. Various RNA isolation kits are available commercially. We used RNA Isolator from Genosys (cat. no. RNA-ISO-050).
2. Isopropanol.
3. Chloroform.

2.3. DNA Isolation

1. Puregene DNA isolation kit (D-5500A) is purchased from Gentra Systems (USA).
2. Isopropanol.

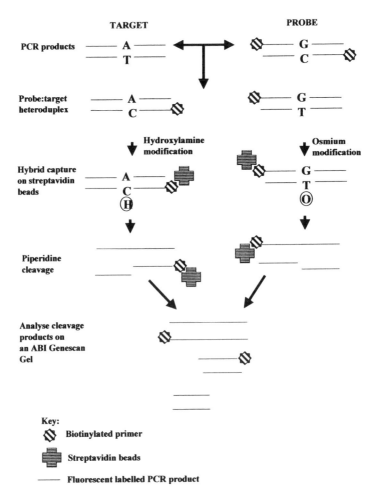

Fig. 2. Diagrammatic representation of steps involved in the solid phase fluorescent chemical cleavage of mismatch method.

2.4. RT-PCR and PCR

1. Primers: The primers used for amplification are listed in **Table 1**. The four RT-PCR fragments are labeled 1, 2, 5, and 6. The outer primers are called A and B whereas the nested ones are given C and D as suffix. Segment 3, 4, P, and T are amplified from DNA and therefore require only one pair of primers each. The nested and the segment 3 and 4 primers are also synthesized with biotin at their 5' end and are marked with an asterisk in the table. A working solution of primers at 100 ng/μL is kept at 4°C and primer stocks are kept at –20°C.
2. Access RT-PCR System (cat. no. A1250) is supplied by Promega, USA. The kit is stored at –20°C.

Table 1
Primers Used for the Amplication of Factor VIII cDNA

1A	GGGAGCTAAAGATATTTTAGAGAAG
1B	CAACAGTGTGTCTCCAACTTCCCCAT
1C*	GAGAAGATTAACCTTTTGCTTCTC
1D*	CCTACCAATCCGCTGAGGGCCATTG
2A	GAAGAAGCGGAAGACTATGATGATG
2B	GCCTAGTGCTACGGTGTCTTGAATTC
2C*	CTGATTCTGAAATGGATGTGGTCAGG
2D*	GGGAGAAGCTTCTTGGTTCAATGGC
3A	AGAGTTCTGTGTCACTATTAAGACCC
3B	TCTGAGGCAAAACTACATTCTCTTGG
4A	CAAAGGACGTAGGACTCAAAGAGATGG
4B	CACCAGAGTAAGAGTTTCAAGACAC
5A	CTTCAG TCAGATCAAGAGGAAATTGAC
5B	GAAGTCTGGCCAGCTTTGGGGCCCAC
5C*	TATGATGATACCATATCAGTTGAAATG
5D*	CTCTAATGTGTCCAGAAGCCATTCCC
6A	TTCATTTCAGTGGACATGTG
6B	CAGGAGGCTTCAAGGCAGTGTCTG
6C*	CAGTGGACATGTGTTCACTGTACGAA ′
6D*	TAGCACAAAGGTAGAAGGCAAGC
PA	GGATGCTCTAGGACCTAGGC
PB	AAGAAGCAGGTGGAGAGCTC
TA	CAAATGTTCATGGAACTAGC
TB	CTGTTCTCCTGGATTGAGGC

*The outer primers for seg 1, 2, 5, and 6 are given A and B as suffix, whereas the nested ones are called C and D. The primers also synthesized with biotin at their 5' end are marked with an asterisk.

3. Fluorescent deoxynucleotides: Fluorescent deoxynucleotides for labeling the PCR products can be obtained from Perkin-Elmer Applied Biosystems. [F]dUTP set (cat. no. P/N 401894) contains 12 nmol of [TAMRA]dUTP, 3 nmol of [R110]dUTP and 3 nmol of [R6G]dUTP. These are stored as 2 μL aliquots at –20°C in the dark to avoid repeated freezing and thawing.

2.5. Solid Phase Fluorescent Chemical Cleavage of Mismatch

1. Hydroxylamine hydrochloride (Sigma, cat no. H 2391): A 4 *M* solution is prepared and titrated to pH 6.0 with diethylamine (Sigma, cat. no. D 3131). Since very little hydroxylamine (20 μL/reaction) is required for the reaction a rough guide to prepare a 4 *M* solution, pH 6.0, is: weight (in mg) of hydroxylamine divided by 0.28 gives the volume (in μL) of water to be added. This value times 0.2 to 0.3 gives the volume of diethylamine (in μL) to be added to the solution to reach pH 6.0. The volume of diethylamine to be added varies from batch to batch

Table 2
Sequence of Primers Used in the Detection of Intron 22 Inversion

Primer	Sequence
P-INV	GCCCTGCCTGTCCATTACACTGATGACATTATGCTGAC
Q-INV	GGCCCTACAACCATTCTGCCTTTCACTTTCAGTGCAATA
A-INV	CACAAGGGGGAAGAGTGTGAGGGTGTGGGATAAGAA
B-INV	CCCCAAACTATAACCAGCACCTTGAACTTCCCCTCTCATA

and should be titrated for every batch. Hydroxylamine is highly toxic and should be handled in fumehood and protective laboratory clothing should be worn.

2. Osmium tetroxide: A 4% solution can be obtained from Sigma (cat. no. O 0631). A working solution of osmium tetroxide is 0.4% osmium tetroxide in 0.2% pyridine (Sigma, cat. no. P 3776). A fresh solution of osmium tetroxide should be made each time it is required as it has a short half life. Fresh stock solution should be purchased every 2–3 mo. Osmium tetroxide is very toxic and should be handled in a fumehood. Protective laboratory clothing should be worn at all times during handling.

3. Formamide loading dye: Deionized formamide containing 10 mg/mL dextran blue. Formamide from Amresco and Sigma give no background fluorescence when used as loading dye on ABI PRISM 377 DNA Sequencer.

4. Piperidine: A 1 M solution of piperidine (Sigma, cat. no. P 5881) is prepared in formamide loading dye.

5. 10X Hybridization buffer: 3 M NaCl in 1 M Tris-HCl, pH 8.0. This is stored at room temperature.

6. 2X Binding buffer: 2 M NaCl, 0.4% Tween 20, 0.1 mM EDTA , 10 mM Tris-HCl, pH 8.0, stored at room temperature.

7. TE 8.0: 10 mM Tris-HCl, pH 8.0, containing 0.1 mM EDTA, autoclaved and stored at room temperature.

8. 6% Acrylamide/Bisacrylamide solution from Severn Biotech Ltd., UK, cat. no. 20-2700-10.

9. Streptavidin coated magnetic beads, Strep Magneshere Paramagnetic Particles, from Promega, USA, cat. no. Z5481.

2.6. Fluorescent Dye Terminator DNA Sequencing

1. ABI PRISM™ DNA Sequencing kit, Big Dye Terminator Cycle Sequencing Ready Reaction (cat. no. 4303152) is supplied by Perkin-Elmer Applied Biosystems and is stored as 4 µL aliquots at –20°.

2. Absolute and 70% ethanol.

3. 3 M Sodium acetate, pH 4.6.

4. AutoMatrix 4.5 (cat. no. EC-854) from National Diagnostics (UK).

5. Sequence Navigator software from ABI.

2.7. Detection of Intron 22 Inversions

1. Primers used for the detection of intron 22 inversion are shown in **Table 2**.
2. Expand Long Template PCR System (cat. no. 1681842) from Boehringer Mannheim, Germany.
3. Deaza dGTP (cat. no. 988537) from Boehringer Mannheim, Germany.

3. Methods
3.1. Isolation of Lymphocytes

1. In a 50 mL polypropylene conical tube, add 10 mL of Histopaque 1077. Gradually overlay equal volume of EDTA anti-coagulated blood. This can be facilitated by tilting the centrifuge tube containing Histopaque so that the blood trickles down the side of the tube. It is essential that Histopaque and the blood are at room temperature (*see* **Note 1**).
2. Centrifuge at 400g at room temperature for 20 min.
3. Transfer the opaque layer of lymphocytes at the interface with a disposable plastic Pasteur pipet into a fresh 50 mL centrifuge tube and add 35 mL of phosphate buffered saline. At this stage, divide the suspension equally between two tubes.
4. Centrifuge at 400g for 10 min at room temperature.
5. In one tube resuspend the pellet in 500 µL of RNA Isolator (Genosys) and store at –70°C until further use. In the other tube add 600 µL of Cell Lysis Solution (Gentra, USA) and store at 4°C for DNA isolation.

3.2. Isolation of DNA and RNA

There are a number of kits available for the isolation of RNA and DNA. We use Puregene DNA isolation kit for DNA isolation and RNA Isolator (Genosys) for RNA isolation.

3.2.1. Isolation of DNA

1. To the 600 µL cell lysate add 300 µL of Protein Precipitation Solution (supplied with the kit). Mix 20 times and centrifuge at 16,000g for 10 min. The precipitated protein forms a tight pellet at the bottom of the tube.
2. Transfer the supernatant into a fresh 2 mL microfuge tube and add 1 mL of 100% isopropanol to it and mix by inverting several times.
3. Centrifuge at 16,000g for 10 min at room temperature.
4. Decant the supernatant and add 70% ethanol and repeat the centrifugation.
5. Air-dry the pellet for 5 min and resuspend in 500 µL of TE, pH 8.0.

3.2.2. Isolation of RNA

1. Thaw the cell lysate in RNA Isolator (Genosys). Add 100 µL chloroform and mix gently 15 times and incubate for 15 min at room temperature.
2. Centrifuge at 16,000g for 15 min at room temperature.

3. Transfer the upper aqueous phase to a fresh 1.5 mL microfuge tube and add 250 µL isopropanol. Mix and incubate for 10 min at room temperature.
4. Centrifuge at 16,000*g* for 10 min at room temperature. At this stage, RNA should be visible as translucent pellet.
5. Wash pellet with 70% ethanol.
6. Air-dry the RNA pellet and dissolve it in RNA Hydration Solution (*see* **Note 2**).

3.3. RT-PCR

For reverse transcription and amplification of four out of the eight segment from factor VIII mRNA, we use Access RT-PCR System. In this kit, the reverse transcription by AMV reverse transcriptase and initial 10 cycles of amplification by Tfl DNA polymerase is performed in a single tube (primary PCR) (*see* **Note 3**). An aliquot is then amplified for another 30 cycles with nested primers (secondary PCR). Two of the four segments (segment 1,5 and 2,6) are multiplexed in the primary PCR so in total we have two primary and four secondary PCRs.

1. Add 5 µL 5X reaction buffer, 0.5µL 10 m*M* dNTP, 3.0 µL 25 m*M* MgSO$_4$, 2.5 µL of 100 ng/µL primer 1A, 1B, 5A, 5B, 0.5 µL AMV reverse transcriptase, 0.5 µL Tfl DNA polymerase, 100–200 ng RNA. Make up the volume with RNAse-free water to 25 µL. Set up a second RT-PCR substituting primer 1A, 1B, 5A, and 5B for 2A, 2B, 6A, and 6B.
2. Place the tubes in controlled temperature block equilibrated at 48°C and incubate for 1 h and proceed immediately to thermal cycling reactions: 93°C for 30 s, 65°C for 30 s, 68°C for 5 min for 10 cycles.
3. To set up secondary PCR, add 2.5 µL of 10X Tfl buffer, 0.5 µL of 10 m*M* dNTP, 1.0 µL of 25 m*M* MgSO$_4$, 2.5 µL of 100 ng/µL primer XC & XD (x is the segment number), 0.5 µL Tfl I, 0.4 µL [F]dUTP, 14.5 µL water, 2.5 µL of primary PCR. This reaction mix is also used for the amplification of segment 3 and 4 from DNA (*see* **Note 4**).
4. Set up the cycling conditions as follows: 94°C for 5 min and 30 cycles of 93°C for 30 s, 65°C for 30 s, 72°C for 3 min.
5. For amplification from control RNA or cloned factor VIII cDNA (referred to as probe from now on) use the same recipe except use biotinylated primers. The products should be gel purified before using in the solid phase fluorescent chemical cleavage of mismatch (*see* **Note 5**).

3.4. Solid Phase Fluorescent Chemical Cleavage of Mismatch

3.4.1. Preparation of Hybrids

1. Set up the hybridization as follows: 3 µL 10X hybridization buffer, 30 ng each of the probe 1, 3 and 6 (the other multiplex will be 2, 4, and 5), 300 ng of target 1, 3, and 6. Make up the volume to 30 µL with TE 8.0.
2. Incubate at 95°C for 5 min and 65°C for 1 h (*see* **Note 6**).

3. 10 µL of Streptavidin coated magnetic beads (a 50% suspension is supplied) are required to bind the biotin tagged products per reaction. Take out the required volume and wash twice with 2X binding buffer. Resuspend in three times the original volume of the beads with 2X binding buffer.
4. Add 30 µL of the washed streptavidin beads to each reaction. Incubate 15 min at room temperature.
5. Place the tube on a magnetic stand to pellet the beads and aspirate the supernatant.

3.4.2. Hybrid Modification and Cleavage

3.4.2.1. HYDROXYLAMINE

1. Resuspend the beads in 20 µL of 4 *M* hydroxylamine, pH 6.0 (*see* **Methods** for preparation procedure).
2. Incubate 2 h at 37°C.
3. Pellet the streptavidin beads on a magnetic stand and wash the beads with TE 8.0 and proceed to **Subheading 3.4.2.3.** below.

3.4.2.2. OSMIUM TETROXIDE

1. Resuspend the beads from **step 5** above in 20 µL 0.4% osmium tetroxide, 0.2% pyridine and incubate for 15 min at 37°C (*see* **Note 7**).
2. Pellet the streptavidin beads on a magnetic stand, wash with TE 8.0 (*see* **Note 8**).

3.4.2.3. CLEAVAGE OF DNA

Add 5 µL of piperidine/ formamide loading dye to the reaction and incubate at 90°C for 30 min (*see* **Note 9**).

3.5. Analysis of the Products from SPFCCM on ABI PRISM 377 DNA Sequencer

3.5.1. Preparation and Loading of Polyacrylamide Gel

1. Clean the glass plates with 3% Alconox. After rinsing them with water wipe them with isopropanol soaked lint free tissue paper.
2. Set up the glass plates on the cassette, with notched plate at the bottom.
3. Prepare the following mix for the acrylamide gel for 12-cm plates (*see* **Note 10**):

6% Acrylamide/Bisacrylamide solution	20 mL
10% APS	47 µL
TEMED	33 µL

4. Pour in between the plates and allow it to set for 1 h. Save the rest of acrylamide solution to check for its polymerization (*see* **Note 11**).
5. Prerun the gel for 15 min at PR 12A-1200.
6. Load 2 µL of the sample from **Subheading 3.4.2.3.** above. Dilute Rox GS-2500 1:5 in formamide loading dye. Heat it at 92°C for 2 min and use 2 µL of this as a marker (*see* **Note 12**).
7. Electrophorese the gel on GS12A-1200 for 5 h.
8. Analyze the gel using Genescan software (**Fig. 3**).

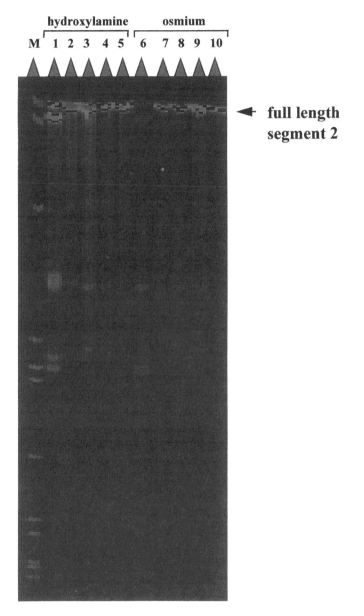

Fig. 3. Gel image of mismatch products obtained by Genescan software on ABI PRISM 377 DNA Sequencer. Lane M, Genescan 2500 Rox marker; lanes 1 and 6, patient UKA162 with G to T mutation at nt 1798 leading to 50 nt deletion in mRNA; lanes 2 and 7, patient UKA160 with A to G mutation at nt 1763 (*see* **Note 13**); lanes 3 and 8, patient UKA144 with A to G mutation at nt 1801. Lanes 4 and 9, patient UKA133 with C to T mutation at nt 1636; lanes 5 and 10 are wild-type controls.

3.6. Analysis of the Promoter and Polyadenylation Signal Region

Samples with mutations in the promoter or polyadenylation signal region of factor VIII may have reduced levels or no transcripts. Therefore, RNA samples that fail to amplify or do not give any mismatch bands (*see* **Note 13**) should be screened for the mutations in these regions. Segment P (promoter region) and segment T (polyadenylation signal region) are amplified in the same way as above (**Subheading 3.3.**, **step 4**) except no FdUTP is added. The primers used for amplification are shown in **Table 1**. The PCR products are gel purified and sequenced with one of the primers used for amplification.

3.7. Sequence Analysis for Complete Mutation Characterization

3.7.1. Setting up the Sequencing Reaction

1. From the size of the mismatch fragment estimate the position of the mutation from either end of the segment.
2. Amplify either the same segment from a new RT-PCR or amplify the relevant exons (*see* **Note 14**).
3. Gel purify the products (*see* **Note 15**).
4. Set up the sequencing reaction as follows: BigDye terminator mix 4 μL, primer 0.8 pmol, DNA (200–400 ng) water to 10 μL.
5. Start the following thermal cycling condition: 96°C for 30 s, 50°C for 15 s, 60°C for 4 min, 25 cycles and hold at 4°C.
6. Transfer the content of the reaction tube to a fresh tube and add 1 μL 3 *M* sodium acetate, pH 4.6, and 25 μL ethanol. Mix and incubate on ice for 10 min.
7. Centrifuge at the maximum speed for 15 min at room temperature.
8. Wash pellet with 70% ethanol. Vortex to resuspend for about 30 s. This is very crucial in removing unincorporated nucleotides.
9. Repeat the centrifugation step (**step 7**).
10. Discard the supernatant and air-dry the pellet for 10 min at room temperature.
11. Resuspend in 2 μL of formamide loading dye. Just before loading heat the sample at 92°C for 2 min.

3.7.2. Preparation of Polyacrylamide Gel

1. Clean and set up the 36-cm glass plates as described in **Subheading 3.5.**
2. Prepare acrylamide mix as follows:

AutoMatrix 4.5	45 mL
10X TBE	5 mL
10% APS	250 μL
TEMED	35 μL
Stir to mix	

3. Immediately pour it between the plates from the top end, pressing the plates to avoid bubbles.

4. Allow it to set for 1 h at room temperature.
5. Prerun the gel for 15 min on Seq PR 36E-1200.
6. Load the samples from **step 11** above (**Subheading 3.7.1.**) and electrophorese for 7 h on Seq Run 36E-1200.

3.7.3. Sequence Analysis

1. To start the sequence analysis program, double click on the Gel file from the sequence run.
2. Define the first and last track on the gel and the click Track and Extract lanes from the Gel menu.
3. Quit sequence analysis program and open Sequence Navigator Program for the analysis of mutations.
4. Import relevant sequences together with the published factor VIII sequence and compare them using Comparative or Clustal program from Align menu.

3.8. Detection of Intron 22 Inversion by Long PCR

Forty five percent of the severe hemophilia A patients have inversions in their factor VIII gene involving intron 22 resulting in the factor VIII coding sequence ending at exon 22 and leading to segment 6 amplification failure *(8)*. The inversions are due to intra-chromatid homologous recombination between a sequence in intron 22 called int22h-1 and either of the two repeats of these sequence (int22h-2 and int22h-3) located 500 and 600 kb more telomerically and in inverted orientation relative to int22h-1 *(9)*. In 1998, Lui et al. *(10)* developed a rapid PCR method for the detection of these inversions. The reaction is performed in a single tube containing four primers. Primers P and Q amplify a 12 kb int22h-1 specific sequence and primers A and B amplify 10 kb of int22h-2 and int22h-3. When an inversion occurs, two new recombined PCR products, PB (11 kb) and AQ (11 kb) are amplified together with the AB (10 kb) from unrecombined int22h-2 or int22h-3 and can be readily distinguished as two bands on a 0.5 % agarose gel (**Fig. 4**).

1. In a 0.2 mL microfuge tube, add 1 µL of 10X Expand buffer 2 (supplied with the enzyme), 1 µL 5 m*M* dNTP (made with 2.5 m*M* deaza dGTP, 2.5 m*M* dGTP and 5 m*M* each of dCTP, dTTP and dATP), 7.5% DMSO, 120 ng of primer P and Q, 50 ng of primer A and B, 0.3 µL of Expand (Boehringer Mannheim, Germany) and 250 ng of DNA, make up the volume to 10 µL with TE 8.0.
2. Place the tube in a controlled temperature block and cycle under following conditions: after 2 min denaturation at 94°, 10 cycles of 94° for 15 s, 68° for 12 min followed by 20 cycles of 94° for 15 s, 68° for 12 min with 20-s increment per cycle.
3. Load the products on 0.5% agarose gel and visualize the band after staining with ethidium bromide.

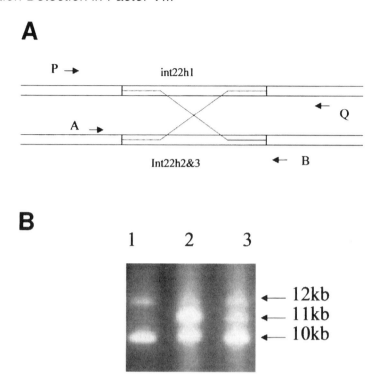

Fig. 4. Schematic diagram depicting the PCR assay for intron 22 inversion in factor VIII gene. (**A**) Diagrammatic representation of the position of the primers around int22h-1 and int22h-2 and -3. (**B**) PCR products (PQ, AB) from a normal male (12 kb and 10 kb, lane 1), PB + AQ, AB from intron 22 inversion male patients (11 kb and 10 kb, lane 2) and a PQ, PB + AQ, AB from female carrier (12, 11 and 10 kb, lane 3).

4. Notes

1. Blood stored for 3 d at room temperature can be used for the isolation of RNA. Although the yield of RNA is low it is usually good enough for the RT-PCR.
2. We normally analyze our RNA on 1% agarose gel before starting the RT-PCR. To 2 μL RNA solution add 2μL bromophenol/glycerol dye (0.1% bromophenol blue in 30% glycerol). Heat it at 65°C for 5 min and load on the gel. Two, sometimes three ribosomal RNA bands are visible under UV light.
3. Some types of mutations, e.g., nonsense or frameshift make the transcript unstable. To compensate for this, increase the number of cycles in primary PCR from 10 to 30.
4. Fluorescent dUTP and dCTP lowers the efficiency of PCR. Avoid using too much fluorescent dUTP in the PCR reaction.

5. We use cloned factor VIII cDNA from Genentech as a probe to amplify segment 1, 2, 5, and 6. Segment 3 and 4 are amplified from normal DNA.
6. Hybrids of target and control DNA can be stored at –20° for up to 3 d.
7. Osmium tetroxide is a strong oxidizing agent, avoid undue exposure to the air.
8. Hydroxylamine and osmium tetroxide reaction can either be performed separately or can be combined. To do that, first do the reaction with hydroxylamine then osmium tetroxide followed by the piperidine cleavage step. Follow the reaction until **step 3 (Subheading 3.4.2.1.)** then start **step 1** of **Subheading 3.4.2.2.** (the osmium tetroxide reaction) as described in the protocol.
9. On a Genescan gel, a couple of fluorescent bands appear around 150–200 bp. These are the degraded fluorescent moieties from the fluorescence dUTPs. They can be removed from the reaction mixture by ethanol precipitation. Perform the cleavage step with aqueous 1 *M* piperidine at 90°C for 30 min. To the supernant add 1/10 vol of 3 *M* sodium acetate, pH 4.6 and 2.5 vol of ethanol. Place the tube in dry ice for 5 min, spin at maximum speed for 10 min in a microfuge. Wash the pellet with 70% ethanol, air-dry and resuspend in 5 µL of formamide loading dye.
10. Twelve centimeter plates give good separation of most of the mismatch bands. However, samples that have large mismatch products are better separated on 36-cm plates.
11. The manufacturer of ABI PRISM 377 DNA sequencer recommend the gel to age for 2 h before loading the samples, However, we find that 15 min setting time is enough. Gels left to set for more 2 h run slower.
12. More accurate sizing of the mismatched products on a Genescan gel will require internal standards. For this purpose, GS 2500-Rox diluted 1:400 in formamide loading dye should be used in place of formamide loading dye at the cleavage step.
13. A T nucleotide mispaired with G is not modified by osmium tetroxide if preceded by G at the 5' end and will only show up in the hydroxylamine reaction *(11)*.
14. Fluorescent DNA sequencing requires more template than ^{35}S dATP sequencing. Don't even attempt to sequence from a poor PCR.
15. For gel purifying the PCR product, remove the relevant band from the low melting agarose gel and add equal volume of water. Heat at 70°C for 10 min, this keeps it in solution. Several matrices are available commercially, e.g., Geneclean (BIO101), PCR purification kit (Promega), centrisep column (Amicon), which can be used for the purification of DNA from this gel.

References

1. Gitschier, J., Wood, W. I., Goralka, T. M., Wion, K. L., Chen, E. Y., Eaton, D. H., et al. (1984) Characterization of the human factor VIII gene. *Nature* **312,** 326–330.
2. Toole, J. J., Knopf, J. L., Wozney, J. M., Sultzman, L. A., Buecker, J. L., Pittman, D. D., et al. (1984) Molecular-cloning of a cDNA-encoding human antihemophilic factor. *Nature* **312,** 342–347.
3. Higuchi, M., Antonarakis, S. E., Kasch, L., Oldenburg, J., Economou-Petersen, E., Olek, K., et al. (1991) Molecular characterisation of mild-to-moderate hemo-

philia A: detection of the mutation in 25 of 29 patients by denaturing gel electrophoresis. *Proc. Natl. Acad. Sci. USA* **88,** 8307–8311.

4. Orita, M., Iwahana, H., Kanazawa, H., Hayashi, K., and Sekiya, T. (1989) Detection of polymorphisms of human DNA by gel-electrophoresis as single-strand conformation polymorphisms. *Proc. Natl. Acad. Sci. USA* **86,** 2766–2770.

5. Oldenburg, J., Ivaskevicius, V., Rost S., Fregin, A., White, K., Holinski-Feder, E., et al. (2001) Evaluation of DHPLC in the analysis of hemophilia A. *J. Biochem. Biophys. Meth.* **47,** 39–51.

6. Waseem, N. H., Bagnall, R., Green, P. M., and Giannelli, F. (1999). Start of UK confidential haemophilia A database: analysis of 142 patients by solid phase fluorescent chemical cleavage of mismatch. *Throm. Haemost.* **81,** 900–905.

7. Naylor, J. A., Green, P. M., Montandon, A. J., Rizza, C. R., and Giannelli, F. (1991) Detection of three novel mutations in two haemophilia A patients by rapid screening of whole essential region of factor VIII gene. *Lancet* **337,** 635–639.

8. Naylor, J. A., Green, P. M., Rizza, C. R., and Giannelli, F. (1992) Factor VIII gene explains all cases of haemophilia A. *Lancet* **340,** 1066–1067.

9. Naylor, J. A., Buck D., Green P., Williamson H., Bentley D., and Giannelli F. (1995) Investigation of the factor VIII intron 22 repeated region (int22h) and the associated inversion junctions. *Hum. Mol. Genet.* **4,** 1217–1224.

10. Liu, Q., Nozari, G., and Sommer S. S. (1998) Single-tube polymerase chain reaction for rapid diagnosis of the inversion hotspot of mutation in hemophilia A. *Blood* **92,** 1458–1459.

11. Forrest, S. M., Dahl, H. H., Howells, D.N., Dianzani, I., and Cotton R. G. H. (1991) Mutation detection in phenylketonuria using the chemical cleavage of mismatch method: Importance of using probes from both normal and patient samples. *Am. J. Hum. Genet.* **49,** 175–183.

14

Denaturing Gradient Gel Electrophoresis

Yvonne Wallis

1. Introduction

Denaturing gradient gel electrophoresis (DGGE) is a powerful mutation detection technique described by Fisher et al. in 1983 *(1)*. It allows the resolution of relatively large DNA fragments (usually polymerase chain reaction [PCR] products) differing by only a single nucleotide. DGGE is most frequently employed for the detection of unknown mutations. A related technique referred to as constant denaturing gradient gel electrophoresis (CDGE) is used for the sensitive detection of known mutations *(2,3)*. More recently a modified version of DGGE known as 2-dimensional DGGE has been employed for the simultaneous analysis of several PCR fragments *(4)*. PCR fragments are initially separated by size on a polyacrylamide gel prior to loading onto a conventional denaturing gradient gel system, as will be described in this chapter.

Denaturing gradient gel electrophoresis offers a number of advantages over other mutation detection techniques, including:

1. It is capable of detecting up to 100% of all single base substitutions in fragments up to 500 bp in length. It has been shown to be more sensitive than many other commonly used mutation detection methods including single-strand conformation polymorphism (SSCP) and heteroduplex analysis *(5,6)*.
2. The availability of user-friendly computer programs reduces the amount of preliminary work required prior to fragment analysis.
3. It uses a nonradioactive method of DNA band detection.
4. The formation of heteroduplexes significantly improves the detection of heterozygotes.
5. Variant DNA bands may be readily isolated from gels for subsequent sequencing analysis.

From: *Methods in Molecular Biology, vol. 187: PCR Mutation Detection Protocols*
Edited by: B. D. M. Theophilus and R. Rapley © Humana Press Inc., Totowa, NJ

DGGE also has a number of limitations, including:

1. Its sensitivity decreases for fragments larger than 500 bp.
2. It requires the use of specialist equipment.
3. The addition of GC clamps makes the purchase of PCR primers expensive.
4. Knowledge of melting behaviour is essential for efficient analysis and therefore preliminary work-ups must be performed before fragment analysis can take place.

DGGE analysis involves the electrophoresis of double-stranded DNA fragments through a polyacrylamide gel containing a linearly increasing concentration of denaturants usually urea and formamide *(1)*. Temperature may also be employed as the denaturant using a similar technique known as temperature denaturing gradient gel electrophoresis (TGGE). During migration discrete regions within the DNA fragments, referred to as melting domains, denature ("melt") at specific positions along the denaturing gradient. The concentration of denaturant at which denaturation of a melting domain occurs is referred to as the melting temperature or T_m. Branched denatured portions of DNA fragments entangle more readily in the polyacrylamide matrix and as a consequence cause an abrupt decrease in the mobility of the whole DNA fragment.

A DNA fragment may contain more than one melting domain with the T_m of each one being strictly dependent upon its nucleotide sequence. Alterations in the nucleotide sequence (e.g., disease-causing mutations) in all but the highest melting temperature domain will alter the melting properties of the whole DNA molecule thereby altering its electrophorectic mobility **(Fig. 1)**.

Base changes in the domain with the highest melting temperature will not be detected due to the loss of sequence dependent migration upon complete strand dissociation. This limitation is easily overcome by the attachment of a highly thermostable segment referred to as a GC clamp to one end of the DNA fragment. The GC clamp is efficiently introduced during the PCR amplification step prior to DGGE analysis by the use of a modified primer containing a GC tail at its 5' end. In general a GC clamp of 40 bp as originally described by Sheffield et al. *(7)* is sufficient (*see* **Note 1**). Longer GC clamps may be required for the sensitive analysis of particularly GC-rich sequences *(8)*. The introduction of a GC clamp increases the percentage of mutations detectable by DGGE close to 100% in fragments up to 500 bp therefore making it more sensitive than other commonly used mutation detection systems.

Successful DGGE requires knowledge of the melting behavior of each DNA fragment to be analyzed. Optimal DGGE results are obtained if the DNA fragment under analysis contains only one or two melting domains (excluding the GC-clamp domain). In fragments containing two melting domains, the GC clamp should be positioned immediately adjacent to the domain with the higher melting temperature. Computational analysis using software such as MELT95

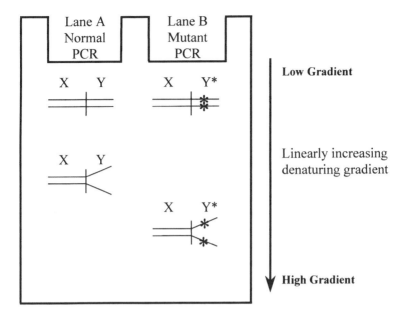

Fig. 1. Schematic diagram showing the electrophoresis of normal (lane A) and mutant (lane B) PCR fragments. The normal PCR product in lane A has two melting domains, X and Y, and the melting temperature of X (T_mX) is higher thatn that of Y (i.e., $T_mX > T_mY$). As the normal fragment migrates through the linearly increasing denaturing gradient gel, domain Y "melts" before domain X and the branched segment of Y causes retardation of the whole PCR fragment. A mutation in domain Y (marked with an *) increases the T_m of domain Y. $Y*$ therefore melts at a higher denaturing concentration and therefore migrates to a position further along the denaturing gradient than its normal counterpart Y. The mutant PCR fragment therefore migrates further along the denaturing gradient.

(9) or MACMELT (commercially available from Bio-Rad Laboratories, UK), allows the melting behavior of each target DNA sequence to be visualized as a meltmap. Using these programs PCR primers can be specifically designed to create fragments containing only one or two melting domains, and GC clamps can be placed to produce products with the most favorable melting behavior.

DGGE is particularly sensitive when screening for heterozygous nucleotide alterations and is used to analyze a number of genes causing dominantly inherited genetic disorders. Examples include the *APC* and the *hMLH1* and *hMSH2* mismatch repair genes *(10–12)*. Mismatches present in heteroduplex molecules produced during the PCR stage significantly reduce the thermal stability of the whole DNA fragment causing them to denature at a lower denaturant concentration than their homoduplex counterparts. Heteroduplex molecules usually

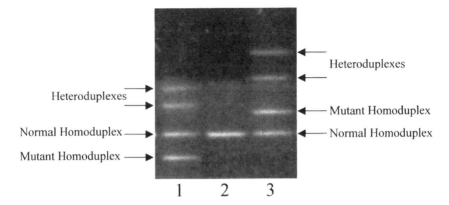

Fig. 2. Photograph of a DGGE gel for the analysis of exon 6 of the *APC* gene. Lane 1 shows an A to G transition at codon 235 and lane 3 shows a C to T transition at codon 232. Lane 2 shows a normal band pattern. Two heteroduplex bands migrate above the normal/mutant homoduplex bands in both cases.

appear as additional bands above the homoduplexes and therefore facilitate detection of heterozygous mutations (**Fig. 2**). Homozygous mutations may also be detected by DGGE if the mutant homoduplex products migrate to a different position along the denaturing gradient than their normal counterparts. If necessary, heteroduplex molecules may be artificially created by mixing normal and patient PCR products prior to DGGE analysis (*see* **Note 2**).

A denaturing gradient linearly increasing from the top to the bottom of a vertical polyacrylamide gel is created using a gradient maker. A "high" denaturant polyacrylamide solution is gradually diluted with a "low" denaturant polyacrylamide solution as it pours into a glass-plate sandwich under gravity (*see* **Fig. 3**).

The choice of denaturant range (and hence the concentration of the "low" and "high" polyacrylamide solutions) is made based on the T_m of the domain of interest (obtained by computational analyses) with a top to bottom difference of 30%. Gels are usually run at 60°C and at this temperature the conversion factor between the T_m of a melting domain and the required percentage of denaturant is derived from the following formula:

$$\% \text{ denaturant} = (3.2 \times T_m) - 182.4$$

Therefore, the starting denaturant range for a particular DNA fragment is usually selected as the percentage of denaturant ±15% of the calculated % denaturant (as derived from the T_m value).

For example, if a PCR fragment has a single melting domain with a T_m of 71°C, the % denaturant required for "melting" at 60°C is calculated to be 45%,

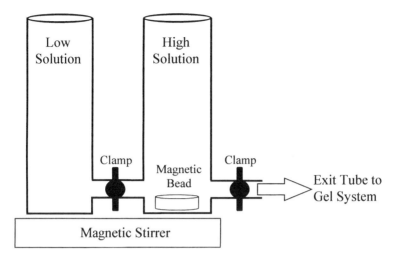

Fig. 3. Schematic diagram of a gradient maker used to generate linearly increasing denaturant gradient gels for DGGE analysis.

and therefore a denaturing gradient of 30–60% should be employed to resolve this fragment. Fragments containing two melting domains should be analyzed on two gradient gels with denaturant gradients specific for each of the two T_m values.

It may be necessary to optimize the denaturing gradient for some PCR products if they migrate either too far along the denaturing gradient or insufficiently so that they come to rest near the top of the polyacrylamide gel (*see* **Note 3**).

2. Materials

The author uses the INGENYphorU DGGE system supplied by Genetic Research Instrumentation Ltd (GRI), and all materials and methods described below are for use with this system. Commercial systems are not essential, however, and "homemade" equipment as originally described by Myers et al. *(13)* may be used if a good electrical workshop is to hand.

2.1. DGGE Equipment

1. DGGE gel system supplied by GRI, comprising of the following components (to make one DGGE gel): two glass plates, one with a large notch and one with a small notch; U-shaped spacer (1 mm thick); 32-well comb; plexi-glass pressure unit.
2. Gradient maker (GRI) with at least a 30 mL capacity per chamber.
3. Electrophoresis unit (GRI): designed to hold two gradient gels. The gels face each other to create the upper buffer reservoir. The unit also incorporates the

cathode and anode platinum electrodes. There is an inlet at the top of the unit that connects to a buffer flow tube. During electrophoresis, buffer is continuously pumped into the upper buffer reservoir and overflows from the chamber via two holes present on each side of the unit positioned above the level of the wells. Circulating buffer maintains a constant pH in the upper reservoir.

4. Buffer tank (GRI): holds 17L TAE (Tris/acetate/EDTA) buffer and incorporates a thermostat to keep the buffer at the designated temperature, as well as a pump to circulate buffer from the tank in to the upper buffer reservoir via a buffer flow tube. A safety mechanism cuts the current off if the tank lid is opened during electrophoresis.

2.2. Reagents and Solutions

1. 20X TAE buffer, pH 8.0: 800 mM Tris base, 20 mM EDTA and 400 mM sodium acetate. For 1 L dissolve the following in 950 mL distilled water: 97 g Tris base, 7.5 g Na$_2$EDTA, and 54.5 g sodium acetate.3H$_2$O

    Adjust to pH 8 with acetic acid (approx 30 mL). Make up to 1 L with distilled water. Store at room temperature.

2. 80% denaturant polyacrylamide stock: 8% polyacrylamide (*see* **Note 4**), 5.6 M urea, and 32% formamide. For 500 mL add together the following: 100 mL 40% polyacrylamide (19:1 acrylamide:bis-acylamide), 170 g urea, 160 mL deionized formamide (*see* **Note 5**), 25 mL 20X TAE buffer, and distilled water up to 500 mL.

    Stir until urea is completely dissolved. Store at 4°C in a dark bottle, stable at 4°C for 3 mo.

3. 0% Denaturant polyacrylamide (8%) stock. For 500 mL combine 100 mL 40% polyacrylamide (19:1 acrylamide:bis-acylamide) and 400 mL distilled water. Store at 4°C, stable at 4°C for 3 mo.

4. 10% Ammonium sulphate (APS). Dissolve 1 g in 10 mL distilled water. Divide into 500 μL aliquots and store at –20°C until required.

5. N,N,N',N'-Tetramethylethylenediamine (TEMED). Store in the dark at room temperature.

6. 6X Gel loading buffer: 0.025% bromophenol blue, 0.025% xylene cyanol, and 20% Ficoll. Dissolve 250 mg bromophenol blue, 250 mg xylene cyanol, and 20 g Ficoll in 100 mL distilled water; can be stored at room temperature.

7. 10 mg/mL Ethidium bromide solution. Dissolve 1 g ethidium bromide in 100 mL distilled water. Store at room temperature in a dark bottle. *Note*: Ethidium bromide is a highly hazardous substance. In addition to standard laboratory safety procedures, always wear appropriate protective equipment when handling ethidium bromide powder (i.e., face mask) and solution (i.e., gloves).

3. Methods
3.1. Preparation of DGGE Gel System

It is important to use thoroughly cleaned glass plates as grease and dust may disrupt the denaturing gradient gel during pouring. Plates should therefore be washed in water immediately after use and cleaned with 100% ethanol before use.

1. Assemble the glass plates and spacer(s) according to manufacturers instructions (GRI or other company). If using a homemade system, use plates that are at least 20 cm by 20 cm and use spacers that are 0.75–1 mm thick.
2. Insert the glass plate sandwich in to the electrophoresis cassette so that the inner smaller plate faces inwards (to create an upper buffer chamber). Insert the plexi-glass pressure unit so that it rests against the larger outer glass plate (this prevents over tightening of the screws and therefore plate cracking during electrophoresis). Lift the U-shaped spacer until it locks into position and tighten all screws to secure the glass plate sandwich into the electrophoresis unit. The design of the U-shaped spacer prevents leakage of the gradient gel during pouring.
3. The comb may be inserted at this point in to the top of the glass plate sandwich.

3.2. Preparation of Denaturing Gradient Gel

1. For each denaturing gradient gel, prepare appropriate "low" and "high" poly-acrylamide solutions, e.g., 30% and 60% denaturant solutions should be prepared to make a 30–60% denaturing gradient gel. The "low" and "high" solutions are prepared by mixing appropriate volumes of 0% and 80% denaturant polyacryla-mide solutions (*see* **Note 6**). The volume of the "low" and "high" solutions is dependent up on the total volume held by the glass plates. The INGENYphorU plates hold a total volume of 55 mL. Therefore, 27.5 mL volumes of "low" and "high" solutions are prepared separately for each denaturing gradient gel. The "low" and "high" solutions should be placed at 4°C for 15 min before pouring.
2. Place the gradient maker on a magnetic stirrer positioned 25–30 cm above the top edge of the glass plate sandwich. The connection between the two chambers should be closed, and the exit tube should be clamped. Insert the end of the exit tube into the top of the glass plate sandwich (this may be facilitated by attaching a yellow tip onto the end of the tubing). Place a magnetic bar into the "high" chamber (*see* **Fig. 3**). The apparatus is now ready for pouring a gradient gel.
3. Add appropriate volumes of 10% APS and TEMED to the "low" and "high" denaturing solutions and mix gently. 140 µL of 10% APS and 14 µL TEMED are added to the 27.5 mL solutions required to make an INGENYphorU gel.
4. Pour the "low" denaturant solution into the "low" chamber. Briefly open the connection between the two chambers to release a small quantity of polyacrylamide into the "high" chamber. This action prevents air bubbles blocking the connection tube.
5. Pour the "high" denaturant solution into the "high" chamber. Activate the magnetic bar to create a vortex, and open the clamped exit tube.
6. By gravity, the "high" denaturant solution will then leave the "high" chamber and pass into the glass plate sandwich via the exit tube. At this point, open the connection between the two chambers. The "low" solution will then flow into the "high" chamber, and mix with the "high" solution.

7. The clamp should be used to adjust the rate of flow of polyacrylamide solution from the "high" chamber through the exit tubing so that a denaturing gradient gel is poured in 5–10 min.

8. Allow the gel to polymerize for between 1–2 h.

9. In the meantime, heat the TAE buffer in the buffer tank to 60°C (*see* **Note 7**). Please note that the TAE buffer may be used three times before replacing.

3.3. Gel Electrophoresis

1. Following polymerization, loosen all screws until they just touch the plexi-glass pressure unit.

2. Place the electrophoresis cassette into the INGENYphorU buffer tank and connect the buffer flow tube. Buffer should overflow from the upper buffer chamber into the main tank through holes in the sides of the cassette. A valve on the outside of the tank controls the rate of buffer flow into the upper buffer reservoir.

3. Carefully remove the comb and push the U-shaped spacer downwards so that the bottom of the polyacrylamide gel makes contact with the buffer. Dislodge any little air bubbles lying below the gel by briefly holding the cassette at a 45° angle.

4. Retighten the upper two screws on each side of the cassette. Connect the electrodes and prerun the gel at 100 V for 10–15 min before loading.

5. During this time, add 6X gel-loading buffer to the PCR samples to be analyzed (*see* **Note 8**). Depending on the yield of the PCR product, load 5–10 μL of sample from a 25 μL PCR reaction (*see* **Note 9**).

6. Before loading, disconnect the power supply and flush out the wells thoroughly with TAE buffer using a needle and syringe. After sample loading reconnect the power supply and electrophorese for the required running time. The optimum running time may be calculated using software mentioned in the introduction. As a guide, the author runs all gels for the analysis of exons 3–14 of the Adenomatous polyposis coli gene at 100 V (for a gel 30 cm in length) for 16 h. Products may be electrophoresed at a higher voltage for a shorter time if required.

3.4. Gel Staining

1. After the required electrophoresis time disconnect the power supply, and switch off the buffer tank. Carefully remove the hot gel electrophoresis unit from the buffer tank.

2. Remove the gel(s) from the unit and separate the glass plates gently to leave the gel on the noneared plate. Remove as appropriate a top gel corner to assign the gel-loading end.

3. Place the gel (still on a noneared plate) into a staining box and stain in 1X TAE buffer (or distilled water) containing 0.5 μg/μL ethidium bromide.

4. Shake gently for 15–30 min.

5. Decant ethidium bromide solution and rinse gel in distilled water and visualize the DNA bands under UV light (254 nm).

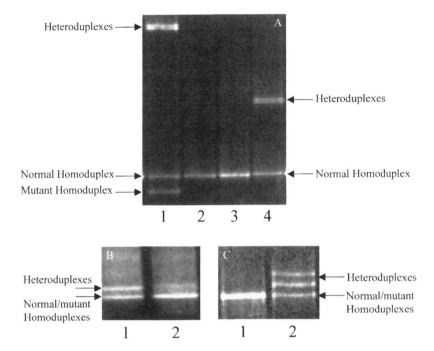

Fig. 4. Photographs of DGGE gels showing examples of variant band patterns detected in exons 10 (panel **A**), 15D (panel **B**) and 5 (panel **C**) of the *APC* gene. Panel A: four bands in lane 1 represent a 5-bp deletion at codon 456 and 3 bands are seen in lane 4 caused by a 2-bp insertion at codon 452. Panel B: a two-band pattern in lane 1 is caused by a 1-bp deletion at codon 964. Panel C: a G to T splice site mutation causes a three-band pattern in lane 2. The presence of heteroduplex bands in lanes 4 of Panel A, lane 1 of Panel B and lane 2 of Panel C facilitates the detection of these variants since the normal and mutant homoduplex bands comigrate.

3.5. Gel Interpretation

A normal PCR fragment (containing no nucleotide alterations) will resolve on a DGGE gel as a single band. Amplification of a DNA fragment containing a heterozygous nucleotide change (whether single base substitution or small insertion/deletion) will result in the formation of both normal and mutant homoduplex molecules as well as two different heteroduplex molecules. Ideally, the mutant homoduplex will be displaced from its normal counterpart. In addition, the two heteroduplexes will migrate to different positions along the denaturing gradient. Following ethidium staining, therefore, a variant band pattern will usually appear as four DNA bands. In some cases, however, variants may manifest as two or even three DNA bands (**Fig. 4**).

4. Notes

1. The 40 bp GC clamp originally described by Sheffield et al. *(7)* has the following sequence: 5'-CGCCCGCCGCGCCCCGCGCCCGTCCCGCCGCCCCCGCCCC-3' and should be attached to the 5' end of either the forward or reverse primer.

2. To make heteroduplex molecules when screening homozygous mutations, mix equal quantities of normal and mutant PCR products and denature at 95°C for 10 min followed by 30 min at the product-specific annealing temperature. Cool to room temperature before loading.

3. Optimization of DNA band migration may be achieved by altering the denaturant gradient range. For example, if fragments migrate too far into the gel before they resolve, the denaturant concentration at the top should be increased. However, if fragments do not migrate sufficiently, the concentration at the top should be lowered.

4. The concentration of polyacrylamide may be altered to suit the fragment size under analysis.

5. To deionize formamide, add 3 g of mixed bed resin to 100 mL formamide and mix for 30 min (in a fume hood). Filter through Whatman paper (again in a fume hood) and store in a dark bottle at 4°C.

6. Appropriate volumes of 0% and 80% denaturant polyacrylamide solutions are mixed to make "low" and "high" gradient solutions. For example, 30% "low" and 60% "high" solutions are made to pour a denaturing gradient gel with a range of 30–60%. The following volumes of 0% and 80% denaturant polyacrylamide solutions are mixed to give 27.5 mL volumes of 30% and 60% solutions:

	30% Low solution	60% High solution
volume 0% solution	17.2 mL	6.9 mL
volume 80% solution	10.3 mL	20.6 mL

7. The temperature of the running buffer is very important, and it must not be allowed to deviate from 60°C, as this is a factor in the derivation of the denaturing gradient. If the temperature is too high, DNA fragments may not migrate far enough into the gel, and if it is too low, migration may be too far.

8. PCR conditions are specific to the fragment under analysis, and therefore have not been discussed in this chapter other than the necessity to incorporate a GC clamp into the product. Reactions however should be free from nonspecific amplification products to avoid confusion caused by extra bands.

9. It is important to avoid overloading the gel as this leads to "fuzzy" bands or smeared lanes making interpretation difficult. "Fuzzy" bands and smeared lanes may also be caused if the samples have not properly focused, a problem that may require either optimization of the denaturing gradient conditions or primer design.

References

1. Fischer, S. G. and Lerman, L. S. (1983) DNA fragments differing by a single base-pair substitution are separated in denaturing gradient gels: correspondence with melting theory. *Proc. Natl. Acad. Sci USA* **80,** 1579–1583.

2. Hovig, E., Smith-Sorenson, B., Brogger, A., and A. L. (1991) Constant denaturant gel electrophoresis, a modification of denaturing gradient gel electrophoresis, in mutation detection. *Mut. Res.* **262,** 63–71.

3. Borresen, A. L., Hovig, E., Smith-Sorenson, B., Malkin, D., Lystad, S., Andersen, T. I., Nesland, J. M., Isselbacher, K. J., and Friend, S. H. (1991) Constant denaturant gel electrophoresis as a rapid screening technique for p53 mutations. *Proc. Natl. Acad. Sci. USA* **88,** 8405–8409.

4. Van Orsouw, N. J., Vijg, J. (1999) Design and application of 2-D DGGE-based gene mutational scanning tests. *Genet Analyt.* **14,** 205–213.

5. Takahashi, N., Hiyami, K., Kodaira, M., and Satoh, C. (1990) An improved method for the detection of genetic variations in DNA with denaturing gradient electrophoresis. *Mut. Res.* **234,** 61–70.

6. Moyret, C., Theillet, C., Puig, P. L., Moles, J. P., Thomas, G., and Hamelin, R. (1994) Relative efficiency of denaturing gradient gel electrophoresis and single strand conformation polymorphism in the detection of mutations in exon 5 to 8 of the *p53* gene. *Oncogene* **9,** 1739–1743.

7. Sheffield, V. C., Cox, D. R., Lerman, L. S., and Myers, R. M. (1989) Attachment of a 40 base pair G+C-rich sequence (GC-clamp) to genomic DNA fragments by the polymerase chain reaction results in improved detection of single base changes. *Proc. Natl. Acad. Sci USA* **86,** 232–236.

8. Olschwang, S., Laurent-Puig, P., Groden, J., White, R., and Thomas, G. (1993) Germ-line mutations in the first 14 exons of the adenomatous polyposis coli (APC) gene. *Am. J. Hum. Genet.* **52,** 273–279.

9. Lerman, L. S. and Silverstein, K. (1987) Computational simulation of DNA melting and its application to denaturing gradient gel electrophoresis, in *Methods in Enzymology Volume 155,* (Wu, R., ed.), Academic Press, NY, pp. 482–501

10. Wallis, Y. L., Morton, D. G., McKeown, C. M., and Macdonald, F. (1999) Molecular analysis of the APC gene in 205 families: extended genotype-phenotype correlations in FAP and evidence for the role of APC amino acid changes in colorectal cancer predisposition. *J. Med. Genet.* **36,** 14–20.

11. Fidalgo, P., Almeida, M. R., West, S., Gaspar, C., Maia, L., Wijnen, J., et al. (2000) Detection of mutations in mismatch repair genes in Portuguese families with hereditary non-polyposis colorectal cancer (HNPCC) by a multi-method approach. *Eur. J. Hum. Genet.* **8,** 49–53.

12. Fodde, R., Vander Luijt, R., Wijnen, J., Tops, C., Van der Klift, H., Van Leeuwen-Cornelisse, I., et al. (1992) Eight novel inactivating germ line mutations of the APC gene identified by denaturing gradient gel electrophoresis. *Genomics* **13,** 1162–1168.

13. Myers, R. M., Maniatis, T. and Lerman, L. S. (1987) Detection and localisation of single base changes by denaturing gradient electrophoresis, in *Methods in Enzymology, Volume 155* (Wu, R., ed), Academic Press, NY, pp. 501–527.

15

Conformation-Sensitive Gel Electrophoresis

Ian J. Williams and Anne C. Goodeve

1. Introduction

Conformation-sensitive gel electrophoresis (CSGE) was first described by Ganguly et al. in 1993 *(1)*. This technique was developed as the result of a study into a rapid, non-radioactive heteroduplex-based detection method for mutation screening. The method relies on the differential migration of DNA heteroduplexes in comparison with homoduplexes during polyacrylamide gel electrophoresis under mildly denaturing conditions. Ethidium bromide staining and visualization under ultraviolet (UV) light determines those samples with aberrant banding patterns resulting from heteroduplexes. These samples are subsequently subjected to DNA sequencing to determine the nature of the nucleotide alteration.

This chapter aims to describe the theory behind CSGE and provide information to establish this method both quickly and effectively in most laboratory situations.

1.1. CSGE Theory

The difference in electrophoretic mobility of DNA homoduplexes from DNA heteroduplexes has been the focal point for the development of rapid screening techniques in the detection of DNA sequence alterations. A DNA homoduplex differs from a DNA heteroduplex in its complete Watson–Crick base pairing of adenine–thymine and guanine–cytosine pairs. A DNA homoduplex consists of entirely complementary DNA and is completely Watson–Crick base-paired. A DNA heteroduplex has incomplete Watson–Crick base-pairing in opposition. Mismatched, noncomplementary base-pairing may involve a single mismatched nucleotide to produce a DNA heteroduplex, detectable by CSGE.

From: *Methods in Molecular Biology, vol. 187: PCR Mutation Detection Protocols*
Edited by: B. D. M. Theophilus and R. Rapley © Humana Press Inc., Totowa, NJ

The CSGE method relies upon mildly denaturing solvents to amplify the conformational changes caused by single-base mismatches during polyacrylamide gel electrophoresis *(1)*. Under nondenaturing conditions, slight conformational changes caused by single-base mismatches in DNA fragments will have an almost identical electrophoretic mobility to wild-type DNA fragments of identical sequence. Under mildly denaturing conditions, one mismatched base becomes rotated out of the double helix, creating a bend or kink in the DNA fragment. The conformational change causes aberrant migration of this fragment compared with wild-type DNA during polyacrylamide gel electrophoresis under mildly denaturing conditions. If the denaturants within the gel matrix are increased above optimal conditions for CSGE, both mismatched bases are rotated out of the double helix and the bend or kink causing the conformational change is eliminated. In this situation, homoduplexed and heteroduplexed DNA will have similar electrophoretic mobilities. Bhattacharrya and Lilley *(2)* proposed that under nondenaturing conditions, DNA fragments with an additional base or bases on one side of the DNA helix, such as a deletion or insertion resulting in a frameshift mutation, exhibit more marked gel retardation than DNA fragments possessing single mismatches, where two non-Watson–Crick bases are in opposition. They also reported an increased gel retardation with an increasing number of mismatched bases. This increased gel retardation with insertion and deletion mutations is also observed under the mildly denaturing conditions of CSGE *(see* **Subheading 3.4.**). Therefore, CSGE aims to optimize conditions for mismatched bases in heteroduplexed DNA to result in a shift in band migration during polyacrylamide gel electrophoresis.

1.2. CSGE Conditions

Optimized conditions for CSGE were described by Ganguly and Prockop *(3)*. Denaturants ethylene glycol and formamide, known to alter the conformation of DNA, were chosen for inclusion into the CSGE gel mix at concentrations of 10% and 15% respectively, because of their compatibility with polyacrylamide gel electrophoresis. A tris-taurine buffer system was employed, replacing a tris-borate buffer, as the tris-taurine buffer is glycerol tolerant *(1)*. Finally, a 10% polyacrylamide gel of 1 mm thickness was used with bisacryloylpiperazine (BAP), replacing bis(*N*,*N*-methylene bisacrylamide) as a crosslinker. These conditions provide strength to CSGE gels and enhance the sieving action of the gel during electrophoresis *(3)*.

1.3. Sensitivity of Technique

Originally, Ganguly et al. tested CSGE on a range of polymerase chain reaction (PCR) products possessing known single-nucleotide changes in amplified

DNA fragments varying in size from 200 to 800 base pairs (bp) *(1)*. Eighteen of twenty-two single-base changes were detected in four different collagen genes. Of the four not detected, three mismatches were located in an isolated high-melting-temperature domain. The remaining undetected mismatch was located 51 bp from one end of the PCR fragment. Subsequently, this mismatch was detected when present 81 bp from the end of a new PCR fragment. A similar observation was made with samples from patients having known *factor IX* gene *(FIX)* mutations, where 31/35 mismatches were detected *(1)*. The four undetected mismatches were located within 50 bp of the end of the PCR fragment. This led to PCR primers being designed further away from the sequence of interest, with 50–100 bp of redundant sequence being incorporated into each end of the region amplified. In the same study, 100% detection of mutations was achieved in a further 11 samples having sequence alterations in the M13 phage and in the elastin gene, giving a total of 60/68 mismatches detected. The mutations not detected were either located in the terminal sequence of the PCR product or located in a high-melting temperature domain. In our laboratory, CSGE was applied to screening the *factor VIII* (*FVIII*) gene of hemophilia A patients with similar success *(4)*. Analysis of the *FVIII* gene of seven patients with hemophilia A revealed seven nucleotide alterations plus an intragenic deletion, detected because of the failure of amplification of exons 23–25. The FIX gene has also been examined by CSGE in our laboratory. CSGE detected nucleotide alterations in 21 of 21 individuals with hemophilia B, 11 of which were previously unknown *(5)*. Recent improvements by Korkko et al. *(6)* yielded 100% detection of mutations by CSGE in 76 different PCR products, ranging in size from 200–450 bp using altered electrophoresis conditions. CSGE has also been adapted to use fluorescent dyes and automated detection on a DNA sequencer for an increased screening throughput *(7)*. In summary, CSGE will detect close to 100% of nucleotide sequence alterations.

1.4. Applications of CSGE

Conformation-sensitive gel electrophoresis has been used for detection of sequence alterations in many different genes. These include inherited defects in several collagen genes *(1,6,8–10)* and genes involved in haemostasis *(4,5,11,12)*, plus acquired mutations in c-kit *(13)* and *BRCA1* and *BRCA2 (7,14)*.

1.5. Variations on CSGE

Many variations on CSGE conditions have been presented by different laboratories. As a standard, we use the PCR sample alone in the heteroduplex reaction, covered by mineral oil to prevent evaporation of the sample, and a 10% acrylamide (99:1 acrylamide:BAP) gel containing 10% ethylene glycol and

15% formamide. This CSGE gel is prerun for 1 h at 750 V and then samples are electrophoresed at 400 V for 16–17 h. Early work involved the addition of ethylene glycol, formamide, xylene cyanol, and bromophenol blue to samples to be heteroduplexed *(1,3,9,10)*. Most laboratories heteroduplex samples without additives. Markoff et al. *(14)* used urea as a denaturant in the gel, at 15%, in place of formamide (15%). Electrophoresis conditions have ranged from 40 W for 6 h *(6)* to a 400 V overnight run *(4)*. Gel documentation has been reported by both ethidium bromide staining and by silver staining *(14)*. A more recent development has also adapted the CSGE mutation scanning assay for use with fluorescent detection (F-CSGE) *(7)*.

1.6. PCR Amplification
1.6.1. Primer Design
Primers for use in PCR amplification of a gene for mutation detection should encompass the entire coding region, including intron/exon boundaries, the promoter region, and the polyadenylation signal. Most primer sets will thus be intronic and should include, for the purpose of CSGE, an additional 80–100 bp at either end of each exon. Larger exons can be split into smaller PCR fragments, with an overlap of 80–100 bp for each fragment.

1.6.2. PCR Conditions
Polymerase chain reaction optimization for mutation screening enhances the quality of CSGE results and reduces PCR product purification steps for subsequent DNA sequencing. Following PCR amplification, samples are electrophoresed on 5% polyacrylamide:bis (*see* **Note 1**) minigels to assess DNA purity and concentration. A single strong PCR product is ideal for CSGE and sequence analysis. Polyacrylamide gel electrophoresis will also provide information relating to the volume of sample to be loaded onto the CSGE gel. If the sample is too concentrated, a smaller volume can be loaded or even diluted in water to maintain the loading volume. An increased volume can also be loaded if the DNA concentration is too low (*see* **Note 2**).

In some situations we have found that the components of the PCR reactions can affect the DNA sample when loaded onto the CSGE gel. In these cases, smearing of the sample has occurred and, instead of a single band, the sample resembles a long streak blending into the DNA front, masking any indication of heteroduplex band separation. This smearing can be easily eliminated by changing PCR buffer components or buffer manufacturer. In our laboratory, Bioline *Taq* polymerase is used with the manufacturers 10X $(NH_4)_2SO_4$ buffer with excellent results. Bioline *Taq* also works reliably with a homemade ammonium-sulfate-based buffer (*see* **Note 3**). Use of commercially supplied buffer ensures standard amplification conditions and eliminates the batch-to-batch variation with homemade buffer.

An incompatibility between PCR amplification buffer and CSGE gel can be identified at an early stage in the procedure by the appearance of smeared bands on CSGE gels, not present on polacrylamide:bis mini gels, even with a dilution of the CSGE sample. Once PCR conditions have been identified that give sharp single bands on CSGE gels, test samples can be amplified, heteroduplexed, and screened by CSGE.

1.7. Heteroduplexing DNA Samples

DNA heteroduplexes are formed between mismatched wild-type and mutant DNA fragments. The heteroduplexing method is a simple two-step process involving an initial denaturing stage, where hydrogen bonds between paired nucleotides are broken, producing single-stranded DNA fragments. This dena-turing step involves heating the sample to 98°C for 5 min. The second step, or annealing stage, involves cooling the sample to 65°C for 30 min (*see* **Note 4**). At this temperature, the DNA can reform back into double-stranded DNA, with the possibility of each sense strand pairing with a different antisense strand (**Fig. 1**). Heteroduplexes are formed when the new paired strands differ in sequence by one or more nucleotides. For inherited disorders, the inheritance pattern of the expected nucleotide change requires consideration. Males posses only one copy of X-linked genes, such as *FVIII*. For their analysis, a PCR sample from a normal individual is mixed with the patient's sample, enabling the two (potentially different) alleles to mix and form heteroduplexes. Five microliters each of patient plus control PCR product are mixed and overlaid with mineral oil to avoid evaporation during heteroduplexing (*see* **Note 5**). In the situation where the subject is expected to be heterozygous for a mutation, such as female carriers of hemophilia, PCR product is heteroduplexed with its own normal allele. This also applies to autosomal dominant inherited disorders such as *von Willebrand disease* (chromosome 12). Therefore, if heterozygosity for a defect is expected, heteroduplexing can be added at the end of the PCR thermocycling and an aliquot loaded directly onto the CSGE gel. Where there is an uncertainty about the nature of a mutation, samples should be analyzed by CSGE following both heteroduplexing against self and against a wild-type sample. The latter technique has been used determine allele frequency in poly-morphism analysis *(4)*.

2. Materials
2.1. Gel Preparation

All chemicals were obtained from Sigma, unless specified.

1. 20X TTE (1.78 *M* Tris, 570 m*M* taurine, 4 m*M* EDTA). A Tris-taurine buffer system was developed for CSGE gels, as it is more alcohol tolerant than the tra-ditional Tris-borate buffer. A 500-mL stock solution of 20X TTE can be used for

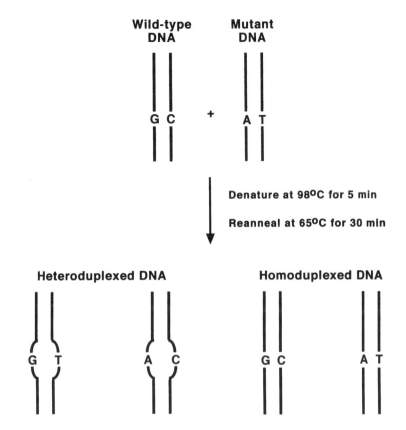

Fig. 1. Illustration of the heteroduplexing reaction between normal or wild-type DNA and mutant DNA. (Top) Prior to heteroduplexing, both DNA strands exhibit Watson–Crick base-pairing. At the mutation site, wild-type pairing is presented as G-C and the mutated nucleotide as A-T. Following the heteroduplexing reaction, both homoduplex and heteroduplex DNA are formed (bottom). In homoduplexed DNA, original base-pairings are reinstated of wild-type G-C and mutant A-T. Heteroduplex DNA is formed where wild-type and mutant DNA strands are paired, which differ by one or more nucleotides. In this example, the two heteroduplexed species formed have non-Watson–Crick base-pairings of G-T and A-C. DNA heteroduplexes such as these can be resolved from DNA homoduplexes, because of their induced conformation under mildly denaturing conditions by CSGE.

   five CSGE gel runs, including subsequent staining procedures, and can be kept at
   room temperature for approx 4 mo.
2. 99:1 Acrylamide:BAP (40%). The crosslinker bisacryloylpiperazine (BAP) has
   been reported to be a more efficient crosslinker than bis(N,N-methylene
   bisacrylamide) in its ability to improve the CSGE gel resolving capacity and to

enhance the physical strength of polyacrylamide gels. A 500 mL gel mix contains 198 g of solid acrylamide (BDH) (*see* **Note 6**) and 2 g of BAP (Fluka). Once dissolved, this stock is kept at 4°C and has a shelf life of 4 mo.

3. Gel loading buffer: 50% glycerol, 0.25% xylene cyanol, 0.25% bromophenol blue. The ingredients are mixed and dissolved in water.

4. 10% Ammonium persulfate (APS). Used in the polymerization of acrylamide gels. A solution of 10% (w/v) APS dissolved in water is made fresh every 2 wk. Reduced APS activity prolongs the polymerization reaction.

5. Ethidium bromide (EtBr, 10 mg/mL). Used in powder form and dissolved in water to yield a concentration of 10 mg/mL. Care must be taken in the preparation, handling, and disposal of this solution (*see* **Note 7**).

6. Gel rigs. Electrophoresis equipment required for running CSGE gels are of a standard design and have few special requirements. Electrophoresis tanks used for manual sequencing can be used for CSGE and make a good starting point to "test" the procedure. Preferably, shorter rigs should be used to minimize wasted gel space. For this reason, we use gel tanks which are 410 mm × 330 mm (Flowgen).

7. Gel plates. The glass plates are a standard plain and lugged glass plate set, dimensions 410 mm × 330 mm, with a siliconized lugged plate. One-millimeter spacers and 1-mm castle combs are used in casting CSGE gels. The thickness of the gel will affect its run time, heat dissipation, and the strength of the gel during the later handling stages. Once the lugged glass plate has been siliconized and both plates cleaned with 70% ethanol, the plates can be sealed using electrical tape (Genetic Research Instrumentation Ltd., code AFT/UT), to prevent leaking of the gel during casting.

8. Combs. One-millimeter castle combs (Web Scientific, special request) are used to cast a CSGE gel with overall comb dimension width 280 mm, depth 35 mm, and thickness 10 mm. Each comb has 40 teeth, but it can be cut in half to give two equal-sized combs of 20 teeth to use with standard sequencing-sized plates. The well dimensions are important to prevent overloading of the well. Well dimensions are 5 mm width, 10 mm depth, and 2 mm well separation.

2.2. Casting the Gel

1. (40%) 99:1 acrylamide:BAP.
2. 20X TTE buffer.
3. Formamide.
4. Ethylene glycol.
5. Deionized water.
6. 10% Ammonium persulfate (APS).
7. TEMED (*N,N,N',N'*-tetramethylethylenediamine).
8. Sealed glass plates (1 mm-thick spacers and comb).
9. Bulldog clips.
10. 50 mL syringe with a 21-gauge needle.

2.3. Loading and Running the Gel

1. One precast CSGE gel, prepared as in **Subheading 3.1.**
2. One gel running rig.
3. One power pack (range 100–1000 V).
4. 2 L of 0.5X TTE buffer (dependent on capacity of gel running rig reservoir).
5. Glycerol loading buffer.
6. Heteroduplexed DNA samples (*see* **Subheading 1.7.**).
7. Positive and negative control samples (*see* **Note 8**).
8. 50 mL syringe with a 21-gauge needle.

2.4. Staining and Viewing the Gel

1. One CSGE gel (loaded and electrophoresed for 16–17 h).
2. Two large staining trays, large enough to accommodate the plain glass plate (e.g., photographic developer tray; one for staining and one for destaining).
3. Plate separator (e.g., plastic wedge tool from Hoeffer).
4. 2 L of 0.5X TTE.
5. 2 L of deionized water.
6. Ethidium bromide (10 mg/mL).
7. Sharp scalpel.
8. Two sheets of 3MM blotting paper (Whatman) (area should be larger than that of the gel).
9. Water bottle containing deionized water.
10. Hand-held UV light (302 nm).
11. UV transilluminator (302 nm).
12. Gel documentation system.

3. Methods

3.1. Gel Preparation

1. For 175 mL of CSGE gel mix (10% 99:1 acrylamide:BAP, 0.5X TTE, 15% formamide, 10% ethylene glycol), mix:

Sterile water	81.38 mL
99:1 acrylamide:BAP	43.75 mL
20X TTE	4.38 mL
Formamide	26.25 mL
Ethylene glycol	17.50 mL

2. To polymerize the gel, add 1.75 mL of 10% APS and 100 µL TEMED (*see* **Note 9**).
3. Mix well and pour into prepared sealed glass plates. (The gel mix can be injected into the space between the plates using the 50-mL syringe. To minimize the occurrence of air bubbles, the gel mix should be introduced down the inside edge of the plate.)
4. Once full, the plate should be laid horizontally with the open end elevated slightly (place top, open end of glass plate on a universal).

5. Insert comb into the mouth of the glass plates at the required well depth. (To prevent the introduction of air bubbles, the gel plates can be overfilled before comb insertion).
6. Clamp the gel plates together with bulldog clips, starting from the bottom of the plates. Allow the gel mix to run out of the mouth of the plates and past the comb.
7. Check that the comb has not been dislodged and clamp firmly into place.
8. Leave the gel to polymerize for at least 1 h in this position, checking regularly for leaks. If leaks do occur, top up gently with the remaining gel mix.

3.2. Loading and Running the Gel

1. Remove comb from precast CSGE gel (*see* **Note 10**).
2. Slit the tape at the bottom of the gel with a scalpel blade to allow buffer access and place gel in electrophoresis apparatus.
3. Add 0.5X TTE to fill reservoirs.
4. Gently clean wells with 0.5X TTE buffer from reservoir using 50 mL syringe and needle (*see* **Note 11**).
5. Prerun gel at 750 V for 1 h.
6. Turn off power and clean out all wells again with 0.5X TTE (*see* **Note 11**).
7. Pipet 2 μL aliquots of loading buffer onto parafilm (parafilm can be temporarily attached to the workbench by wetting with a small amount of water).
8. Mix 2–8 μL of heteroduplexed sample with 2 μL loading buffer and load onto the CSGE gel (*see* **Notes 2** and **5**).
9. Depending on the speed of sample loading, wells should be washed out every four to five wells with the syringe containing 0.5X TTE buffer. Care must be taken not to elute samples already loaded onto the gel.
10. Once all samples have been loaded, electrophorese at 400 V (10 V/cm) for 16–17 h at room temperature (the second dye front, xylene cyanol, will run at about 200 bp).

3.3. Staining and Viewing the Gel

1. Make space on a workbench close to the gel documentation system (e.g., darkroom).
2. Fill staining tray with 2 L of 0.5X TTE and add 200 μL of EtBr (10 mg/mL) and mix well.
3. Add 2 L of water to the destaining tray.
4. Turn off electrophoresis power supply from gel rig and take CSGE gel to the designated staining area.
5. Place gel face up on the work surface and remove sealing tape. The CSGE gel can be exposed by removing the top, siliconized, lugged plate with a plate separator. The CSGE gel should stick to the plain glass plate (*see* **Notes 12** and **13**).
6. Carefully lower gel and plate into staining tray (gel face up) and leave for 10 min.
7. Drain off excess buffer and lower gel into destaining tray for 10–15 min (gel face up).
8. After destaining, carefully drain off excess water and place plate and gel on work surface (gel face up).
9. The CSGE gel can be transferred from the glass plate to 3MM blotting paper by covering the gel with two sheets of 3MM paper. Apply firm pressure to obtain good contact between the gel and blotting paper.

10. Carefully peel back the blotting paper, ensuring that the CSGE gel is firmly attached.
11. Place gel face up on the work surface.
12. Locate bands by visualizing the gel with a hand-held UV light (302 nm) and excise relevant portions of DNA band containing gel using the scalpel blade (*see* **Notes 13** and **14**).
13. Wet the surface of the UV transilluminator with deionized water from the water bottle.
14. Place the excised gel section face down on the wet transilluminator.
15. Wet the 3MM paper with water from the water bottle and peel off the 3MM paper, leaving the excised gel section on the UV transilluminator (*see* **Note 15**).
16. Visualize bands in the gel using the UV transilluminator and document results (*see* **Note 16**).

3.4. Gel Banding Patterns

The CSGE results often reflect the nature of the nucleotide change. The variation in enhancement or retardation of DNA migration of heteroduplexes from homoduplexes will mainly depend on the type or nature of the nucleotide change. The overall fragment size, sequence composition, and flanking nucleotide sequence at the mutation site will also play a role in the degree of DNA band separation during polyacrylamide gel electrophoresis.

Small insertions or deletions of one or a few base pairs produce the largest band separations, because of an increased conformational change from wild-type DNA. In these cases, all four possible reannealled conformations can often be seen on CSGE gels (**Figs. 2** and **3**). Single-nucleotide substitutions produce less marked band separations, because of the small conformational change induced. Results obtained from CSGE gels detecting different nucleotide substitutions, insertions, or deletions will vary dramatically.

Most banding patterns produced by single-nucleotide substitutions can be easily identified when compared to wild-type homoduplex DNA. Some heteroduplexes only induce a slight retardation or enhancement of migration during polyacrylamide gel electrophoresis. These CSGE patterns consist of a single band, which is slightly thicker in width than wild-type DNA. These "fatter" bands can mimic an overloaded well and the nucleotide substitution can be overlooked. An extensive visual examination of the CSGE gel is essential for comparing all samples with normal or wild-type DNA. After gel documentation, all altered banding patterns must be investigated, either by a repeat CSGE, possibly with loading less DNA, or direct sequencing of the sample. All samples displaying altered CSGE migration should then be sequenced to determine the nature of the nucleotide change (*see* Chapter 10).

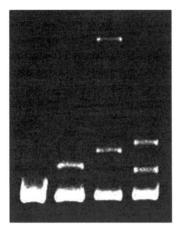

Fig 2. *FVIII* gene, exon 6 (423 bp): Lane 1, negative control, heteroduplexed against self; lane 2, heteroduplexed sample with an A to G substitution; lane 3, heteroduplexed sample with a 2-bp insertion of CC; lane 4, heteroduplexed sample with a single-base deletion of a T.

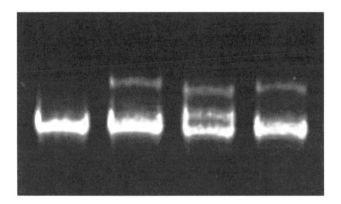

Fig 3. *FVIII* gene, exon 8 (547 bp): Lane 1, negative control heteroduplexed against self; lane 2, heteroduplexed sample with an A to G substitution; lane 3, heteroduplexed sample with an A to G substitution and a C to T substitution, hence the different banding pattern; lane 4, heteroduplexed sample with an A to G substitution.

Figs. 2 and 3. CSGE gel stained with ethidium bromide and visualized under UV light. All samples have been mixed with a negative control or wild-type sample of equal concentration and subjected to the heteroduplexing reaction, 98°C for 5 min, 65°C for 30 min). Each lane contains a total of 5 µL heteroduplexed PCR product mixed with 2 µL dye and electrophoresed at 400 V for 17 h on a 10% polyacrylamide get (99:1, acrylamide:BAP), 0.5X TTE, 15% formamide, 10% ethylene glycol).

4. Notes

1. Stock of 40% (w/v) acrylamide: 2.105% (w/v) *N,N*-methylene bisacrylamide ratio 19:1.

2. Both well preparation and sample loading are an essential part in obtaining reproducible results. When loading the gel, keep the sample volume to a minimum and allow it to cover the bottom few millimeters of the well.

3. Homemade PCR buffer (10X) is 166 mM (NH$_4$)$_2$SO$_4$, 670 mM Tris-HCl, pH 8.8, 67 mM MgCl$_2$, 2% β-Mercaptoethanol, and 1 mg/mL bovine serum albumin (BSA). The latter two ingredients are added immediately prior to use. Commercial PCR buffer (1X) is 16 mM (NH$_4$)$_2$SO$_4$, 67 mM Tris-HCl, pH 8.8, 0.1% Tween-20.

4. DNA thermocyclers, hot blocks, hot ovens, and water baths can all be used for heteroduplexing PCR product.

5. This will give a total sample volume of 10 μL, of which 2–8 μL can be loaded onto the CSGE gel. Where PCR product volume is limited, each PCR product can be reduced to, for example, 3 μL, providing a total volume to load of 5 μL. Where one PCR sample is at a lower concentration, volumes mixed should be adjusted to produce equal DNA concentrations to be mixed together.

6. Care must be taken when handling solid acrylamide. Always read safety guidelines before attempting to make solutions.

7. Ethidium bromide is mutagenic. Care must be taken in its use and disposal.

8. A "heterozygous" positive control sample should be loaded with each CSGE run to ensure that the gel has run properly and is capable of heteroduplex detection. Samples with a known polymorphism or mutation genotype can be used. A negative control must also be included where wild-type DNA is heteroduplexed to self. For samples that require mixing with wild-type DNA, the same wild-type DNA should also be used in the negative control. A negative control should be incorporated into the CSGE screening procedure for each DNA fragment studied, for comparison with band characteristics of heteroduplexed samples.

9. The CSGE gels should be prepared, polymerized, and electrophoresed on the same day. Keeping the CSGE gel overnight, even in a cold room, may affect results.

10. Combs must be removed from the gel without disfiguring the wells. The loaded sample will take up the shape of the well, so avoid loading misshapen wells. These can be marked on the glass plates with a marker pen prior to loading and then avoided during loading. Vacuum pressure may also cause wells to collapse when the comb is removed. Two methods are available to remove combs; in both instances, unpolymerized liquid acrylamide is first removed by blotting the area around the comb with blue roll or tissue, enabling air to enter the well space more freely. The first method involves carefully coaxing the comb out of the well space at the same time as allowing air to enter the well space. The second method involves the separation of the lugged glass plate from the cast gel. With the gel laid flat and open face up on the bench, a sharp object, such as the point of a pair of scissors, can be inserted in the center of the gel plate between the comb and the

plate (in a gap between the teeth of the comb) and rotated slightly. The gel will be seen to come away from the glass plate and the comb can be removed with ease. This will not affect the running of the gel.

11. The syringe can be filled from the top reservoir, providing that there is buffer remaining to cover the wells. Take care when placing the needle onto the syringe. With constant pressure, the needle tip can be inserted into each well, but avoiding touching the well base. Salts collecting at the base of the well will be seen as a viscous liquid.

12. The CSGE gels can stretch when mishandled, mainly at the edges where the gel has come away from the glass plate. Care must be taken to avoid agitating the gel when staining and destaining. A CSGE gel that has become unstuck from the plain glass plate will grow a few centimeters in all directions. Therefore, stretching the gel can affect the results and must be avoided.

13. Make a note of the orientation of the gel. Cut a corner off each segment of the gel to indicate the location of the first sample loaded.

14. To help to locate the DNA bands when using the hand-held UV light, samples of similar size should be loaded in clusters. The detection of minor changes is improved when samples are loaded adjacent to a wild-type sample or negative control. Samples will not be visualized with a hand-held UV light before the gel is transferred from the glass plate to the blotting paper.

15. The procedure of transferring the CSGE gel with 3MM paper can be repeated, if required, to reposition the gel for accurate documentation.

16. It is important to document all CSGE results. Banding patterns can vary from large band separations to slightly thicker bands when compared with the negative control. All banding patterns that differ from the negative control must be treated as a positive result until proven otherwise.

References

1. Ganguly, A., Rock, M. J., and Prockop, D. J. (1993) Conformation-sensitive gel electrophoresis for rapid detection of single-base differences in double-stranded PCR products and DNA fragments: evidence for solvent induced bends in DNA heteroduplexes. *Proc. Natl. Acad. Sci. USA* **90,** 10,325–10,329.

2. Bhattacharyya, A. and Lilley, D. M. J. (1989) The contrasting structures of mismatched DNA sequences containing looped-out bases (bulges) and multiple mismatches (bubbles). *Nucl. Acids Res.* **17,** 6821–6840.

3. Ganguly, A. and Prockop, D. J. (1995) Detection of mismatched bases in double stranded DNA by gel electrophoresis. *Electrophoresis* **16,** 1830–1835.

4. Williams, I. J. Abuzenadah, A. Winship, P. R. Preston, F. E. Dolan, G. Wright, J. et al. (1998) Precise carrier diagnosis in families with haemophilia A: use of conformation sensitive gel electrophoresis for mutation screening and polymorphism analysis. *Thromb. Haemost.* **79,** 723–726.

5. Hinks, J. L., Winship, P. R., Makris, M., Preston, F. E., Peake, I. R., and Goodeve, A. C. (1999) Conformation sensitive gel electrophoresis for precise haemophilia B carrier analysis. *Br. J. Haemotol.* **104,** 915–918.

6. Körkkö, J., Annunen, S., Pihlajamaa, T., Prockop, D. J., and Ala-Kokko, L. (1998) Conformation sensitive gel electrophoresis for simple and accurate detection of mutations: Comparison with denaturing gradient gel electrophoresis and nucleotide sequencing. *Proc. Natl. Acad. Sci. USA* **95,** 1681–1685.

7. Ganguly, G., Dhulipala, R., Godmilow, L., and Ganguly, A. (1998) High throughput fluorescence-based conformation-sensitive gel electrophoresis (F-CSGE) identifies six unique BRCA2 mutations and an overall low incidence of BRCA2 mutations in high-risk BRCA1-negative breast cancer families. *Hum. Genet.* **102,** 549–556.

8. Ganguly, A. and Williams, C. (1997) Detection of mutations in multi-exon genes: Comparison of conformation sensitive gel electrophoresis and sequencing strategies with respect to cost and time for finding mutations. *Hum. Mut.* **9,** 339–343.

9. Williams, C. J., Rock, M., Considine, E., McCarron, S., Gow, P., Ladda, R., et al. (1995) Three new point mutations in type II procollagen (COL2A1) and identification of a fourth family with the COL2A1 Arg519(Cys base substitution using conformation sensitive gel electrophoresis. *Hum. Mol. Genet.* **4,** 309–312.

10. Williams, C. J., Ganguly, A., Considine, E., McCarron, S., Prockop, D. J., Walsh-Vockley, C., et al. (1996) A-2(G Transition at the 3' acceptor splice site of IVS17 characterises the COL2A1 gene mutation in the original stickler syndrome kindred. *Am. J. Med. Genet.* **63,** 461–467.

11. Beauchamp, N. J., Daly, M. E., Makris, M., Preston, F. E., and Peake, I. R. (1998) A novel mutation in intron K of the PROS1 gene causes aberrant RNA splicing and is a common cause of protein S deficiency in a UK thrombophilia cohort. *Thromb. Haemost.* **79,** 1086–1091.

12. Abuzenadah, A. M., Gursel, T., Ingerslev, J., Nesbitt, I. M., Peake, I. R., and Goodeve, A. C. (1999) Mutational analysis of the von Willebrand disease. *Thromb. Haemost.* (Suppl. A) Abstract 887.

13. Gari, M., Goodeve, A. C., Wilson, G., Winship, P. R., Langabeer, S., Linch, D., et al. (1999) c-kit proto-oncogene exon 8 in frame deletion plus insertion mutations in acute myeloid leukaemia. *Br. J. Haematol.* **10,** 894–900.

14. Markoff, A., Sormbroen, H., Bogdanova, N., Priesler-Adams, S., Ganev, V. Dworniczak, B., et al. (1998) Comparison of conformation-sensitive gel electrophoresis and single-strand conformation polymorphism analysis for detection of mutations in the BRCA1 gene using optimised conformation analysis protocols. *Eur. J. Hum. Genet.* **6,** 145–150.

16

SSCP/Heteroduplex Analysis

Andrew J. Wallace

1. Introduction

Single-strand conformation (SSCP) and heteroduplex analysis are separate mutation scanning methods in their own right. They are, however, unusual in that they can be carried out simultaneously on a single gel.

The technique of SSCP analysis was originally described in 1989 (*1*). It involves the heat denaturation of polymerase chain reaction (PCR) amplified DNA followed by electrophoresis under nondenaturing conditions. The fragments in the original protocol were visualized by radiolabeling and autoradiography, although a variety of nonisotopic methods are now available, including silver staining (*2*), fluorescent labels (*3*), and ethidium bromide staining (*4*). The SSCP technique relies on the propensity for single-stranded DNA (ssDNA) in nondenaturing conditions to take on a three-dimensional, or secondary, structure that is highly sequence dependent. Consequently, sequence differences can cause alterations to the DNA's secondary structure. Because the electrophoretic mobility of DNA under nondenaturing conditions is dependent on its shape as well as other factors like charge, point mutations can give rise to mobility shifts. The gels used for SSCP are usually native acrylamide, typically with a low level of crosslinking (49:1), although there is a great deal of variability between published protocols. The low level of crosslinking gives a large pore size, thus permitting efficient separation of the bulky structures that ssDNA forms under these conditions. The detection efficiency of SSCP is highly variable, the most important parameter to consider is fragment size. The optimum sensitivity is with fragments as small as 150 bp, where under a single condition, 90% of mutations are detected (*5*).

Heteroduplexes are hybrid DNA molecules that, although largely matched, have one or more mismatched base pairs. Heteroduplexes have been used as a

From: *Methods in Molecular Biology, vol. 187: PCR Mutation Detection Protocols*
Edited by: B. D. M. Theophilus and R. Rapley © Humana Press Inc., Totowa, NJ

tool to scan for point mutations since 1992 *(6)*. They typically appear on native polyacrylamide gels as one or two bands of reduced mobility relative to the homoduplex DNA. The mismatched bases present in heteroduplexes are thought to affect electrophoretic mobility by inducing bends in the DNA *(7)*. Because two different DNA sequence variants must be present to form heteroduplexes, they may need to be created for some types of analysis. For example, in order to analyze male samples for loci on the X chromosome by heteroduplex analysis, heteroduplexes can be created by mixing, denaturing, and annealing the test sample PCR amplification with a known normal control amplification. In heterozygotes, however, heteroduplexes form as a natural by product of PCR reactions. During the latter stages of PCR amplification, when the polymerase activity is limiting, some of the denatured ssDNA can spontaneously reanneal without primer extension with an opposite strand from the other allele, thus creating heteroduplex DNA.

Heteroduplex analysis is carried out by electrophoresis of the fragment of interest on long (usually polyacrylamide) gels with low ratios of crosslinking. Heteroduplexes have been visualized using radioisotopes *(6)*, silver staining *(2)*, and ethidium bromide staining *(8)*. Fluorescent labeling should also be theoretically possible. The detection efficiency of heteroduplex analysis has been reported to approach 90% under ideal conditions *(7)* and the optimum size, 250–500 bp, is not as tightly defined as SSCP.

Combined SSCP/heteroduplex analysis exploits the tendency for a proportion of the DNA denatured during sample preparation for SSCP to spontaneously reanneal to form dsDNA and, hence, heteroduplexes when there are sequence differences in the sample. The gel conditions for both SSCP and heteroduplex analysis are compatible and so it is possible to get "two techniques for the price of one." There are, however, some limitations to the technique; foremost among these is that the dsDNA, with which the heteroduplexes are associated, have a much higher mobility than the SSCPs formed by the ssDNA. This limits the electrophoresis time in order to retain the dsDNA on the gel, reducing the resolution of the SSCPs. In practice, the loss of sensitivity is more than made up for by the complementarity between the two techniques, a point evinced by the observation that every one of 134 different cystic fibrosis transmembrane conductance regulator (CFTR) mutations are detectable by a combined SSCP/heteroduplex strategy *(9)*. Consequently, a combined SSCP/heteroduplex approach is now used more frequently by laboratories than either technique alone.

2. Materials

2.1. Preparation and Electrophoresis of SSCP/Heteroduplex Gels

1. S2 or SA32 sequencing gel system (Gibco-BRL; cat. nos. S2–21105-036 and SA32–31096-027) (*see* **Note 1**).

2. 1-mm-thick combs and spacers. These are a nonstandard thickness and need to be custom-made to order.

3. Acrylamide powder preweighed (49:1 acrylamide:bis-acrylamide) (e.g., Sigma; cat. no. A0924). Dissolve according to the manufacturer's instructions with deionized water to make a 40% stock solution and store at 4°C. Use within 1 mo (*see* **Notes 2** and **3**).

4. TEMED (*N,N,N',N'*-tetramethylethylenediamine) (e.g., Sigma; cat. no. T7024).

5. 10% Ammonium persulfate solution (10% AMPS) (e.g., Sigma; cat. no. A9164). Make up fresh on the day of use.

6. 10X TBE electrophoresis buffer (e.g., Gibco-BRL; cat. no. 15581-036). Use within 1 mo of opening.

7. Formamide loading buffer: 10 mL deionized formamide, 200 μL of 0.5 *M* EDTA, pH 8.0, 15 mg xylene cyanol, 3 mg bromophenol blue. Store at room temperature.

8. Electrophoresis power pack capable of maintaining 600 V and with a voltage preset.

9. Access to a 4°C cold room or cold cabinet with power supply (*see* **Note 4**).

10. Flat gel loading tips (e.g., Life Sciences International; cat. no. PP000-0GEL-F01).

2.2. Silver Staining and Drying of SSCP/Heteroduplex Gels

1. Plastic photographic style staining trays (e.g., Jencons Scientific; part no. 682-172).

2. Orbital shaker.

3. Silver staining solution 1: 10% industrial methylated spirit, 0.5% glacial acetic acid. Store at room temperature. Solution 1 may be recycled up to 10 times.

4. Silver staining solution 2: 0.1% $AgNO_3$. Prepare as a 10X stock solution (1% $AgNO_3$). Store at room temperature in a brown bottle. Prepare and store the 1X working solution in a clear bottle so that any precipitate is clearly visible, an indication that the solution should not be used. Both the 10X stock and the 1X working solution should be stored out of direct sunlight. The 1X working solution may be reused to stain up to three gels.

5. Silver staining solution 3: 1.5% NaOH, 0.15% formaldehyde (*see* **Note 5**). This solution is labile and the formaldehyde should only be added immediately prior to use. The 1.5% NaOH may be made up in bulk and stored at room temperature.

6. Silver staining solution 4: 0.75% Na_2CO_3. Make as a 10X stock and dilute down as necessary;, store the 10X stock at room temperature.

7. Gel drying frame and platform (e.g., Pharmacia Biotech; part no. 80-6122-37 and 80-6122-94).

8. Cellophane sheets (e.g., Pharmacia Biotech; part no. 80-6121-99).

3. Methods

3.1. Preparation and Electrophoresis of SSCP/Heteroduplex Gels

1. Wash a suitable comb, pair of glass plates, and spacers thoroughly with warm water and a household detergent. Rinse with deionized water and dry with disposable tissues.

2. Lay the glass plates on a clean section of the bench and wipe with a disposable tissue soaked in 100% ethanol.

3. Smear the spacers with Pritt™ or an equivalent paper adhesive to prevent them from slipping while assembling and pouring the gel. Assemble the glass plates and spacers as shown in **Fig. 1A,B** using four strong binder clips to to hold the plates together.

4. Lay the gel plates, short plate uppermost, on a box or upturned Eppendorf rack on a flat section of benching.

5. For an 8% gel for the SA32 system, place the following reagents in a clean dry beaker:

    15 mL 40% (49:1) acrylamide solution;

    7.5 mL of 10X TBE solution;

    52 mL dH$_2$O;

    90 μL TEMED.

    For the S2 system all these volumes should be doubled (*see* **Notes 2**, **4**, and **6**).

6. Add 500 μL (for SA32 gel) or 1 mL (for S2 gel) of freshly prepared 10% AMPS solution to the acrylamide solution and mix thoroughly by gently swirling.

7. Carefully draw the acrylamide solution into a 50-mL disposable syringe, avoiding introducing air bubbles.

8. Rest the nozzle of the syringe on the protruding portion of the long glass plate, about 2 mm from the edge of the short plate, and slowly expel the contents of the syringe. The acrylamide solution should run down between the plates quite evenly by capillary action (**Fig. 1A**). The syringe will need to be refilled three to four times for the S2 system. Gently tapping the glass plates just ahead of the acrylamide solution can help to prevent the formation of trapped air bubbles.

9. Once the acrylamide has completely filled the space between the plates, any air bubbles can be removed using a hook-shaped "bubble catcher." The comb should then be carefully inserted while avoiding the introduction of further air bubbles. The gel should be left to polymerize at room temperature for at least 1 h preferably 2 h.

10. Remove the comb and straighten any uneven wells using an old gel loading tip.

11. Place the gel in the electrophoresis apparatus and firmly tighten the four locking nuts. Close the drain tap for the upper buffer chamber and fill the upper buffer chamber with 1X TBE. Check for leaks by leaving the gel for 10 min and inspecting the lower buffer chamber.

12. Fill the lower buffer chamber with 1X TBE and leave the whole apparatus to equilibrate for at least 4 h at 4°C (*see* **Notes 4** and **7**).

13. Combine the PCR amplification with an equal volume of formamide loading buffer (typically 10 μL) and mix well (*see* **Notes 8–10**).

14. Place the samples on a heated block or thermal cycler set at 94°C for 3 min then snap-chill in a bath of crushed ice.

15. Load between 6 and 12 μL of each sample (the optimum volumes depend on the size of well and amplification efficiency) in each well using flat gel loading tips.

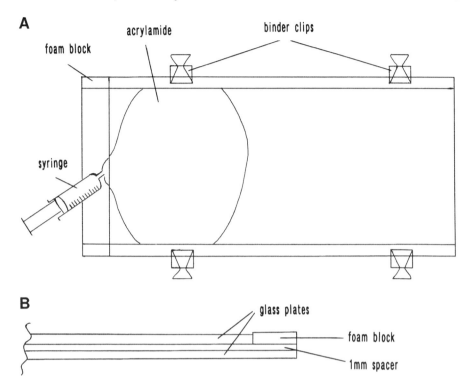

Fig. 1. Assembly of glass plates for S2/SA32 sequencing system and pouring without tape: (**A**) view from above to illustrate the pouring of the gel on a horizontal bench using a syringe; (**B**) profile view to illustrate the position of the foam block against the short upper plate, which forms a watertight seal against the gel tank gasket.

16. Electrophorese for typically 16 h at a constant voltage of 370 V (for SA32 gel) or 450 V (for S2 gel).

17. After 16 h the xylene cyanol should have run to the end of the aluminum cooling plate on an 8% gel. Double-stranded DNA (heteroduplexes) of about 180 bp comigrates with the xylene cyanol on an 8% gel, although there can be considerable sequence-dependent variability. It is best to determine optimum electrophoresis times empirically.

3.2. Silver Staining and Drying of SSCP/Heteroduplex Gels

1. Slide one of the spacers out from between the glass plates and gently prise apart with a plastic spatula or other nonmetal instrument. Ensure that the gel is attached to the lowermost plate before completely removing the upper plate. Mark the gel orientation by removing a corner adjacent to lane 1. For S2 gels, cut the gel vertically in half for staining by applying pressure on the gel with the edge of a ruler.

2. Carefully lift the bottom edge of the gel and fold over a 10-cm length. Repeat this action until the gel is completely rolled up; then, place the plate in a clean staining tray and dislodge or lift the gel into the tray. For S2 gels, place each half in a separate tray.

3. Pour on 400 mL of silver staining solution 1 and leave on an orbital shaker for about 5 min. In the specified trays, the gel should unfurl to a degree but remain folded in half during the staining process.

4. Pour off solution 1 (*see* **Note 11**) and save for reuse. Pour on 500 mL of silver staining solution 2 and shake gently for 15 min (*see* **Note 12**).

5. Pour off solution 2, remembering to reuse for up to three gels. Add 400 mL of freshly prepared silver staining solution 3 (*see* **Note 5**). Place the tray in a fume cabinet and leave for 20 min, shaking occasionally. A small amount of powdery black precipitate should be observed when solution 3 is added and the bands slowly appear on the gel during this stage.

6. Pour off solution 3 and rinse the gel twice with deionized water. Add 400 mL of silver staining solution 4 and leave for 15 min (*see* **Notes 13–15**).

7. Take two sheets of precut cellophane and soak in a sink of tap water. The sheets become pliable when wet; ensure that the whole sheet is adequately soaked.

8. Place a drying frame inner section over the platform, ensuring that the inner section is in the correct orientation. Place a moistened sheet of cellophane over the inner section, ensuring that the whole of the inner section is covered.

9. Gently slide an old sheet of X-ray film under the gel while still in solution 4. Carefully pour off solution 4 and rinse the gel briefly twice with tap water. Lift the gel out of the staining tray using the X-ray film as a support onto a clean area of benching or plastic sheet. The gel can now be carefully unfolded and unwanted areas of the gel trimmed away by pressing with the edge of a ruler. If the whole gel is to be dried down, then the gel will have to be cut in half and dried down in two separate frames.

10. Transfer the gel onto the cellophane sheet placed over the drying frame and wet the gel surface with a few milliliters of deionized water from a wash bottle.

11. Carefully lay the second sheet of moistened cellophane over the top of the gel and frame, taking care not to trap air bubbles between the sheets. Place the outer section of the drying frame firmly over the inner section's rubber gasket. The cellophane should now be drawn taut and the frame can be carefully lifted off the platorm.

12. Turn the retaining screws 90° to retain the inner frame in position and trim any surplus cellophane away with a pair of scissors. Wipe away excess water with a disposable tissue and leave the frame to dry either on a warm shelf for 24 h or in a 37°C incubator for 6 h (*see* **Note 16**).

13. Once the gel is thoroughly dry, dismantle the frame and cut excess cellophane away from the dried gel with scissors. Seal the cellophane by folding strips of adhesive tape around all of the edges. This prevents the cellophane from peeling apart.

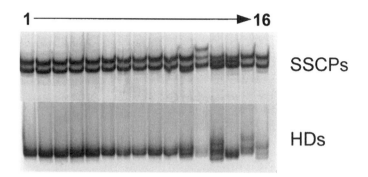

Fig. 2. SSCP/heteroduplex (HD) analysis of CFTR exon 3 (309 bp). Lanes 1–10 are test samples; lanes 11–16 are heterozygote mutation controls. The mutant controls are as follows: lane 11—P67L(nt332c→t); lane 12—R75X(nt355c→t); lane 13—G85E(nt386g→a); lane 14—L88S(nt395t→c); lane 15—E60X(nt310g→t). Lane 16 is from a heterozygote for the polymorphism R75Q(nt356g→a). Exon 3 gives a typical SSCP pattern of two discrete bands corresponding to the forward and reverse strands. Note how the patterns of shifts are different for each mutation (i.e., the SSCP pattern of lanes 13 and 14 are quite similar, but their heteroduplex mobility shifts are completely different). Also note how both lanes 15 and 16 are not discernible by SSCP analysis alone, but they give rise to clear and characteristic heteroduplex mobility shifts.

14. Store the dried down gels flat and away from moisture. Avoid bending and folding the gels because they are very brittle (*see* **Note 17**).

3.3. Interpretation

1. SSCP/heteroduplex gels are straightforward to interpret. Provided that a control normal sample is loaded onto each gel, an SSCP shift or heteroduplex of differing mobility is indicative of the presence of a sequence difference. Common polymorphisms may complicate interpretation, but the use of controls of known polymorphism genotype and experience allows the most complex of combinations to be successfully interpreted (*see* **Figs. 2–4** for examples of typical data).
2. ssDNA has a much lower mobility than dsDNA in the native acrylamide gels used for SSCP/heteroduplex analysis. Consequently, the limiting factor for resolution of SSCP shifts is the need to retain the dsDNA on the gel. For maximum resolution, the dsDNA should be run as close to the end of the gel as possible. The ssDNA (SSCPs) tends to stain a reddish brown color, whereas the dsDNA (heteroduplexes) usually stain a dark gray or black color.
3. A homozygote will typically give rise to two distinct SSCP bands corresponding to the two complementary strands of DNA. Quite often, more than two SSCP bands are present, some of which stain more weakly than the others. Assuming that the PCR is optimized and that no background amplification has taken place, the subsidiary bands are the result of alternative stable conformations.

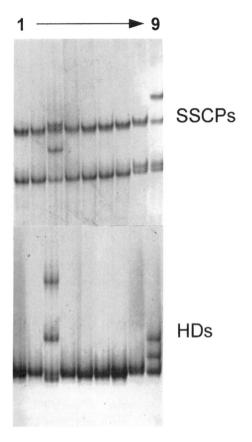

Fig. 3. SSCP/heteroduplex analysis of CFTR exon 23 (223 bp). Lanes 1–7 are test samples; lanes 8 and 9 are control heterozygotes for the mutations Q1412X(nt4366c→t) and 4279insA, respectively. Note how lane 3 clearly has a mobility shift different from the two controls. Direct sequencing revealed that this sample was heterozygous for the 4326ΔTC mutation. Also note how Q1412X gives rise only to a SSCP shift.

4. Occasionally, a fragment only gives rise to a single SSCP band; this is due to both complementary strands having identical mobilities, and, generally, does not lead to loss of sensitivity.
5. Rarely, there is no distinct SSCP band present for a given fragment. Closer inspection of the gel usually reveals the presence of a faint in track smear. This appears to be caused by the fragment adopting a whole range of conformations, each having only a slight mobility difference from the next. The presence of a mutation seems to destroy this balance, leading to the creation of a typical SSCP band.
6. When a genuine SSCP shift is present, the relative staining of the normal bands in the sample will be reduced relative to those of other normal samples on the

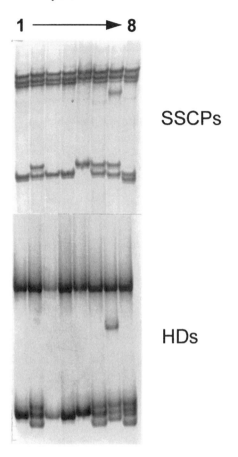

Fig. 4. SSCP/heteroduplex analysis of CFTR Exon 14a. This fragment at 511 bp is too large to analyze effectively as one fragment and so has been digested with *Rsa*I to give rise to a 277-bp and 234-bp fragment *(10)*. There is a frequent polymorphism in exon 14a (nt2694t→g) that is visible as both SSCP and heteroduplex shifts. Lanes 2 and 6 are heterozygotes for this polymorphism. Lanes 1, 3, and 4 are homozygotes for the faster migrating allele, whereas lane 5 is a homozygote for the slower migrating allele. Lanes 7 and 8 are heterozygotes for the polymorphism but also are heterozygotes for the mutations 2711ΔT and W846X1(nt2669g→a), respectively. Note how the presence of these mutations within the same fragment as the polymorphism modifies the appearance of the polymorphism, allowing them to be clearly differentiated.

same gel that have amplified with the same efficiency. This is the result of the ssDNA being spread over a larger gel volume. This effect can be used to filter out false positives resulting from spurious amplification.

7. Excess primers left over in unpurified PCR reactions reannealing to the ssDNA after gel loading can lead to a third set of bands intermediate in mobility to the

ssDNA and dsDNA. Although they usually show a similar pattern to the standard SSCPs, they can occasionally highlight differences that are not detected elsewhere.

8. Heteroduplexes usually show as bands of reduced mobility relative to the homoduplex dsDNA. They are often more weakly staining than the homoduplex DNA (*see* **Note 8**).

9. Occasionally, the homoduplex DNA itself displays mobility shifts. This may be the result of the presence of an insertion/deletion mutation leading to an alteration in molecular weight. Some single-base substitutions can also give the same effect. This is thought to be caused by the base substitution causing the DNA to bend to a greater or lesser degree, thus affecting mobility.

10. Frequently, homozygous normal samples will give rise to two homoduplex bands; the reason for this is not known.

4. Notes

1. Fluorescent SSCP/heteroduplex analysis is possible. Both the Perkin-Elmer 377 fluorescent analyzer and the 310 capillary electrophoresis instrument have temperature control and are thus suitable for this technique. There are several advantages to this approach. Fluorescent analysis permits multiplexing by dye color, thus increasing productivity. The gel images are also analyzed and stored electronically, thus avoiding the need to dry down and store gels. The internal size standards commonly used in fluorescent analysis also makes comparisons between samples more objective *(11)*. However, the major drawback is one of cost, not only in capital equipment and software but also the fluorescently labeled primers.

2. There are commercial gel matrices specifically marketed for SSCP and heteroduplex analysis, such as Hydrolink and MDE™. They do have different characteristics from the 49:1 acrylamide recommended here. They tend to work out to be quite expensive for routine use.

3. Acrylamide is a cumulative neurotoxin. Great care should be taken when handling acrylamide powder, solution, and gels. Gloves should be worn at all times when handling acrylamide. A face mask and fume cabinet should also be used when adding water to preweighed bottles of acrylamide powder.

4. Gel additives are recommended in many SSCP protocols to improve detection efficiency. The addition of glycerol in particular is reported to enhance the sensitivity of SSCP analysis *(9)*. Gels containing glycerol should also be run at room temperature, thus offering an alternative method to workers without access to a cold room or cabinet. We have found that the detection rate of heteroduplex analysis is compromised in the presence of glycerol, so we can only recommend its use as an additional condition to enhance the detection rate when this needs to be maximized. The addition of 10% sucrose to SSCP gels has also been used to enhance the sensitivity of SSCP *(9)*.

5. Formaldehyde is usually sold as a 37% solution; remember to take this into account when calculating volumes for solution 3.

6. There is some evidence that the addition of mild denaturants like 10% urea *(6)* or 10% ethanediol and 15% formamide *(7)* can enhance the efficiency of heteroduplex analysis. We have found that these additives reduce the efficiency of SSCP analysis and are not recommended for combined SSCP/heteroduplex analysis.

7. Replicate gels can vary markedly in appearance; the major causes of variability are gel quality and environmental factors. Take great care when measuring reagents and always use fresh 10% AMPS solution. Try to get into a routine so that gels are always poured at a certain time of the day and left to polymerize and then equilibrate at 4°C for the same time. Your results will become more reproducible, although some variability should still be expected.

8. The formation of heteroduplex DNA is dependent on spontanous reannealing in the formamide loading buffer or during the earliest phase of electrophoresis. Some fragments reanneal very poorly, preventing efficient heteroduplex analysis and thus reducing the mutation detection rate. This problem can be overcome simply by preloading the gel with a proportion of the sample in formamide loading buffer (usually 25% of the final volume to be loaded) before heat denaturation of the sample. The remainder of the sample can then be loaded after denaturation in the same wells and electrophoresis commenced.

9. Larger fragments can be analyzed without loss of detection efficiency by cutting the sample with an appropriate restriction enzyme to yield fragments of optimum size (200–250 bp) (**Fig. 4**) *(10)*. Running both restricted and unrestricted samples in the same lane can increase efficiency still further because some loss of heteroduplex detection efficiency has been noted within 50 bp of the end of fragments *(7)*.

10. If the gene under analysis is X-linked or mitochondrial in origin, or if the mutation could be homozygous or hemizygous, the samples should be mixed with a known normal control DNA and then subjected to a single round of denaturation and renaturation in order to encourage the formation of heteroduplexes prior to loading.

11. Pouring solutions off can be difficult, the gels are large and can easily tear. The talc on some latex gloves can also mark the gels. Placing a piece of old X-ray film over the gel while pouring off the solutions helps to support the gel, preventing it from tearing. It also prevents gloves from marking the gels.

12. Contaminating salts or alkali can make silver salts precipitate out in solution 2, causing the gel to become milky white in appearance. If you continue to stain in solution 3 once this has happened, these areas will turn black and the gel will be unreadable. The gel can often be saved by washing twice in dH_2O and then immersing in 2.5% ammonium hydroxide for 15 min followed by two further washes in dH_2O. Silver staining can then be recommenced with solution 1. However, prevention is better than cure, the most common causes of this are dirty staining trays and contact with gloves that have been used to prepare solution 3. Always wash trays thoroughly before use and change gloves after preparing solution 3 if you need to subsequently pour off solution 2.

13. Sometimes, we have observed gels with bands that fade and disappear while drying down; this is due to solution 4 being made up incorrectly.

14. It is not possible to isolate and reamplify from silver stained bands with the given protocol because of the inhibitory effect of residual silver ions on PCR. However, if solution 4 is substituted for a 50 mM solution of EDTA (pH 8.0), then this chelates the remaining silver ions and permits reamplification from the band of interest. We have found this to be particularly useful for isolating minority alleles in mosaic samples.
15. Silver-stained gels can be temporarily stored by heat sealing in plastic. However, the gels are delicate and easily crushed and tend to dry out after 2–3 yr.
16. High-percentage acrylamide gels (i.e., over 9%) are prone to cracking while drying down. Try to dry these slowly at room temperature rather than in an incubator, as this minimizes cracking.
17. Sometimes, a salty deposit builds up on the surface of old gels, making them difficult to read. This can be removed by spraying the gel with a little household polish and then rubbing the gel firmly with a soft cloth.

References

1. Orita, M., Iwahana, H., Kanazawa, H., Hayashi, K., and Sekiya, T. (1989) Detection of polymorphisms of human DNA by gel electrophoresis as single-strand conformation polymorphisms. *Proc. Natl. Acad. Sci. USA* **86,** 2766–2770.
2. Tassabehji, M., Newton, V. E., Turnbull, K., Seemanova, E., Kunze, J., Sperling, K., et al. (1994) PAX3 gene structure and mutations: close analogies between Waardenburg syndrome and the *Splotch* mouse. *Hum. Mol. Genet.* **7,** 1069–1074.
3. Makino, R., Yazyu, H., Kishimoto, Y., Sekiya, T., Hayashi, K. (1992) F-SSCP: fluorescence-based polymerase chain reaction single-strand conformation polymorphism (PCR-SSCP) analysis. *PCR Meth. Applic.* **2,** 10–13.
4. Hongyo, T., Buzard, G. S., Calvert, R. J., and Weghorst, C. M. (1993) 'Cold SSCP': a simple, rapid and non-radioactive method for optimized single-strand conformation polymorphism analyses. *Nucl. Acids Res.* **21,** 3637–3642.
5. Sheffield, V. C., Beck, J. S., Kwitek, A. E., Sandstrom, D. W., and Stone, E. M. (1993) The sensitivity of single-strand conformation polymorphism analysis for the detection of single base substitutions. *Genomics* **16,** 325–332.
6. White, M. B., Carvalho, M., Derse, D., O'Brien, S. J., and Dean, M. (1992) Detecting single base substitutions as heteroduplex polymorphisms. *Genomics* **12,** 301–306.
7. Ganguly, A., Rock, M. J., and Prockop, D. J. (1993) Conformation-sensitive gel electrophoresis for rapid detection of single-base differences in double-stranded PCR products and DNA fragments: evidence for solvent-induced bends in DNA heteroduplexes. *Proc. Natl. Acad. Sci. USA* **90,** 10,325–10,329.
8. Keen, J., Lester, D., Inglehearn, C, Curtis, A., and Bhattacharya, S. (1991) Rapid detection of single base mismatches as heteroduplexes on Hydrolink gels. *Trends Genet.* **7,** 5.
9. Ravnik-Glavac, M., Glavac, D., and Dean, M. (1994) Sensitivity of single-strand conformation polymorphism and heteroduplex method for mutation detection in the cystic fibrosis gene. *Hum. Mol. Genet.* **3,** 801–807.

10. Lee, H.-H., Lo, W.-J., and Choo, K.-B. (1992) Mutational analysis by a combined application of the multiple restriction fragment-single strand conformation polymorphism and the direct linear amplification sequencing protocols. *Anal. Biochem.* **205,** 289–293.

11. Inazuka, M., Wenz, H. M., Sakabe, M., Tahira, T., and Hayashi, K. (1997) A streamlined mutation detection sysytem: multicolor post-PCR fluorescence labeling and single-strand conformational polymorphism analysis by capillary electrophoresis. *Genome Res.* **7,** 1094–1103.

17

Cleavase® Fragment Length Polymorphism Analysis for Genotyping and Mutation Detection

Laura Heisler and Chao-Hung Lee

1. Introduction

DNA sequencing is the gold standard in DNA diagnostics and is the only absolute means of establishing the identity of a new mutation. However, the clinical cost of obtaining this information is often prohibitive, particularly when large DNA fragments are interrogated for the presence of any of a number of either known or previously undescribed genetic alterations (*1*). Instead, several mutation scanning methods have been developed to eliminate the need to sequence every nucleotide when it is only the precise identity of one or a few nucleotides that is clinically significant. Until now such methods have provided only a "yes" or "no" answer in determining whether a test sample differs from a known reference. Relatively few methods have been proven capable of unambiguously identifying unique nucleic acid variants, particularly when multiple sequence changes occur (*2*). Consequently, the majority of existing mutation scanning methods are unsuitable for PCR-based genotyping applications in which regions of sequence variability are used to categorize isolates for their similarities to known variants.

Third Wave Technologies has pioneered a novel mutation and polymorphism screening method that accurately and precisely distinguishes nucleic acid variants (*3*). This approach relies on enzymatic cleavage of characteristic structures formed by single-stranded nucleic acids. On sequential denaturation and renaturation, both single-stranded DNA and RNA molecules assume three-dimensional conformations that are a precise reflection of their nucleic acid sequences (*4*). This principle is the foundation of several mutation scanning techniques, such as single-strand conformation polymorphism (SSCP) and dideoxy fingerprinting (*5,6*). Instead of relying on direct observation of these

From: *Methods in Molecular Biology, vol. 187: PCR Mutation Detection Protocols*
Edited by: B. D. M. Theophilus and R. Rapley © Humana Press Inc., Totowa, NJ

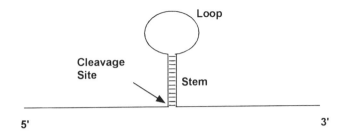

Fig. 1. Structures recognized by the Cleavase I enzyme. The Cleavase I enzyme is a structure-specific nuclease that recognizes the junctions between single- and double-stranded regions of nucleic acids, i.e., so-called hairpins or stem loops. Cleavage occurs on the 5' side of such structures (*see* arrow). These intrastrand structures are formed when nucleic acid molecules are denatured and then allowed to cool.

structures, e.g., by noting subtle differences in how different DNA strands migrate through nondenaturing gels, the Third Wave Technologies' enzyme-based approach uses a structure-specific endonuclease engineered from the nuclease domain of Taq DNA polymerase, dubbed the Cleavase® I enzyme, to cut DNA strands wherever these structures occur *(3)* (**Fig. 1**). By analogy to restriction fragment length polymorphism analysis, Third Wave has named their method Cleavase Fragment Length Polymorphism (CFLP®) analysis.

The Cleavase I enzyme rapidly and specifically cleaves these structures, many of which are formed on a given DNA fragment, albeit transiently, in equilibrium with alternative, mutually exclusive structures. The CFLP method is thus able to elucidate a considerable amount of information about the sequence content of a DNA fragment without relying on cumbersome high-resolution analysis of each base. Each unique DNA sequence produces a distinctive collection of structures that, in turn, results in the generation of a singular fingerprint for that sequence. This capability makes the CFLP technology suitable for diverse mutation scanning applications, including genotyping *(1,3,7–14)*. Furthermore, the CFLP reaction is unaffected by the length of the DNA fragment and can be used to analyze far longer stretches of DNA than is currently possible with other methods, up to at least 2.7 kb (unpublished data).

1.2. Visualizing Sequence Differences in CFLP Fingerprints

The CFLP method comprises the steps of separation of DNA strands by heating, formation of intrastrand structures on cooling with rapid enzymatic cleavage of these structures before they are disrupted by reannealing of the complementary strands, and separation and visualization of the resulting "structural fingerprint" (**Fig. 2**). When closely related DNA fragments, such as a wild-type and a mutant version of a gene, are compared, the CFLP fingerprints

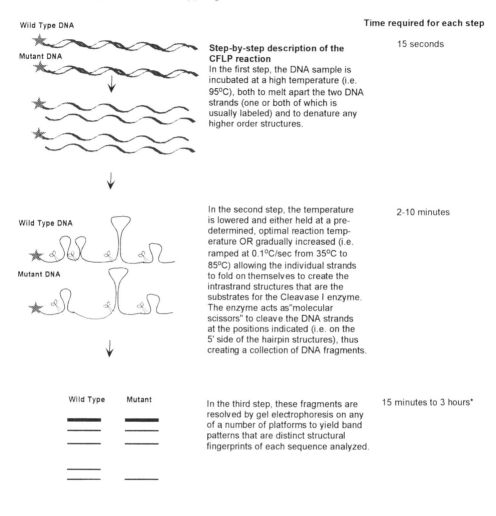

Step-by-step description of the CFLP reaction

In the first step, the DNA sample is incubated at a high temperature (i.e. 95°C), both to melt apart the two DNA strands (one or both of which is usually labeled) and to denature any higher order structures.

Time required for each step

15 seconds

In the second step, the temperature is lowered and either held at a predetermined, optimal reaction temperature OR gradually increased (i.e. ramped at 0.1°C/sec from 35°C to 85°C) allowing the individual strands to fold on themselves to create the intrastrand structures that are the substrates for the Cleavase I enzyme. The enzyme acts as "molecular scissors" to cleave the DNA strands at the positions indicated (i.e. on the 5' side of the hairpin structures), thus creating a collection of DNA fragments.

2-10 minutes

In the third step, these fragments are resolved by gel electrophoresis on any of a number of platforms to yield band patterns that are distinct structural fingerprints of each sequence analyzed.

15 minutes to 3 hours*

* The time is dependent on the gel-based instrument, which varies from the traditional vertical apparatus to fluorescence sequencers with fragment analysis software.

Fig. 2. CFLP reaction. The CFLP reaction itself is a simple three-step procedure that relies on the use of temperature to change some of the physical characteristics of DNA molecules.

exhibit strong familial resemblance to one another such that they share the majority of bands produced. The sequence differences are revealed as changes in one or several bands. These unique band changes are visualized as mobility shifts, the appearance or disappearance of bands, and/or significant differences in band intensity (**Fig. 3**).

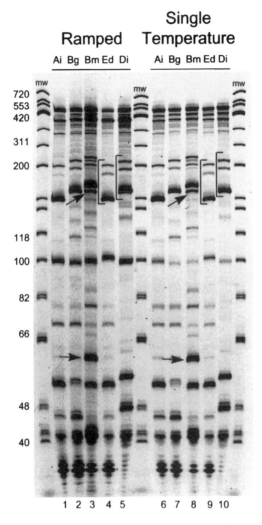

Fig. 3. CFLP analysis of the ITS regions of *P. carinii*. PCR products spanning the ITS1 and ITS2 regions of *P. carinii*, 534 bp long and labeled on the 5' ends of both strands with TET, were subjected to CFLP analysis. Approximately 250 fmol of labeled PCR product was analyzed in the "ramped" reactions and approx 150 fmol in the single temperature reactions. The DNA aliquots were supplemented with DNA dilution buffer. The ramping assay was performed as described in **Subheading 3.2.** The genotypes of the samples from which the DNA was amplified are indicated above the lanes. Lanes marked "mw" contain molecular weight markets with fragment sizes as indicated. The gel was electrophoresed at constant wattage (20 W) until a bromophenol blue market dye, loaded in a far lane (not shown), reached the bottom of the gel. The gel cassette was scanned on a Hitachi FMBIO-100 fluorescence imager with a 585-nm emission filter.

CFLP analysis is exquisitely sensitive to the presence of minor sequence variations and can detect changes involving one or more bases, including missense mutations, with >95% sensitivity and 100% specificity. Because the CFLP method results in an easily examined pattern, rather than base-by-base analysis of each sequence, the value of this approach may be even more pronounced in genotyping applications. In these cases, what is sought is the rapid identification of compound sequence variations occurring throughout an amplified fragment. Rapid inspection of the patterns generated by CFLP analysis of fragments containing multiple, dispersed base changes has proven to be an effective approach to classifying bacterial and viral sequences according to genotype *(1,3,11)*.

Pneumocystis carinii f. sp. *hominis* is the leading cause of pneumonia and the most commonly transmitted life-threatening opportunistic infection among AIDS patients *(15)*. To establish the origin of particular infections, verify localized outbreaks, and determine whether an individual has sustained multiple, coincident infections, researchers have attempted to classify individual *P. carinii* strains based on sequence variability among isolates *(16)*. Sequence variation in the internal transcribed spacer (ITS) regions of the rRNA genes of *P. carinii* can be used for such genotypic identification *(17)*. The region located between the 18S and 5.8S rRNA genes is called ITS1, and that between the 5.8S and 26S rRNA genes is ITS2. Among the two regions, approx 60 different ITS sequences have been characterized by direct DNA sequencing *(18)*. Sequence variation occurs throughout these 161- and 192-bp regions, respectively, and the majority of sequence changes within each ITS have been determined to be significant in establishing type *(18)*.

The suitability of the CFLP scanning method for differentiating sequences in the ITS region of five cloned *P. carinii* sequences belonging to different types was investigated. The ITS region was amplified by polymerase chain reaction (PCR), and the 5' ends of both strands were labeled by using tetrachlorofluorescein (TET) sense strand labeled primers (*see* **Note 1**). The amplified products were purified and then partially digested with the Cleavase I enzyme. The samples were analyzed in duplicate sets, one of which was subjected to CFLP digestion at a predetermined, optimized reaction temperature, whereas the other was digested under conditions in which the temperature was continually increased, or "ramped" (*see* **Subheading 3.2.** and **Note 2**).

The results indicate that the CFLP method is highly effective in reproducibly distinguishing different *P. carinii* types (**Fig. 3**). An inspection of the fingerprints generated from these samples reveals a high degree of similarity overall, indicative of the fact that only a few bases are altered in the variants, with some marked differences that reflect those base changes. In **Fig. 3**, there are several examples of bands that appear in some lanes but that are absent in others, as well as bands that appear shifted in some lanes relative to others. Unique bands, indicated by arrows, are apparent, e.g., in lanes Bm. In other cases, the most notable difference is a composite shifting downward of a substantial portion of the pattern, indicative of a small deletion, such that the fragments are shortened relative to the labeled 5′ ends (e.g., lanes Ed and Di, as indicated by brackets).

Note that the patterns generated by the ramping procedure appear to be enhanced relative to the single temperature procedure in several places. In particular, note the appearance of additional bands between 82 and 118 bp. This improvement in the richness of the patterns is likely due to the fact that certain substrate structures may not be as favored at a single temperature, as is used in the conventional approach, but rather emerge as the temperature changes over the course of the ramping reaction. In some cases the ramping approach not only eliminates the need for preliminary optimization steps but may also serve to improve the sensitivity of the CFLP method.

2. Materials

2.1. Preparation of End-Labeled PCR-Amplified Fragments

2.1.1. PCR Reagents

1. Sterile double-distilled H_2O.
2. 10X PCR buffer: 500 mM KCl, 100 mM Tris-HCl, pH 9.0, 15 mM $MgCl_2$.
3. dNTP (deoxynucleotide) mix: 0.2 mM each dNTP in sterile double-distilled H_2O.
4. Mineral oil or wax overlay.
5. Oligonucleotide primers at a concentration of 10 µM, at least one of which is labeled with a fluorescent dye (e.g., tetrachlorofluorescein, fluorescein) or a moiety detectable by chemiluminescence (e.g., biotin, digoxyigenin).

2.1.2. Post-PCR Fragment Purification

1. Exonuclease I: Available at 10 U/µL from Amersham Pharmacia Biotech (Arlington Heights, IL), cat. no. E70073Z, or at 20 U/µL from Epicentre Technologies (Madison, WI), cat. no. X40505K.
2. High Pure PCR Product Purification Kit (HPPPPK), available from Roche Molecular Biochemicals (Indianapolis, IN), cat. no. 1732668.
3. Sterile double-distilled H_2O or $T_{10}E_{0.1}$: 10 mM Tris-HCl, pH 8.0, 0.1 mM EDTA, pH 8.0.

2.2. CFLP Analysis

1. Cleavase I enzyme (25 U/µL) in Cleavase enzyme dilution buffer: 20 mM Tris-HCl, pH 8.0, 50 mM KCl, 0.05% Tween® 20, 0.05% Nonidet™ P 40, 100 µg/mL bovine serum albumin, and 50% (v/v) glycerol.
2. DNA dilution buffer: 5 mM MOPS, pH 7.5.

3. 10X CFLP buffer: 100 mM MOPS, pH 7.5, 0.5% Tween 20, 0.5% Nonidet P 40.
4. 2 mM MnCl$_2$.
5. 10 mM MgCl$_2$.
6. Stop solution (for nonfluorescent gel-based detection): 95% formamide, 10 mM EDTA, pH 8.0, 0.05% xylene cylanol, 0.05% bromophenol blue (*see* **Note 3**).
7. Stop solution (for fluorescent gel-based detection): 95% formamide, 10 mM EDTA, pH 8.0, 0.05% crystal violet (*see* **Note 3**).
8. Sterile double-distilled H$_2$O.
9. Thin-walled reaction tubes (200 or 500 µL).

2.3. Gel Electrophoresis

1. Gel solution: 6–10% acrylamide:bis (19:1) solution, 7 M urea, 0.5X Tris-borate EDTA (TBE) buffer.
2. 0.5X TBE gel running buffer (pH 8.3): 44.5 mM Tris, 44.5 mM borate, 1 mM EDTA, pH 8.0.
3. Ammonium persulfate (10% [w/v]).
4. TEMED.
5. Teflon flat-bottomed combs and spacers (0.5 mM thick) *(2)*.
6. Glass plates (20 × 20 cm), nonfluorescing for use with fluorescence imager or standard for chemiluminescence detection.
7. Gel electrophoresis support.
8. Power supply capable of supplying up to 2000 V.

2.4. Visualization of CFLP Patterns

2.4.1. Fluorescence Detection

1. Hitachi FMBIO®-100 Fluorescent Method Bio-Image Analyzer (Hitachi Software, San Bruno, CA) or Molecular Dynamics 595 FluorImager™ (Molecular Dynamics, Sunnyvale, CA).
2. Lint-free laboratory wipes.
3. Lens paper.
4. Nonfluorescing detergent, e.g., RBS 35 Detergent Concentrate (Pierce, Rockford, IL).

2.4.2. Chemiluminescence Detection

1. 10X SAAP: 1 M NaCl, 0.5 M Tris-base, pH 10.0.
2. 1X SAAP, 0.1% sodium dodecyl sulfate (SDS): 100 mM NaCl, 50 mM Tris-Base, pH 10.0, 0.1% SDS (w/v).
3. 1X SAAP, 1 mM MgCl$_2$: 100 mM NaCl, 50 mM Tris-base, pH 10.0, 1 mM MgCl$_2$.
4. Sequenase Images™ 5X Blocking Buffer (cat. no. US75354; Amersham Pharmacia Biotech).
5. Streptavidin-Alkaline-Phosphatase Conjugate (cat. no. US11687; Amersham Pharmacia Biotech).
6. CDP-*Star*™ substrate (cat. no. MS250R; Tropix, Bedford, MA).

7. Isopropanol.
8. Latex gloves (powder free).
9. X-ray film.
10. Positively charged nylon membrane, pore size 0.2 μm (e.g., Nytran® Plus Membrane, Schleicher and Schuell, Keene, NH).
11. Blotting paper (20 × 20 cm) (cat. no. 28303-100; VWR Scientific).
12. Sealable plastic bags.
13. Forceps.
14. Small plastic containers for processing membranes.
15. Darkroom/film-developing facilities.
16. Permanent laboratory marker.

3. Methods

3.1. Purification of PCR-Generated Fragments (see Note 1)

PCR amplification should be performed according to established protocols for the particular locus in question. When PCR products are visualized by gel electrophoresis followed by sensitive detection of the label to be used to visualize CFLP products, contamination by labeled primers and prematurely truncated single-stranded PCR products is evident. These contaminating DNA species are effectively removed by the procedures noted. In particular, the HPPPPK procedure has been proven effective for eliminating lower molecular weight (i.e., >100 bp) DNA species, whereas Exonuclease I digestion is effective for removing larger DNA species. An alternative to the HPPPPK columns in conjunction with Exonuclease I digestion is to precipitate with 1 vol of isopropanol following Exonuclease I digestion.

If a single, labeled product is detected following PCR and HPPPPK and Exonuclease I digestion, then proceed with CFLP analysis. If more than one product is detected, then optimization of the PCR reaction or gel isolation of the desired product is necessary. The following protocol describes the method for Exonuclease I digestion:

1. Following PCR amplification, incubate the reaction mixture at 70°C for 10 min.
2. Bring the reaction mixture to 37°C.
3. Add 1 U of Exonuclease I/μL of original PCR reaction mixture (e.g., 100 U to a 100-μL reaction mixture).
4. Incubate for 30 min at 37°C.
5. Inactivate the reaction by heating at 70°C for 30 min.
6. Following Exonuclease I digestion, apply the reaction mixtures to the HPPPPK spin columns according to the manufacturer's suggested procedures. The supplied elution buffer should be replaced with either sterile double-distilled H_2O or $T_{10}E_{0.1}$, pH 8.0.

3.2. Preparation and Performance of CFLP Reactions

Prior to performing CFLP analysis, it is strongly recommended that the quality and quantity of the PCR-generated fragments following purification be checked. This can be done by visualizing the label used (i.e., by fluorescence analysis or chemiluminescence detection) on an aliquot of the DNA in a small denaturing polyacrylamide gel.

As seen in **Fig. 3**, there are two alternative approaches to be taken in configuring the CFLP reaction. The initial configuration of the assay involves performing the reaction under an abbreviated matrix of reaction times and temperatures in order to identify the optimal conditions for generation of a pattern with a broad spectrum of evenly distributed bands (temperature/time optimization). Alternatively, recent studies have demonstrated that the use of a programmable thermal cycler enables informative patterns to be generated by increasing the reaction temperature throughout the course of the reaction, specifically from 25 to 85°C at a rate of 0.1°C/s for a total ramping time of 10 min. In some genetic systems, such as *P. carinii*, the ramping approach appears to generate somewhat more even distributions of fragments and has improved mutation detection sensitivity. Furthermore, provided suitable thermal cyclers are available, the ramping approach is simpler and requires less DNA, since optimization reactions need not be run prior to analysis of test samples. The following protocol describes the method of performing CFLP analysis utilizing either the single temperature or ramping procedure:

1. Aliquot the desired amount of end-labeled DNA (approx 100–200 fmol) into a thin-walled reaction tube (200 or 500 μL, depending on the capacity of the thermal cycler). Bring the DNA to a final volume of 13 μL with DNA dilution buffer, if necessary.

2. In a separate tube, prepare an enzyme master mix that contains the following for each reaction: 2 μL of 10X CFLP buffer, 2 μL 2 m*M* MnCl$_2$, 1 μL of Cleavase I enzyme, 2 μL of 10 m*M* MgCl$_2$ (optional, *see* **Note 4**), and DNA dilution buffer to a final volume of 7 μL (if needed).

3. To denature samples, place tubes containing DNA in a programmable thermal cycler (or heat block) and incubate at 95°C for 15 s. If the single temperature method is used proceed to **step 4**. If the ramping method is to be used proceed to **step 5**.

4. Temperature/time optimization: After the 15-s denaturation step, set the thermal cycler to the desired reaction temperature (or place the tubes in a heat block held at reaction temperature if no thermal cycler is available). Optimal times and temperatures can be determined by examining matrices of different reaction times (e.g., 1, 3, and 5 min) and temperatures (40, 50, and 55°C). Choose the conditions that yield the richest and most even pattern (*see* **Note 5**). Incubate the CFLP reactions for the amount of time determined to be optimal, holding the

temperature constant. After the incubation period, stop the reactions with 16 μL of stop solution. Proceed to **Subheading 3.3.**

5. Ramping: After the 15-s denaturation step, set the thermal cycler to 35°C. As soon as the thermal cycler reaches 35°C, add 7 μL of the enzyme/buffer mixture. Mix well by pipetting up and down several times. Incubate the CFLP reactions for 15 s at 35°C. Program the thermal cycler to increase in temperature at a rate of 0.1°C/s to 85°C, or set to ramp for an 8-min period from 35 to 85°C. On reaching 85°C, stop the reactions with 16 μL stop solution (*see* **Note 3**).

3.3. Separation of CFLP Fragments

1. Prepare a denaturing polyacrylamide gel, choosing a percentage of acrylamide (19:1) appropriate for the size of the fragment being analyzed (*see* **Note 6**).
2. Prerun the gel for approx 30 min before loading the samples at sufficient wattage to warm the gel (e.g., 18–20 W).
3. Heat denature the CFLP reactions at 80°C for 2 min immediately prior to loading onto the gel. The best resolution is achieved when the samples are fully denatured.
4. Load 5–10 μL of the appropriate CFLP reaction per well. The remainder of the reactions can be stored at 4°C or –20°C for later analysis.
5. Continue electrophoresis until sufficient separation of the CFLP fragments is achieved (the time will depend on the fragment size and the percentage of acrylamide used).

3.4. Visualization of CFLP Patterns

3.4.1. Fluorescence Imaging of CFLP Patterns

1. Following gel electrophoresis, thoroughly wash the outside of the gel plates using nonfluorescent soap.
2. Dry and wipe clear with lens paper to remove residual debris.
3. Place the gel carefully in the fluorescence scanning unit (Hitachi FMBIO-100 or Molecular Dynamics 595).
4. Scan using the correct wavelength or filter for the fluorescent group to be detected.

3.4.2. Chemiluminescence Detection of CFLP Patterns

1. After electrophoresis, wearing powder-free latex gloves that have been washed with isopropanol (*see* **Note 7**), carefully separate the glass plates to expose the acrylamide gel.
2. Cut a piece of Nytran Plus membrane (Schleicher and Schuell) to fit the gel size and moisten by applying 5–10 mL of 0.5X TBE.
3. Carefully place the moistened membrane onto the gel, avoiding lifting and repositioning the membrane, and smooth out air bubbles with a clean pipet. Transfer starts immediately, so the membrane should not be picked up and repositioned once it has come into contact with the gel.
4. Cover the membrane with two pieces of precut blotting paper, cover with a glass plate, and place a binder clip on each side of the sandwiched gel. Alternatively,

for large gels (i.e., 20 × 20 cm or larger), place an approx 2-kg weight on top of the sandwich.

5. Allow the DNA to transfer onto the membrane for 4–16 h (e.g., overnight, if convenient) at room temperature.

6. After the transfer, disassemble the sandwiched gel and remove the membrane by carefully moistening it with distilled water. Mark the DNA side (i.e., the side touching the gel during transfer) using a permanent laboratory marker.

7. Rinse a dish thoroughly with isopropanol (*see* **Note 7**) and fill with 0.2 mL/cm² of 1X blocking buffer (e.g., 100 mL for a 20 × 20 cm membrane).

8. Transfer the membrane to the dish containing the blocking buffer and allow to rock gently for 15 min.

9. Repeat the 15-min wash with fresh blocking buffer and discard the buffer.

10. Add 2 μL of Streptavidin-Alkaline-Phosphatase Conjugate to 50 mL of fresh blocking buffer (or add conjugate to the blocking buffer at a volume ratio of 1 : 4000).

11. Pour the conjugate/blocking buffer mixture onto the blocked membrane and rock gently for 15 min.

12. Remove the conjugate and rinse for 5 min with 0.1% SDS/1X SAAP buffer, 0.5 mL/cm² each (200 mL for a 20 × 20 cm membrane). Repeat twice, for a total of three washes.

13. Remove the SDS and rinse 5 min with 0.5 mL/cm² 1 mM MgCl₂/1X SAAP buffer. Repeat twice, for a total of three washes.

14. Place the membrane in a sealable bag and add 4 mL of CDP-*Star* (or 0.01 mL/cm²).

15. Seal the bag and spread the CDP-*Star* gently over the membrane for 3–5 min.

16. Completely remove the CDP-*Star* and any air bubbles. Transfer the membrane while still in the bag to a film exposure cassette.

17. In the darkroom, expose the membrane to X-ray film. Initially expose for 30 min. For subsequent exposures, adjust the time for clarity and intensity (*see* **Note 8**).

18. Develop the film.

4. Notes

1. Depending on the objective of the analysis in question, the DNA can be labeled on either strand or on both strands using this approach. Single end labeling, i.e., of the sense or the antisense strand, permits some degree of localization of the base change(s) corresponding to the observed pattern changes (*3*). The sensitivity of the CFLP method is approx 90% for single-stranded analysis and >95% for two-strand analysis. While double end labeling precludes this localization, it affords more sensitive mutational analysis.

2. It has been determined empirically that samples analyzed according to the ramping procedure require approximately twofold more DNA than do those analyzed by the conventional method. This is likely because in the ramping procedure, digestion occurs throughout the course of the temperature increase and optimally cleaves different hairpins at different temperatures.

3. The choice of dyes used in the stop solution depends on the system used to visualize the CFLP patterns. If chemiluminescence detection is used, then the stop

solution should include 0.05% bromophenol blue and 0.05% xylene cyanol (**Subheading 2.2., item 6**). If fluorescent scanning is used, then a dye that migrates with opposite polarity, such as crystal violet (0.05%), is preferable, because dyes that migrate into the gel fluoresce at the wavelengths used to detect the fluorescent dyes, thereby obscuring a portion of the CFLP pattern. Note that when a dye with opposite polarity is used, it is advisable to load 3–5 µL of stop solution containing bromophenol blue and xylene cyanol in a lane that does not contain CFLP reactions in order to monitor the progress of gel electrophoresis.

4. $MgCl_2$ dramatically reduces the rate of cleavage in the CFLP reaction. When $MgCl_2$ is added to a final concentration of 1 mM in the presence of standard $MnCl_2$ concentrations of 0.2 mM, the rate of cleavage is slowed by as much as 10-fold. This reduced reaction rate can be useful for analysis of DNA fragments that assume highly favored secondary structures that are rapidly cleaved in the CFLP reaction. The presence of such structures is readily identified by the appearance of a structural fingerprint comprising one or two prominent bands. When 1 mM $MgCl_2$ is added, the optimal time and temperature of digestion should be reevaluated to reflect the reduced rate of cleavage (*see* **Note 5**).

5. The structural fingerprint produced by CFLP digestion is a collection of fragments resulting from partial digestion, usually of 5' end labeled fragments. Because the CFLP reaction is a partial digestion and because the formation of the substrate secondary structures depends on reaction temperature, it is possible to modulate the extent of digestion through variations in the duration and temperature of the reaction. Specifically, lower temperatures stabilize secondary structure formation whereas higher temperatures reduce the number of structures formed by a given molecule. Similarly, longer reaction times lead to increased accumulation of smaller cleavage products. The most informative fingerprints are those that contain a relatively even distribution of low and high molecular weight products, including a fraction of full-length, uncut DNA. Ensuring that the entire size distribution of cleavage products is visible increases the likelihood of detecting the products that reflect the presence of a polymorphism.

6. The percentage of polyacrylamide to be used is dictated by the size of the PCR fragment being analyzed. Appropriate percentages of polyacrylamide for various size ranges are well established *(19)*.

7. The objective of this step is to minimize carryover of alkaline phosphatase from previous reactions and from exogenous sources (e.g., skin). Throughout this procedure, it is of paramount importance to minimize contamination by this ubiquitous enzyme.

8. An alkaline phosphatase reaction with the chemiluminescence substrate produces a long-lived signal, especially on membranes. Light emission increases of >300-fold are seen in the first 2 h on application of the substrate onto nylon membranes, with the chemiluminescence signal persisting up to several days. Because film exposure times range from minutes to several hours, multiple images may be acquired. Varying film exposure times enables the user to optimize signal to noise.

Acknowledgments

We wish to acknowledge the efforts of Mary Oldenburg, Senior Technical Scientist, Third Wave Technologies, in performing much of the CFLP reaction optimization as well as in providing critical commentary on the manuscript.

References

1. Sreevatsan, S., Bookout, J. B., Ringpis, F. M., Pottathil, M. R., Marshall, D. J., de Arruda, M., Murvine, C., Fors, L., Pottathil, R. M., and Barathur, R. R. (1998) Algorithmic approach to high-throughput molecular screening for alpha interferon-resistant genotypes in hepatitis C patients. *J. Clin. Microbiol.* **36,** 1895–1901.
2. Cotton, R. G. H. (1997) Slowly but surely towards better scanning for mutations. *Trends Genet.* **13,** 43–46.
3. Brow, M. A., Oldenberg, M., Lyamichev, V., Heisler, L., Lyamicheva, N., Hall, J., Eagan, N., Olive, D. M., Smith, L., Fors, L., and Dahlberg, J. (1996) Differentiation of bacterial 16S rRNA genes and intergenic regions and Mycobacterium tuberculosis katG genes by structure-specific endonuclease cleavage. *J. Clin. Microbiol.* **34,** 3129–3137.
4. Orita, M., Suzuki, Y., Sekiya, T., and Hayashi, K. (1989) Rapid and sensitive detection of point mutations and DNA polymorphisms using the polymerase chain reaction. *Genomics* **5,** 874–879.
5. Hayashi, K. (1991) PCR-SSCP: a simple and sensitive method for detection of mutations in the genomic DNA. *PCR Meth. Applica.* **1,** 34–38.
6. Sarkar, G., Yoon, H., and Sommer, S. S. (1992) Dideoxy fingerprinting (ddE): a rapid and efficient screen for the presence of mutations. *Genomics* **13,** 441–443.
7. Sreevatsan, S., Bookout, J. B., Ringpis, F. M., Mogazeh, S. L., Kreiswirth, B. N., Pottathil, R. R., and Barathur, R. R. (1998) Comparative evaluation of cleavase fragment length polymorphism with PCR-SSCP and PCR-RFLP to detect antimicrobial agent resistance in Mycobacterium tuberculosis. *Mol. Diagn.* **3,** 81–91.
8. Tahar, R. and Basco, L. K. (1997) Analysis of Plasmodium falciparum multidrug-resistance (pfmdr1) gene mutation by hairpin-dependent cleavage fragment length polymorphism. *Mol. Biochem. Parasitol.* **88,** 243–247.
9. Wartiovaara, K., Hytonen, M., Vuori, M., Paulin, L., Rinne, J., and Sariola, H. (1998) Mutation analysis of the glial cell line-derived neurotrophic factor gene in Parkinson's disease. *Exp. Neurol.* **152,** 307–309.
10. Schlamp, C., Poulsen, G. L., Nork, M., and Nickells, R. W. (1997) Nuclear exclusion of wild-type p53 in immortalized human retinoblastoma cells. *J. Natl. Cancer Inst.* **89,** 1530–1536.
11. Marshall, D. J., Heisler, L. M., Lyamichev, V., Murvine, C., Olive, D. M., Ehrlich, G. D., Neri, B. P., and de Arruda, M. (1997) Determination of hepatitis C virus genotypes in the United States by Cleavase Fragment Length Polymorphism analysis. *J. Clin. Microbiol.* **35,** 3156–3162.

12. Rainaldi, G., Marchetti, S., Capecchi, B., Meneveri, R., Piras, A., and Leuzzi, R. (1998) Absence of mutations in the highest mutability region of the p53 gene in tumour-derived CHEF18 Chinese hamster cells. *Mutagenesis* **13,** 153–155.

13. Rossetti, S., Englisch, S., Bresin, E., Pignatti, P. F., and Turco, A. E. (1997) Detection of mutations in human genes by a new rapid method: cleavage fragment length polymorphism analysis (CFLPA). *Mol. Cell. Probes* **11,** 155–160.

14. Eisinger, F., Jacquemier, J., Charpin, C., Stoppa-Lyonnet, D., Bressac-de Paillerets, B., Peyrat, J.-P., Longy, M., Guinebretiere, J.-M., Sauvan, R., Noguichi, T., Birnbaum, D., and Sobol, H. (1998) Mutations at BRCA1: the medullary breast carcinoma revisited. *Cancer Res.* **58,** 1588–1592.

15. Centers for Disease Control and Prevention (1989) AIDS Weekly Surveillance Report, Centers for Disease Control and Prevention, Atlanta, GA.

16. Latouche, S., Ortona, E., Masers, E., Margutti, P., Tamburrini, E., Siracusano, A., Guyot, K., Nigou, M., and Roux, P. (1997) Biodiversity of Pneumocystis carinii hominis: typing with different DNA regions. *J. Clin. Microbiol.* **35,** 383–387.

17. Lu, J.-J., Bartlett, M. S., Shaw, M. M., Queener, S. F., Smith, J. W., Ortiz-Rivera, M., Leibowitz, M. J., and Lee, C.-H. (1994) Typing of Pneumocystis carinii strains that infect humans based on nucleotide sequence variations of internal transcribed spacers of rRNA genes. *J. Clin. Microbiol.* **32,** 2904–2912.

18. Lee, C.-H., Tang, X., Jin, S., Li, B., Bartlett, M. S., Helweg-Larsen, J., Olsson, M., Lucas, S. B., Roux, P., Cargnel, A., Atzori, C., Matos, O., and Smith, J. W. (1998) Update on Pneumocystis carinii f. sp. hominis typing based on nucleotide sequence variations in internal transcribed spacer regions of rRNA genes. *J. Clin. Microbiol.* **36,** 734–741.

19. Sambrook, J., Fritsch, E. F., and Maniatis, T. (eds.) (1989) *Molecular Cloning: A Laboratory Manual,* Cold Spring Harbor Laboratory Press, Cold Spring Harbor, NY.

18

Automated Genotyping Using the DNA MassArray™ Technology

Christian Jurinke, Dirk van den Boom, Charles R. Cantor, and Hubert Köster

1. Introduction

1.1. Markers Used for Genetic Analysis

The ongoing progress in establishing a reference sequence as part of the Human Genome Project (*1*) has revealed a new challenge: the large-scale identification and detection of intraspecies sequence variations, either between individuals or populations. The information drawn from those studies will lead to a detailed understanding of genetic and environmental contributions to the etiology of complex diseases.

The development of markers to detect intraspecies sequence variations has evolved from the use of restriction fragment length polymorphisms (RFLPs) to microsatellites (short tandem repeats [STRs]) and very recently to single nucleotide polymorphisms (SNPs).

Although RFLP markers (*2*) are useful in many applications, they are often of poor information content, and their analysis is cumbersome to automate. STR markers (*3*), by contrast, are fairly highly informative (through their highly polymorphic number of repeats) and easy to prepare using polymerase chain reaction (PCR)-based assays with a considerable potential for automation. However, using conventional gel electrophoresis-based analysis, typing of large numbers of individuals for hundreds of markers still remains a challenging task.

Within the last few years, much attention has been paid to discovery and typing (scoring) of SNPs and their use for gene tracking (*4,5*). SNPs are biallelic single-base variations, occurring with a frequency of at least 1 SNP/ 1000 bp within the 3 billion bp of the human genome. Recently, a study on the

From: *Methods in Molecular Biology, vol. 187: PCR Mutation Detection Protocols*
Edited by: B. D. M. Theophilus and R. Rapley © Humana Press Inc., Totowa, NJ

sequence diversity in the human lipoprotein lipase gene suggested that the frequency of SNPs might be much higher *(6)*. The diversity in plant DNA, which would be relevant for agricultural applications, is five to seven times larger than in human DNA *(7)*.

Even though the use of SNPs as genetic markers seems to share the same limitations as relatively uninformative RFLPs, when used with modern scoring technologies, SNPs exhibit several advantages. Most interesting for gene tracking is that SNPs exist in the direct neighborhood of genes and also within genes. Roughly 200,000 SNPs are expected *(4)* in protein coding regions (so-called cSNPs) of the human genome. Furthermore, SNPs occur much more frequently than STRs and offer superior potential for automated assays.

1.2. Demand for Industrial Genomics

1.2.1. Genetics

The efforts of many researchers are dedicated to the exploration of the genetic bases of complex inherited diseases or disease predispositions. Studies are performed to identify candidate or target genes that may confer a predisposition for a certain disease *(8)*. Linkage analysis can be done as a genome wide screening of families; association or linkage disequilibrium analysis can be done with populations. Either approach can use STR or SNP markers. Once a potential candidate gene is discovered, a particular set of markers is compared between affected and unaffected individuals to try to identify functional allelic variations. To understand the genotype-to-phenotype correlation of complex diseases, several hundred markers need to be compared among several hundred individuals *(9,10)*. To get an impression of the complexity of the data produced in such projects, imagine a certain multifactorial disease in which predisposition is linked to, e.g., 12 genes. Consider that each of those 12 genes can be present in just two different alleles. The resulting number of possible genotypes (2 homozygotes and 1 heterozygote = 3 for each gene) is $3^{12} = 531,441$.

The whole process of drug development, including hunting for new target genes and especially the subsequent validation (significant link to a certain disease), will benefit from high-throughput, high-accuracy genomic analysis methods. Validated target genes can also be used for a more rational drug development in combination with genetic profiling of study populations during clinical trials.

1.2.2. Pharmacogenetics

Traits within populations, such as the ABO blood groups, are phenotypic expressions of genetic polymorphism. This is also the case for variations in response to drug therapy. When taken by poor metabolizers, some drugs cause exaggerated pharmacological response and adverse drug reactions. For example, tricyclic antidepressants exhibit order of magnitude differences in blood concentrations depending on the enzyme status of patients *(11)*. Pharmacogenetics is the study of genetic polymorphism in drug metabolism. Today, pharmaceutical companies screen individuals for specific genetic polymorphisms before entry into clinical trials to ensure that the study population is both relevant and representative. Targets for such screenings are cytochrome P450 enzymes or *N*-acetyltransferase isoenzymes (NAT1 and NAT2). Potential drug candidates affected by polymorphic metabolism include antidepressants, antipsychotics, and cardiovascular drugs.

1.2.3. Current Technologies

In addition to candidate gene validation and pharmacogenetics, many other applications such as clinical diagnostics, forensics, as well as the human sequence diversity program *(12)* are dealing with SNP scoring. In agricultural approaches, quantitative trait loci can be explored, resulting in significant breeding advances. Methods are required that provide high-throughput, parallel sample processing; flexibility; accuracy; and cost-effectiveness to match the different needs and sample volumes of such efforts.

Large-scale hybridization assays performed on microarrays have enabled relatively high-throughput profiling of gene expression patterns *(13)*. However, several issues must be considered in attempting to adapt this approach for the large-scale genotyping of populations of several hundred individuals. Hybridization chips for SNP scoring can potentially analyze in parallel several hundred SNPs per chip—with DNA from one individual. Therefore, several hundred hybridization chips would be needed for projects with larger populations. If during the course of a study an assay needs to be modified or new assays have to be added, all chips might have to be completely remanufactured.

Also, note that DNA hybridization lacks 100% specificity. Therefore, highly redundant assays have to be performed, providing a statistical result with a false-negative error rate of up to 10% for heterozygotes *(14)*. Finally, because of the inherent properties of repeated sequences, hybridization approaches are hardly applicable to STR analysis.

1.3. DNA MassArray Technology

Within the last decade, mass spectrometry (MS) has been developed to a powerful tool no longer restricted to the analysis of small compounds (some hundred Daltons) but also applicable to the analysis of large biomolecules (some hundred thousand Daltons). This improvement is mainly based on the invention of soft ionization techniques. A prominent example is matrix assisted laser desorption/ionization (MALDI) time-of-flight (TOF) MS, developed in the late 1980s by Karas and Hillenkamp *(15)*.

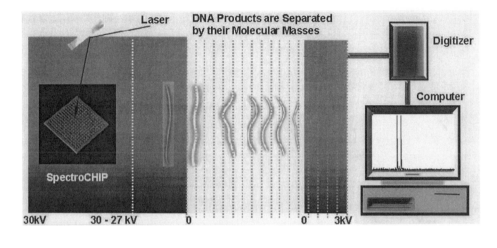

Fig. 1. Schematic drawing of the MALDI-TOF MS process, as used in the DNA MassArray method.

The general principle of MS is to produce, separate, and detect gas-phase ions. Traditionally, thermal vaporization methods are used to transfer molecules into the gas phase. Most biomolecules, however, undergo decomposition under these conditions. Briefly, in MALDI MS, the sample is embedded in the crystalline structures of small organic compounds (called matrix), and the cocrystals are irradiated with a nanosecond ultraviolet-laser beam. Laser energy causes structural decomposition of the irradiated crystal and generates a particle cloud from which ions are extracted by an electric field. After acceleration, the ions drift through a field-free path (usually 1 m long) and finally reach the detector (e.g., a secondary electron multiplier) (*see* **Fig. 1**). Ion masses (mass-to-charge ratios, m/z) are typically calculated by measuring their TOF, which is longer for larger molecules than for smaller ones (provided their initial energies are identical). Because predominantly single-charged nonfragmented ions are generated, parent ion masses can easily be determined from the spectrum without the need for complex data processing and are accessible as numerical data for direct processing.

The quality of the spectra, which is reflected in terms of resolution, mass accuracy, and also sensitivity, is highly dependent on sample preparation and the choice of matrix compound. For this reason, the early applications of MALDI-TOF MS were mostly for analyzing peptides and proteins. The discovery of new matrix compounds for nucleic acid analysis *(16)* and the development of solid-phase sample conditioning formats *(17,18)* enabled the analysis of nucleic acid reaction products generated in ligase chain reaction or PCR *(19)*.

The more demanding DNA sequence determination with MALDI-TOF MS can be addressed using exonucleolytic digestion *(20)*, Sanger sequencing *(21)*, or solid-phase Sanger sequencing approaches *(22)*. These approaches are currently restricted to comparative sequencing, and the read length is limited to about 100 bases. Further improvements in reaction design and instrumentation *(23)* will surely lead to enhanced efficacy and longer read length. For genotyping applications, this limitation is not relevant because scientists at Sequenom (San Diego, CA) developed the primer oligo base extension (PROBE) reaction especially for the purpose of assessing genetic polymorphism by MS *(24)*. The PROBE assay format can be used for the analysis of deletion, insertion, or point mutations, and STR, and SNP analysis, and it allows the detection of compound heterozygotes. The PROBE process comprises a postPCR solid-phase primer extension reaction carried out in the presence of one or more dideoxynucleotides (ddNTPs) and generates allele-specific terminated extension fragments (*see* **Fig. 2**). In the case of SNP analysis, the PROBE primer binding site is placed adjacent to the polymorphic position. Depending on the nucleotide status of the SNP, a shorter or a longer extension product is generated. In the case of heterozygosity, both products are generated. After completion of the reaction, the products are denatured from the solid phase and analyzed by MALDI-TOF MS. In the example given in **Fig. 2**, the elongation products are expected to differ in mass by one nucleotide. **Figure 3** presents raw data for a heterozygous DNA sample analyzed by this PROBE assay. The two SNP alleles appear as two distinct mass signals. Careful assay design makes a high-level multiplexing of PROBE reactions possible.

In the case of STR analysis, a ddNTP composition is chosen that terminates the polymerase extension at the first nucleotide not present within the repeat *(25)*. For length determination of a CA repeat, a ddG or ddT termination mix is used. Even imperfect repeats harboring insertion or deletion mutations can be analyzed with this approach. **Figure 4** displays raw data from the analysis of a human STR marker in a heterozygous DNA sample. Both alleles differ by four CA repeats. The DNA polymerase slippage during amplification generates a pattern of "stutter fragments" (marked with an asterisk in **Fig. 4**). In the case of heterozygotes that differ in just one repeat, the smaller allele has higher intensities than the larger allele, because allelic and stutter signals are added together. A DNA MassArray compatible STR portfolio with a 5-cM intermarker distance is currently under development at Sequenom.

When compared to the analysis of hybridization events by detecting labels, even on arrays, the DNA MassArray approach differs significantly. The PROBE assay is designed to give only the relevant information. The mass spec-

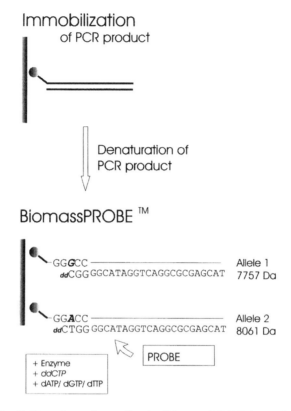

Fig. 2. Reaction scheme for the BiomassPROBE reaction.

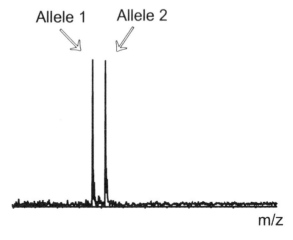

Fig. 3. Raw data of SNP analysis (heterozygous sample) using the BiomassPROBE reaction.

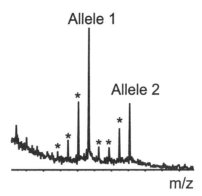

Fig. 4. Raw data of microsatellite analysis (heterozygous sample) using the BiomassPROBE reaction. Signals marked with an asterisk are stutter fragments (*see* **Subheading 1.3.**).

trometric approach enables direct analyte detection with 100% specificity and needs no redundancy. This accuracy and efficacy is combined with sample miniaturization, bioinformatics, and chip-based technologies for parallel processing of numerous samples.

Now, the use of an advanced nanoliquid handling system based on piezoelectric pipets combined with surface-modified silicon chips permits an automated scanning of 96 samples in about 10 min. Currently, up to 10 SpectroCHIPs (960 samples) can be analyzed in one automated run using a Bruker/Sequenom SpectroSCAN mass spectrometer (*see* **Fig. 5**). The SpectroSCAN mass spectrometer addresses each position of the chip sequentially, collects the sum of 10 laser shots, processes and stores the data, and proceeds to the next spot of the chip. In **Fig. 6**, 96 raw data spectra from a heterozygous sample are depicted resulting from a SpectroCHIP with one sample spotted 96 times. Using a proprietary algorithm, masses as well as signal intensities are automatically analyzed and interpreted. After completion of analysis, the results are transferred to a database and stored as accessible genetic information (*see* **Fig. 7**). The database also provides a tool for visual control and comparison of spectra with theoretically expected results (*see* **Fig. 8**).

The DNA MassArray throughput in terms of genetic information output depends on the chosen scale. Using microtiter plates and 8-channel pipets, the analysis of 192 genotypes (two 96-well microtiter plates) a day is routine work. With the use of automated liquid handling stations, the throughput can be increased by a factor of about four. An automated process line was been developed during the last year to increase the throughput to an industrial scale. The automated process line integrates biochemical reactions including PCR setup, immobilization, PROBE reaction sample conditioning, and recovery

Fig. 5. Sample holder for 10 SpectroCHIPs for use in the SpectroSCAN mass spectrometer.

from the solid-phase into a fully automated process with a throughput of about 10,000 samples per day.

2. Materials

2.1. PCR and PROBE Reaction

1. Dynabeads M-280 Streptavidin (Dynal, Oslo, Norway).
2. Separate PROBE stops mixes for ddA, ddC, ddG, and ddT (500 μ*M* of the respective ddNTP and 500 μ*M* of all dNTPs not present as dideoxynucleotides) (MassArray Kit; Sequenom).
3. 2X B/W buffer: 10 m*M* Tris-HCl, pH 7.5, 1 m*M* EDTA, 2 *M* NaCl (all components from Merck, Darmstadt, Germany).
4. 25% Aqueous NH₄OH (Merck, Darmstadt, Germany).
5. 10 m*M* Tris-HCl, pH 8.0 (Merck).
6. AmpliTaq Gold (Perkin-Elmer, Foster City, CA).
7. AmpliTaq FS (Perkin-Elmer).
8. Magnetic particle concentrator for microtiter plate or tubes (Dynal).
9. Specific PCR and PROBE primer (*see* **Note 1**).

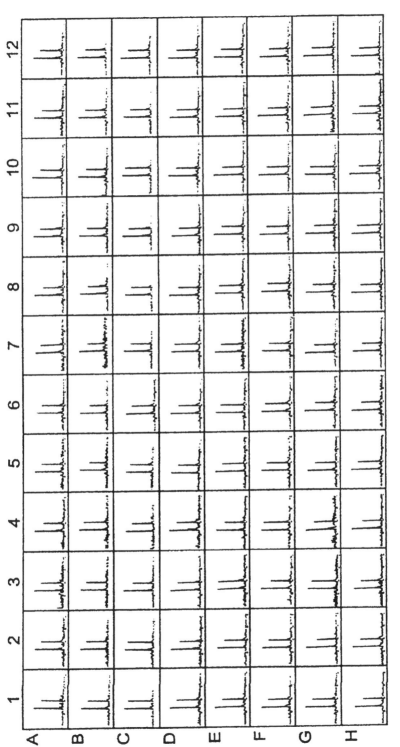

Fig. 6. Raw data generated during the analysis of one sample spotted 96 times on a SpectroCHIP.

Reaction_Details				Sample_Details				Assay_Details			
PlateNo.	PlateID	Well	SampleNo.	SampleID	PlateNo.	PlateID	Well	AssayNo.	Name	Result	Spectrum
1	11S7872	A1	1	14G88	1	23R902	A1	1	AMG	Male	11S7872_A1.sq
1	11S7872	A2	2	14G89	1	23R902	A2	1	AMG	Male	11S7872_A2.sq
1	11S7872	A3	3	14G90	1	23R902	A3	1	AMG	Female	11S7872_A3.sq
1	11S7872	A4	4	14G91	1	23R902	A4	1	AMG	Female	11S7872_A4.sq
1	11S7872	A5	5	14G92	1	23R902	A5	1	AMG	Male	11S7872_A5.sq
1	11S7872	A6	6	14G93	1	23R902	A6	1	AMG	Female	11S7872_A6.sq
1	11S7872	A7	7	14G94	1	23R902	A7	1	AMG	Male	11S7872_A7.sq
1	11S7872	A8	8	14G95	1	23R902	A8	1	AMG	Female	11S7872_A8.sq
1	11S7872	A9	9	14G96	1	23R902	A9	1	AMG	Male	11S7872_A9.sq
1	11S7872	A10	10	14G97	1	23R902	A10	1	AMG	Male	11S7872_A10.sq
1	11S7872	A11	11	14G98	1	23R902	A11	1	AMG	Female	11S7872_A11.sq
1	11S7872	A12	12	14G99	1	23R902	A12	1	AMG	Female	11S7872_A12.sq
1	11S7872	B1	13	14G100	1	23R902	B1	1	AMG	Male	11S7872_B1.sq
1	11S7872	B2	14	14G101	1	23R902	B2	1	AMG	Female	11S7872_B2.sq

Fig. 7. Sequenom data analysis software reports for automated sex typing using the DNA MassArray.

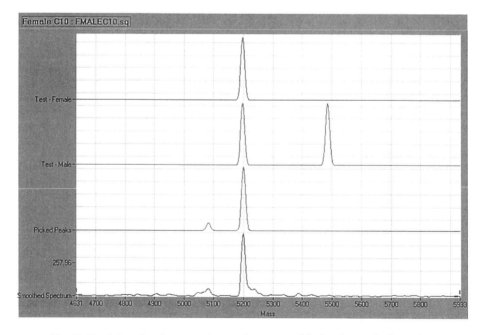

Fig. 8. Tool for visual comparison of spectra with the theoretical results.

2.2. Nanoliquid Handling and SpectroCHIPs

1. SpectroCHIP (Sequenom).
2. SpectroJET (Sequenom).

2.3. SpectroCHIP Analysis

1. SpectroSCAN (Sequenom).
2. SpectroTYPER (Sequenom).

3. Methods

3.1. PCR and PROBE Reaction

The following steps can be performed either in microtiter plates using multichannel manual pipettors or automated pipetting systems or on the single-tube scale.

3.1.1. Preparation of PCR

Perform one 50-µL PCR per PROBE reaction with 10 pmol of biotinylated primer and 25 pmol of nonbiotinylated primer (*see* **Note 2**).

3.1.2. Immobilization of Amplified Product

1. For each PCR use 15 µL of streptavidin Dynabeads (10 mg/mL).
2. Prewash the beads twice with 50 µL of 1X B/W buffer using the magnetic rack.
3. Resuspend the washed beads in 50 µL of 2X B/W buffer and add to 50 µL of PCR mix.
4. Incubate for 15 min at room temperature. Keep the beads resuspended by gentle rotation.

3.1.3. Denaturation of DNA Duplex

1. Remove the supernatant by magnetic separation.
2. Resuspend the beads in 50 µL of 100 mM NaOH (freshly prepared).
3. Incubate for 5 min at room temperature.
4. Remove and discard the NaOH supernatant by magnetic separation.
5. Wash three times with 50 µL of 10 mM Tris-HCl, pH 8.0.

3.1.4. PROBE Reaction

1. Remove the supernatant by magnetic separation, and add the following PROBE mix: 3 µL of 5X reaction buffer, 2 µL of PROBE nucleotide mix (ddA, ddC, ddG, or ddT with the respective dNTPs), 2 µL of PROBE primer (20 pmol), 7.6 µL of H$_2$O, 0.4 µL of enzyme (2.5 U).
2. The PROBE temperature profile comprises 1 min at 80°C, 3 min at 55°C, followed by 4 min at 72°C. Cool slowly to room temperature. Keep the beads resuspended by gentle rotation (*see* **Notes 3** and **4**).

3.1.5. Recovery of PROBE Products

1. After the reaction is completed, remove the supernatant by magnetic separation.
2. Wash the beads twice with 50 µL of 10 mM Tris-HCl, pH 8.0.
3. Resuspend the beads in 5 µL of 50 mM NH$_4$OH (freshly aliquoted from 25% stock solution).
4. Incubate for 4 min at 60°C.
5. Transfer the supernatant to a microtiter plate, and discard (or store) the beads.

3.2. SpectroCHIP Loading (see Note 5)

1. Fill containers with ultrapure water.
2. Initialize the nanoplotter.
3. Place the SpectroCHIP and microtiter plate on the nanoplotter (*see* **Note 6**).
4. Start the sample spotting program.

3.3. SpectroCHIP Scanning

1. Place the loaded SpectroCHIP on the sample holder.
2. Insert the sample holder into the SpectroSCAN.
3. Define which spots or chips have to be analyzed.
4. Choose analysis method and start the automated run.
5. Transfer the data to the processing server.

4. Notes

1. For PCR as well as PROBE primers it is useful to verify the masses before use. Primers that are not completely deprotected (mass shift to higher masses) or mixed with n-1 synthesis products should not be used.
2. Use asymmetric primer concentrations in PCR, with the nonbiotinylated primer in excess.
3. The length of the PROBE primer should not exceed 20-25 bases; try to have C or G at the 3' end, and avoid mismatches, especially at the 3' end.
4. The second temperature step in the PROBE program (55°C) depends on the primer length.
5. After the reaction, the beads can be stored for further reactions in Tris-HCl buffer at 4°C.
6. Be sure to handle SpectroCHIPs with gloves and avoid any contact with moisture.

References

1. Collins, F. S., Patrinos, A., Jordan, E., Chakravarti, A., Gesteland, R., Walters, L., and the members of DOE and NIH planning groups. (1998) New goals for the U.S. human genome project: 1998-2003. *Science* **282,** 682–689.
2. Botstein, D., White, D. L., Skolnick, M., and Davis, R. W. (1980) Construction of a genetic linkage map in man using restriction fragment length polymorphisms. *Am. J. Hum. Genet.* **32,** 314–331.

3. Weber, J. L. and May, P. E. (1989) Abundant class of human DNA polymorphisms which can be typed using the polymerase chain reaction. *Am. J. Hum. Genet.* **44,** 388–396.
4. Collins, F. S., Guyer, M. S., and Chakravarti, A. (1997) Variations on a theme: Cataloging human DNA sequence variation. *Science* **278,** 1580–1581.
5. Kruglyak, L. (1997) The use of a genetic map of biallelic markers in linkage studies. *Nat. Genet.* **17,** 21–24.
6. Nickerson, D. A., Taylor, S. L., Weiss, K. M., Clark, A. G., Hutchinson, R. G., Stengard, J., Salomaa, V., Vartiainen, E., Boerwinkle, E., Sing, C.F. (1998) DNA sequence diversity in a 9.7-kb region of the human lipoprotein lipase gene. *Nature Genet.* **19,** 233–240.
7. Sun, G. L., Diaz, O., Salomon, B., von Bothmer, R. (1999) Genetic diversity in Elymus caninus as revealed by isozyme, RAPD, and microsatellite markers. *Genome* **42,** 420–431.
8. Gusella, J. F., Wexler, N. S., Conneally, P. M., Naylor, S. L., Anderson, M. A., Tanzi, R. E., Watkins, P. C., Ottina, K., Wallace, M. R., and Sakaguchi, A. Y. (1983) A polymorphic DNA marker genetically linked to Huntigton's disease. *Nature* **306,** 234–238.
9. Risch, N. and Merikangas, K. (1996) The future of genetic studies of complex human diseases. *Science* **273,** 1516–1517.
10. Risch, N. and Teng, J. (1998) The relative power of family-based and case-control designs for linkage disequilibrium studies of complex human diseases. *Genome Res.* **8,** 1273–1288.
11. Larrey, D., Berson, A., Habersetzer, F., Tinel, M., Castot, A., Babany, G., Letteron, P., Freneaux, E., Loeper, J., and Dansette, P. (1989) Genetic predisposition to drug hepatotoxicity: role in hepatitis caused by amineptine, a tricyclic antidepressant. *Hepatology* **10,** 168–173.
12. Collins, F. S., Brooks, L. D., and Chakravarti, A. (1998) A DNA polymorphism discovery resource for research on human genetic variation. *Genome Res.* **8,** 1229–1231.
13. Christopoulos, T. K. (1999) Nucleic acid analysis. *Anal. Chem.* **71,** 425R–438R.
14. Hacia, J. G. (1999) Resequencing and mutational analysis using oligonucleotide microarrays. *Nat. Genetics Suppl.* **21,** 42–47.
15. Karas, M. and Hillenkamp, F. (1988) Laser desorption ionization of proteins with molecular masses exceeding 10,000 daltons. *Anal. Chem.* **60,** 2299–2301.
16. Wu, K. J., Steding, A., and Becker, C. H. (1993) Matrix-assisted laser desorption time-of-flight mass spectrometry of oligonucleotides using 3-Hydroxypicolinic acid as an ultraviolet-sensitive matrix. *Rapid Commun. Mass Spectrom.* **7,** 142–146.
17. Tang, K., Fu, D., Kötter, S., Cotter, R. J., Cantor, C. R., and Köster, H. (1995) Matrix-assisted laser desorption/ionization mass spectrometry of immobilized duplex DNA probes. *Nucleic Acids Res.* **23,** 3126–3131.
18. Jurinke, C., van den Boom, D., Jacob, A., Tang, K., Wörl, R., and Köster, H. (1996) Analysis of ligase chain reaction products via matrix assisted laser desorption/ionization time-of-flight mass spectrometry. *Anal. Biochem.* **237,** 174–181.

19. Jurinke, C., Zöllner, B., Feucht, H.-H., Jacob, A., Kirchhübel, J., Lüchow, A., van den Boom, D., Laufs, R., and Köster, H. (1996) Detection of Hepatitis B virus DNA in serum samples via nested PCR and MALDI-TOF mass spectrometry. *Genet. Anal.* **13,** 67–71.

20. Pieles, U., Zurcher, W., Schar, M., and Moser, H. E. (1993) Matrix-assisted laser desorption ionization time-of-flight mass spectrometry: a powerful tool for the mass and sequence analysis of natural and modified oligonucleotides. *Nucl. Acids Res.* **21,** 3191–3196.

21. Köster, H., Tang, K., Fu, D. J., Braun, A., van den Boom, D., Smith, C. L., Cotter, R. J., and Cantor, C. R. (1996) A strategy for rapid and efficient DNA sequencing by mass spectrometry. *Nat. Biotechnol.* **14,** 1123–1129.

22. Fu, D. J., Tang, K., Braun, A., Reuter, D., Darnhofer-Demar, B., Little, D. P., O'Donnell, M. J., Cantor, C. R., and Köster, H. (1998) sequencing exons 5 to 8 of the p53 gene by MALDI-TOF mass spectrometry. *Nature Biotechnol.* **16,** 381–384.

23. Berkenkamp, S., Kirpekar, F., and Hillenkamp, F. (1998) Infrared MALDI mass spectrometry of large nucleic acids. *Science* **281,** 260–262.

24. Braun, A., Little, D. P., and Köster, H. (1997) Detecting CFTR gene mutations by using primer oligo base extension and mass spectrometry. *Clin. Chem.* **43,** 1151–1158.

25. Braun, A., Little, D. P., Reuter, D., Muller-Mysock, B., and Köster, H. (1997) Improved analysis of microsatellites using mass spectrometry. *Genomics* **46,** 18–23.

19

An Introduction to Bioinformatics

Henry Brzeski

The purpose of this chapter is to illustrate how to obtain information on DNA and protein sequences from databases. This is most conveniently achieved using a Web browser (Netscape or Internet Explorer). This chapter is loosely based on a course given by the author at the University of Hertfordshire from a set of Web pages that facilitate Internet navigation by using hyperlinks and allow sequences to be copied from the Web page and pasted into the relevant search engine. (The online version of this information can be found at http://www.herts.ac.uk/natsci/Bio/informatics.htm.)

Many simple queries about protein and DNA sequences can now be answered using a Web browser. The purpose of this chapter is to give you a flavor of the sorts of things which are now possible, but first it is necessary to explain some of the jargon. If you know all about the World Wide Web (WWW or "the Web"), URLs, and hyperlinks then you can bypass the introductory information.

1. Introduction
1.1. The World Wide Web

The World Wide Web and the Internet were not invented by Bill Gates! It was originally put together in the 1960s by, primarily, the US Department of Defense (DOD) to ensure continuity of communication during a war. The DOD relied on the fact that every computer connected to its lines could communicate with any other computer. However, to make the system impregnable to attack, there was not a straightforward connection between each computer. Instead, multiple computers were connected to each other. As a result of this form of connection, there was always more than one way to connect between A and J. It may have been A-B-C-G-J, it may have been A-C-H-I-J, or it may have been, A-G-C-I-J, etc. This provided the resilience to possible attempts at dis-

From: *Methods in Molecular Biology, vol. 187: PCR Mutation Detection Protocols*
Edited by: B. D. M. Theophilus and R. Rapley © Humana Press Inc., Totowa, NJ

ruption. The original work was done by the Advanced Research Projects Agency, and the network was known as *ARPAnet*. Later on, academic institutions saw this as a wonderful way for researchers to communicate and so they started to connect their computers to this international network or "Internet." This gave rise to the ability to communicate via e-mail and also allowed collaborators to share large amounts of data by transferring the files rapidly via the Internet rather than by mailing a pile of disks or tapes. Now the Internet is being used for advertising and other commercial purposes.

1.2. Navigating on the Web

The files on computers scattered around the world must all have a unique name so that you can access each one specifically. This means that their names sometimes can be rather complex. Initially, it is necessary to tell your Web browser where to go to pick up a specific page. (The address of this page is given a jargon name; it is a "URL," which stands for Universal Resource Locator.) However, navigating (surfing) on the Internet would be tedious if surfers had to type in these names continually. They would soon get bored with typing "http://www.expasy.ch/prosite/," one of the addresses we will used later.) For this reason navigation is accomplished by using "hyperlinks" displayed in the now ubiquitous Web browsers. Hyperlinks can readily be identified on Web pages, because they are usually represented as underlined text in color, or as a button that is 'pressed' by clicking on it with the computer mouse and cursor. If the pointer is positioned over the hyperlink the address or URL it represents will appear in the status bar at the bottom of the browser window. Single click on the hyperlink and the Web browser will load the page at the new address. Hyperlinks can refer to different places in the same document or to totally new addresses. Hyperlinks should be traversed with care, as it is all to easy to follow links without thinking and end up miles from home, both figuratively and literally. It is usually possible to retrace the original path by pressing the **Back** button, but take care, this does not always work. All Web browsers record a history of traversed links, which can be used to connect rapidly to recently visited sites.

1.3. Databases and the Web

Since the early days of DNA and protein sequencing, such information has been deposited in computer databases so that many individuals could access this information. When the World Wide Web greatly expanded the reach of networked computers, it was not long before the Web browser became the interface between a very widely scattered population of researchers and the programs that could sift through the large amounts of data that were being accumulated.

2. DNA Databases

In the following sections, I will introduce the reader to a few of the programs available via the Web for finding and analyzing biochemical information.

2.1. Using Entrez to Search for Relevant Database Entries

Sooner or later a project reaches the point when it is necessary to devise primers to amplify known sequences from cells. This section describes how to obtain sequences for known genes/mRNAs, making it possible to devise primers to characterize genes/mRNAs. The National Center for Biotechnology Information (NCBI) has a very powerful computer with an easy-to-use Web-based interface for accessing sequence information. Follow these instructions to find sequences of particular genes/mRNAs.

1. Start a Web browser (*Netscape* or *Internet Explorer*) by clicking on its icon.
2. Go to **File/Open** in the menubar, enter http://www.ncbi.nlm.nih.gov/Entrez/ in the dialog box and press the **OK** button. This *will not* load the file into a new window.
3. This will load the *Entrez* page which allows users to quiz the databases available at NCBI for textual information connected with any required topic.
4. Click on the **Nucleotide** hyperlink.
5. In the text box enter 'p53' and press the **Search** button.
6. This very simple search will find many records (note the button with **Retrieve** *N* **records** at the top right hand side of the page).
7. These matches or 'hits' will consist of database entries containing partial and complete genomic or cDNA sequences from *Homo sapiens* and many other species.
8. Press the **Back** button on the browser.
9. Change the query to p53 & human (the "&" tells the server to find all records which contain both the word "p53" **and** the word "human"—an example of Boolean logic) and press the **Search** button. (You can find out more about Boolean expressions by clicking on **Detailed help** on the *Entrez* page: scroll to the top of the page, find the section labeled **For Experts Only**, and click on **Entering Complex Boolean Expressions**.)
10. This query finds fewer records. You can adjust your query using various required words and Boolean operators. By adding extra keywords, e.g., "complete," "mRNA," etc., you can fine tune your search and hit fewer documents.
11. Once the list is manageable, press the **Retrieve** *N* **records** to receive the first summary page of hits.
12. Each hit contains a checkbox, an accession number (the ID of the record), a brief summary of the entry (taken from the file), and various related links.
13. Check the boxes of the relevant hits and then press the **Display** button to retrieve the actual record(s).
14. Each record contains a number of fields that describe the sequence, e.g., the organism, whether the sequence is genomic or derived from mRNA, or relevant

published information. Depending on the sequence, the record may contain information on biologically important areas of the sequence, e.g., promoters, start AUG, introns, etc.) followed by the final, and most important, part of the record, the sequence itself.

2.2. Searching for Database Entries That Match a Sequence

The first step of the human genome sequencing project has involved identifying those DNA sequences most important for a cell. These code for proteins synthesized by the cell, which defines the cell's enzymatic complement and therefore its function. For this reason, people have been isolating the mRNAs expressed in cells, converting them to DNA (cDNA), cloning and sequencing them in their thousands. These expressed sequences are given the jargon name of *Expressed Sequence Tag* (EST) and will define the proteins made by a cell. As the sequencing of the human genome progresses, the function of more and more DNA/protein sequences will be identified. It is now routine to generate many ESTs and then to compare them with sequences in the databases to determine their function. A number of such ESTs are given in **Table 1** (2–8) along with one bacterial gene (1). The next procedure illustrates how to compare these sequences against the DNA databanks using *Basic Local Alignment Search Tool* (*BLAST*) to find out what they code for.

1. Identify a sequence to use (this can be an in house sequence or one of those provided in **Table 1**). Copy the sequence to your clipboard.
2. Open your Web browser and go to http://www.ncbi.nlm.nih.gov/BLAST/. Click on **Basic *BLAST* Search** to load a page containing the search form.
3. **Paste** the copy on the clipboard into the **search window**.
4. Use the default conditions for the search, i.e., *blastn*.
5. Press the **Search** button and wait while the sequence is compared to the databases and the matches displayed. (There is now a formal queuing system at NCBI, and you will wait for your results as explained on the page.)

The Washington University-Merck collaboration for EST sequencing (http://genome.wustl.edu/est/esthmpg.html) generates a large amount of sequencing information, and pictures of every sequencing gel are available at this site.

The Washington University Medical School Genome Sequencing Centre (http://genome.wustl.edu/gsc/index.shtml) is also involved in sequencing the human genome and information can be found here.

2.2.1. Interpreting the Results

The results from the NCBI *BLAST* server are presented both graphically and textually. The graphical view shows the query sequence as a thick red line with base numbers attached to it. Below this are a series of thin lines which represent matches to the query sequence. The length of the line indicates that part of

Table 1

Number	Sequence
	GGAAAGAAATGCATAAGCTTTTGCCATTC
	TCACCGGATTCAGTCGTCACTCATGGTGATT
	TCTCACTTGATAACCTTATTTTTGACGAGG
	GGAAATTAATAGGTTGTATTGATGTTGGAC
	GAGTCGGAATCGCAGACCGATACCAGGATC
1	TTGCCATCCTATGGAACTGCCTCGGTGAGT
	TTTCTCCTTCATTACAGAAACGGCTTTTTCA
	AAAATATGGTATTGATAATCCTGATATGA
	ATAAATTGCAGTTTCATTTGATGCTCGATG
	AGTTTTTCTAATCAGAATTGGTTAATTGGT
	TGTAACACTGGCAGAGCATTACGCTGACT
	TGACGGGACGGCGGCTTTGTT
2	TCCTGGNTCTGTTCTTCATCTTCACCTACTTCAAAGTTCCTGAGACTAAA
3	GGCCAAATTTGAAGAGCTCAACATGGATCTGTTCCGGTCTACTATGAAGC
4	GATGTCCAGAAGAATATTCAGGACTTAACGGCTNCAGGNTTTTAACAAAA
5	ATTGGCAGCCACACGGTGCTGGAGCTGCTGGAGGCTGGCTACTTGCCTGT
6	CATCGTGGAGAAGCCCTTCGGGAGGGACCTGCAGAGCTCTGACCGGCTGT
7	GCCCTGTCGAGACACTTGCCTTCTTCACCCAGCTAATCTGTAGGGCTGGA
8	TACATAATGTATTTATATATTTTTTGTATAATCACTATCTTTGTATTTAC

the query sequence which matches the hit sequence. The color represents the quality of the match.

Below the picture is a list of files which correspond to these matches sorted in match-quality order. The first hyperlink is to the file containing the entire sequence. This is followed by a very brief description of the file. The next number is a numerical score which represents how good the match was. This score is hyperlinked to the actual match found between your query sequence and the match itself. Finally, the last number gives the statistical significance of the match (the *E* value) and the chances of finding this match by chance.

2.3. Designing PCR Primers

2.3.1. Designing Primers for PCR Using xprimer

xprimer is a Web-based primer design package. Go to http://alces.med. umn.edu/xprimerinfo.html to see a detailed explanation of the various conditions that the primers must fulfil.

The purpose of this exercise is to learn how to design primers using *xprimer* and the sequence in **Table 2** (insert.seq) as the template. This sequence represents an insert in a plasmid plus 50–100 bp of vector sequence on either side.

Table 2

insert.seq	TGTGAGCGGATAACAATTTCACACAGGAAACAGCTATGACCATGAT
	TACGAAAGGTGCTTTTGGGGGCCGTCAGGGTCGAGGGTTCCTATTT
	CCTGGTCTATGGGGTCCCCGGCTTCGGGAAAGATAATGAAAGCCT
	CATCAGCAGGGAGGAGTTTTTAGGGGGGGTCCGCATGGGGGTCCC
	CCAAGCGACCGAATTGGCGGCTGAGGCCGTGGTGCTTCATTACAC
	CGATTTTCGAGCTCGGTACCCGGGGATCCTCTAGAGTCGACCTGC
	AGGCATGCAAGCTTGGCACTGGCCGTCGTTTTACAACGTCGTGAC
	TGGGAAAACCCTGGCGTTACCCAACTTAATCGC

1. Start a Web browser (*Netscape* or *Internet Explorer*) by clicking on the relevant icon.
2. Go to **File/Open** in the menubar, enter http://alces.med.umn.edu/webprimers.html in the dialog box and press the **OK** button.
3. After connection the Web browser will open the **Primer selection (image)** window.
4. Click in the **Query sequence**: text box (towards the bottom of the page).
5. Paste or type the sequence into this text. *Don't worry about any spaces which might appear.*
6. It is possible to fine tune the search parameters, i.e., primer length or Tm difference, by altering the values in the various list boxes but for the moment use the suggested defaults.
7. Press the **Submit** button.
8. After a few seconds/minutes the results will be returned in the form of a GIF file (it has a '.gif' file extension). This is a format for displaying images on a computer, and it is not possible to copy and paste primer sequences from here! If you want to copy and paste sequences then use the **Text** version of *xprimer* available from the **Primer selection (image)** window.
9. Compare these sequences with those of the M13 forward (GTTTTCCCAGT-CACGAC) and reverse (GGAAACAGCTATGACCATG) primers. *Note*: The terms forward and reverse used for M13 primers are not the same as the terms used for forward and reverse primers.
10. Do they match?
11. Can you find these sequences in insert.seq? Remember that these are PCR primer sequences and will be given in a 5' to 3' direction for *each* strand. Remember insert.seq is single stranded and does not include the complementary strand.
12. Do they match?

2.3.2. Checking the Suitability of Your Primers

The final part of primer design is to ensure that the chosen sequences will be specific for the required DNA target. This is achieved by repeating the *BLAST* search performed earlier. However, in this instance the two PCR primers should be used as the queries to ensure that there are no other sequences in the database which might be amplified along with your own sequence.

This is obviously not a guarantee against mis-primes, but it will help avoid the more obvious problems, such as unknowingly including regions containing repeated sequences.

2.4. The Human Genome Project

The chromosomal location of completed human DNA sequences can be found at (http://www.ncbi.nlm.nih.gov/genemap/). This Web address contains data on all chromosomes. Selection of a chromosome number will display a figure which contains three parts. First, there are three different ways of displaying the mapping data: two RH (radiation hybrid G3 and GB4) and one genetic map. Second, a drawing of the gene density on this particular chromosome is shown, and, third, the chromosome is drawn as an ideogram. Below this is a wealth of information on what has been sequenced and its relevance, if known. This site is continually updated.

Clicking on a region on the GB4 or G3 map will display the available sequencing information below the mapping data. All this information contains hypertext links to the actual sequencing data itself.

2.4.1. Genes Associated with Human Diseases

The Online Mendelian Inheritance in Man (OMIM) Web site, edited by Dr. Victor A. McKusick and his colleagues at Johns Hopkins University, and elsewhere, contains information on a large number (10,000) of diseases that have been identified as being linked with particular genes. This site can be accessed at http://www3.ncbi.nlm.nih.gov/omim and provides not only a description of the clinical symptoms of the disease but also the genetic lesion that gives rise to it.

It is possible to display the results from a search of this site in two different ways.

1. Searching the **Gene map** (http://www3.ncbi.nlm.nih.gov/Omim/searchmap. html) accesses the database using the name of the disease of interest and will display the result in the order in which the genes are found on the chromosome.
2. Searching the **Morbid map** (http://www3.ncbi.nlm.nih.gov/Omim/searchmorbid. html) will allow a search of the database using the name of the disease of interest (or a general descriptive term such as anemia) and will display, in alphabetical order, a list of diseases found that contain the keyword(s). The list contains information on the chromosome location and details of the genetics and clinical symptoms of the disease.

2.5. Sequencing Genomes

The genomes of the following species have been or are being sequenced, and data can be found at the given Web site.

- Human (http://www.ornl.gov/TechResources/Human_Genome/home.html).
- Mouse (http://www.informatics.jax.org).
- *Escherrichia coli* (http://www.genetics.wisc.edu).
- *Haemophilus influenzae* (http://www.tigr.org/tdb/mdb/hidb/hidb.html).
- *Caenorhabditis elegans* (http://www.sanger.ac.uk/Projects/C_elegans).
- *Arabidopsis thaliana* (http://genome-www.stanford.edu/Arabidopsis).
- Rice (http://www.dna.affrc.go.jp:82).
- Yeast (http://genome-www.stanford.edu/Saccharomyces).

A more exhaustive list can be found at http://www.ncbi.nlm.nih.gov/Entrez/Genome/org.html.

3. Protein Databases

3.1. The Databases

There are essentially three databases: the *Protein Information Resource* (PIR®; http://pir.georgetown.edu), *SWISS-PROT®* (http://www.expasy.ch/prosite) and *OWL* (http://www.biochem.ucl.ac.uk/bsm/dbbrowser/OWL/OWL.html).

3.2. The Sequence Retrieval System (SRS)

SRSWWW is a World Wide Web interface to the *Sequence Retrieval System* (*SRS*). It can be accessed at a number of different Web sites. *SRSWWW* is widely used because of the simplicity of Web browsers as an interface.

Detailed instructions on how to use *SRS* can be found in the SRS online manual (http://www.expasy.ch/srs5/man/srsman.html). Here I will introduce the basics of the program. In essence, *SRS* will allow the construction of a query that will look for the requested information in a number of databases. This is not as straightforward as it may seem because different databases organize the data into different fields, so it is necessary to construct the query with care. *SRS* ensures that this query construction is as simple as possible.

There are a number of *SRS* Web sites at which users can search various databases. Compare the *SRS* page at Heidelberg (http://www.embl-heidelberg.de/srs5; set up to find nucleic acid and protein database entries) with the version of *SRS* at *SWISS-PROT* (http://www.expasy.ch/srs5).

1. Go to the *SWISS-PROT SRS* page (http://www.expasy.ch/srs5).
2. Press the **Start** button to **Start a new SRS session**.
3. Click on the **TREMBL check box** to deselect it.
4. A detailed explanation of the databases are available by clicking on the hyperlinked database name.
5. Press the **Continue** button.
6. Type oxygen in the first field (leave the default **All text** in the drop-down list box) in the **SRS: Query Form Page**.

7. Press the **Do query** button.
8. This search will find over 1000 entries.
9. Press the **Back** button on your Web browser to return to the **SRS: Query Form Page**.
10. Click on the drop down list box to the left of the first text field and select **Description**.
11. Press the **Do query** button.
12. This search will find about 100 entries.
13. Press the **Back** button to return to the **SRS: Query Form Page**.
14. On the next line change **All text** to **Organism**.
15. Type *Homo sapiens* in the adjacent text box.
16. Press the **Do query** button.
17. This search will find just less than 10 entries.
18. This query is looking for database entries in *SWISS-PROT* that contain only the word "oxygen" in the **Description field** and "*Homo sapien*" in the **Organism field**. (It is possible to change this to an OR search in the drop down list box adjacent to the **Do query** button.)
19. Using this search technique it is possible to find entries from one or many databases using only one set of search parameters.

3.3. Searching for Database Entries That Match a Sequence

The first step of the human genome sequencing project has involved identifying those DNA sequences most important for a cell: the ones coding for proteins synthesized by the cell and, hence, which define the cell's enzymatic complement and thus its function. For this reason, people have been isolating the mRNAs expressed in cells, converting them to DNA (cDNA), cloning and sequencing them in their thousands. These expressed sequences (cDNAs) define those proteins made by a cell and are given the jargon name *Expressed Sequence Tag* (EST). As the sequencing of the human genome progresses, the functions of more and more DNA/protein sequences are identified. It is now routine to generate many ESTs, which are then sequenced and translated into proteins that can be compared with the protein databases to determine their function. You will find a number of such protein sequences derived from ESTs in **Table 3**. Compare these against the protein databanks using *Basic Local Alignment Search Tool* (*BLAST*) to find out what they code for:

1. Choose a sequence to use (this can be an in-house sequence or one of those provided in **Table 3**).
2. Copy the sequence to your clipboard.
3. Now click on **Basic *BLAST* SEARCH** at (http://www.ncbi.nlm.nih.gov/BLAST/) to load a page containing the search form.
4. **Paste** the copy on the clipboard into the search window.

Table 3

Number	Sequence
1	GDAAKNQLTSNPENTVFDAKRLI
2	EKASGKKIPYKVVARREGDVAACY
3	KLGKSFEMLILGRFIIGVYCGL
4	KGRTFDEIASGFRQGGASQSDKTPEELFHP
5	DDERNGWPVEQVWKEMHKLLPFSPDSVV
6	WRIFTPLLHQIELEKPKPIPYIYGSRG
7	PGAPGGGGGMYPPLIPTRVPTPSNGAPEIP
8	AVFYYSTSIFEKAGVQQPVYATIG

5. Click in the **Program** drop down list box, which at present says *blastn* (*blast nucleic acid*), and choose *blastp* to carry out a blast search on the protein databases. Now press the **Search** button and wait while the chosen sequence is compared to the databases and the matches displayed.

3.3.1. Interpreting the Results

The results from the NCBI *BLAST* server are presented both graphically and textually. The **graphical view** shows the query sequence as a thick red line with base numbers attached to it. Below this are a series of thin lines which represent matches to the query sequence. The length of the line indicates which part of the query sequence matches the hit sequence. The color represents the quality of the match.

Below the picture is a list of files corresponding to these matches and sorted in match-quality order. The first hyperlink (blue and underlined) is to the file containing the entire sequence. This is followed by a very brief description of the file. A numerical score represents how good the match was. This score is hyperlinked to the actual match found between your query sequence and the match **itself**. Finally, the last number gives the statistical significance of the match (the E value) and the chances of finding this match by chance.

Note that the color of the line represents a hit of poor quality. However, clicking on the hyperlinked colored line will display the matching sequence, which will be very similar. The reason for the apparently poor match is that the chance of finding such a short sequence match is high and so the score will be correspondingly low.

3.4. Aligning Protein Sequences Using CINEMA

There are two commonly used programs for sequence alignment: *CLUSTALW* and *pileup*. One of these programs (*CLUSTALW*) can be accessed using a Web browser and a Java Applet called *CINEMA* (Colour Interactive

Editor for <u>M</u>ultiple <u>A</u>lignment). This applet will access local or database sequences over the Internet and then, once they have been retrieved, align them. The alignments are color coded. If you wish to modify the alignment, it is possible using the *CINEMA* interface.

The purpose of an alignment is to compare two sequences and align the related regions to identify conserved and non-conserved regions. This alignment is built on the underlying assumption that the two sequences being aligned have evolved from a common precursor. If this evolutionary relationship is, in fact, true, then substitution, addition, or deletion of amino acids will be a rare occurrence because of the evolutionary constraints on biological function; the relationship is scored by assigning positive and negative values to matches and mismatches. However, not all amino acid changes are necessarily equally disadvantageous (for instance, the substitution of one hydrophobic amino acid with another is less likely to cause dramatic changes in protein structure than substituting a hydrophobic amino acid with a polar one). Each substitution has a "cost" associated with it and this cost is contained in tables which have such names as PAM-30, PAM-70, BLOSUM-80, and BLOSUM-62. This concept of "cost" is also true for the introduction of gaps into either sequence (addition or deletion of amino acids). The introduction of gaps is an undesirable event and so the introduction of gaps carries penalties in the summation of the final score.

There are penalties to pay for mismatches and introduction of gaps into an alignment. Depending on the specific aims of your own particular alignment, you might want to change these penalties (press the **Advanced button** in the *CLUSTALW* interface window). If you do so, then the results you obtain will probably be different. The default values suggested by the program are a good starting point. Don't be afraid to experiment with these penalties to look for less obvious similarities.

The *CINEMA* home page can be found at http://www.biochem.ucl.ac.uk/ bsm/dbbrowser/CINEMA2.1. Alternatively, if this server is proving slow, then try one of the mirror sites at Venus Internet (http://www.venus.co.uk/cinema) or The Weizmann Institute (http://bioinformatics.weizmann.ac.il/CINEMA).

1. Go to the *CINEMA* home page (http://www.biochem.ucl.ac.uk/bsm/dbbrowser/ CINEMA2.1) directly into a Web browser.
2. In the top window click on **Applet here** under the *CINEMA* logo.
3. This will load a separate window entitled *CINEMA* which contains multiple color sequences.
4. Select **File/Clear all** in the menu bar to give a clean starting window.
5. Select **Pluglets/Load pluglets** in the menu bar.
6. This opens the **Load Pluglet** window, and you should select **Clustal** and then press the **Load Pluglets** button. **AutoAlign** will appear in the menu bar.

7. Press the **Close** button in the **Load Pluglet** window.
8. Use **SRS** at *SWISS-PROT* to find human globin sequences (**All text** = globin, **Organism** = Homo sapiens) and note the SWISS-PROT file names, e.g., HBA HUMAN.
9. In the *CINEMA* window.

   EITHER

   a. Press the **DB seq** button
   b. In the **Load database sequence** window, delete Enter ID code here and then

   EITHER

   i. Enter the *OWL* code (use the protein ID code, e.g., opsd_sheep, not accession number).

   OR

   i. Change the **Database name** to *SWISSPROT* (or *PIR*) in the drop down list box at the top of the window.
   ii. Enter the *SWISS-PROT (PIR)* code, e.g., hba_human (this is *not* case sensitive).

   c. Press the **Get sequence** button.
   d. The requested sequence will be loaded into the *CINEMA* window.
   e. Repeat this procedure until you have accumulated all your sequences.

   OR (if the *OWL* server is down)

   a. Load a database file in the browser window (e.g., from *SWISS-PROT*) by clicking on the relevant hyperlink.
   b. Find the protein sequence at the end of the file.
   c. Drag the mouse across the sequence to select it (you will know it is selected because the sequence is now seen as white text on a black background; don't worry if you have selected numbers).
   d. Choose **Edit/Copy** from the menu bar
   e. Click on the button labeled *CINEMA* in the task bar at the foot of the screen.
   f. Click on the **Seq Editor** button.
   g. Give your sequence a title in the **Accession name** window.
   h. Click in the large text window at the bottom and

   EITHER

   i. Press **Ctrl-V** to paste your sequence from the clipboard

   OR

   i. Right click in the window then select **Paste** from the **Shortcut** menu.
   ii. Press the **Add sequence** button.
   iii. Press the **Clear All** button then repeat this procedure until you have included all the required sequences.

10. Select **AutoAlign/Clustal interface** in the menu bar.
11. In the *CLUSTALW* interface window note that you have a series of dark gray buttons with white text on them which will indicate the progress of the alignment.
12. In the *CLUSTALW* interface window enter a name for your job and then press the **Submit job** button.

a. You can fine tune your search by pressing the **Advanced** button. This will give you the opportunity of using a different scoring matrix (default is PAM 250) or by altering the criteria used for the alignment, e.g., by changing the penalties for opening or extending a gap.

13. When the alignment has been performed the dark gray buttons will change to light gray and the text will become blue.
14. The alignment will be loaded into the *CINEMA* window so you should clear this first by switching to the *CINEMA* window and choosing **File/Clear all** from the menu bar.
15. Move back to the *CLUSTALW* interface window and view your alignment by pressing the **Load Alignment** button.
16. Note that

a. The amino acids are color coded according to their properties, e.g., Polar positive—H, K, R (Blue); Polar negative—D, E (Red); Polar neutral—S, T, N, Q (Green); Non-polar aliphatic-A, V, L, I, M (White); Non-polar aromatic-F, Y, W (Purple); P, G (Brown); and C (Yellow). Colors can be viewed and modified by pressing the **COLORS** button.
b. Where gaps have been introduced this is indicated with a - (dash).
c. You can move through the aligned sequences using the scroll bars on the *CINEMA* window.

3.5. Comparing 2D Gels in Databases from Different Tissues

Identifying differences between two 2D gels is not always easy. Using the *Flicker* program it is possible to compare two 2D gels on the screen at the same time, and this program will make differences between the two gels more obvious by making the unique spots flicker.

1. Go to the *Flicker* Web site, http://www-lmmb.ncifcrf.gov/flicker.
2. Scroll down the page until you see **C) Lists of 2D PAGE gel images—you pick two from each list to compare** (in Section 1.1), then locate **6.T-lymphocyte phosphoproteins from IL-2/IL-4 dependent cell line 2D gel studies** and click on this. The reason for choosing these gels are that they contain only phosphoproteins; this makes the patterns simpler and so differences are more easily seen.
3. Click on the hyperlink **Select two gels and Flicker Compare them** and then accept the default choices, i.e., **lymphocyte-T_mouse_32P_59g-PPDB - G1-phase** in the upper window and **lymphocyte-T_mouse_32P_59h-PPDB - G2/M-phase** in the lower window.
4. Press the **Go Flicker** button.
5. This may take some time to display the gels but you will eventually see a new page with two 2D gels, one from each stage of the cell cycle, at the bottom of the window.
6. Each gel image will have a set of cross hairs in blue. Look at the two gels and

a. Decide on a common spot,

b. Press and hold down the **Control key (Ctl)**, and

c. Click on the common spot with the mouse in each gel.

7. This will move the gel image and position the cross hairs onto the spot. If you are not happy with the position then repeat this procedure.

8. Scroll back up the screen and click in the ***Flicker*** box to check it. The two images will be viewed in quick succession so that small changes can be easily seen.

3.6. Comparing 2D Gels in House from Different Tissues (Flicker)

This is relatively easy to accomplish as long as you can download your 2D gels onto a Web server that can be accessed using a normal Web addressing system.

1. Log onto the *Flicker* home page, http://www-lmmb.ncifcrf.gov/flicker.

2. Once you have connected with the *Flicker* page you should scroll down the page until you see **4. Flicker compare images from any two URLs**, then scroll further until you get to **Enter two images URLs:**.

3. Click in the **Left image** box and type http://www.herts.ac.uk/natsci/Bio/2Dimages/Image10alt.gif (or select, copy, and paste!).

   EITHER

   a. Click in the **Right image** box and type in the gel address, i.e., http://www. herts. ac.uk/natsci/Bio/2Dimages/Image12alt.gif (or select, copy, and paste!)

   OR

   a. Drag the mouse cursor across the whole of the address you have just typed in **step 3** to highlight it.

   b. Hold down **(Ctrl)**.

   c. Press and release the **C key** to copy the address to the clipboard.

   d. Click in the **Right image** box.

   e. Hold down **Ctrl**.

   f. Press and release the **V key** to paste the address into the box.

   g. Edit the address to give http://www.herts.ac.uk/natsci/Bio/2Dimages/Image12alt.gif (or select, copy, and paste!).

4. Now press the **Go Flicker** button.

5. The relevant gels and the program to view them will now be downloaded. This may take some time but you will eventually see a new page with your two 2D gels. These will be found at the bottom of the window, and if you can't see them then scroll down the window.

6. Each gel image will have a set of cross hairs in blue. Look at the two gels and decide on a common spot, press and hold down **Ctl**, and click on the common spot with the mouse in each gel. This will move the gel image and position the cross hairs on the spot. If you are not happy with the position then repeat this procedure.

7. Now scroll back up the screen and click in the **Flicker box** to check it. The two images will be viewed in quick succession so that small changes can be easily seen.

4. URLs Cited

Chromosomal location of genes, http://www.ncbi.nlm.nih.gov/genemap

Chromosomal location of completed human DNA sequences, http://www.ncbi.nlm.nih.gov/genemap

CINEMA, http://www.biochem.ucl.ac.uk/bsm/dbbrowser/CINEMA2.1

Entrez, NCBI http://www.ncbi.nlm.nih.gov/Entrez

fFlicker (comparing two images), http://www-lmmb.ncifcrf.gov/flicker

Genes associated with human diseases, http://www3.ncbi.nlm.nih.gov/omim

Genomes which have been or are being sequenced:

> Human, http://www.ornl.gov/TechResources/Human Genome/home.html
>
> Mouse, http://www.informatics.jax.org
>
> *E. coli*, http://www.genetics.wisc.edu
>
> *Haemophilus influenzae*, http://www.tigr.org/tdb/mdb/hidb/hidb.html
>
> *Caenorhabditis elegans*, http://www.sanger.ac.uk/Projects/C elegans
>
> *Arabidopsis thaliana*, http://genome-www.stanford.edu/Arabidopsis
>
> Rice, http://www.dna.affrc.go.jp:82
>
> Yeast, http://genome-www.stanford.edu/Saccharomyces

A list can be found at http://www.ncbi.nlm.nih.gov/Entrez/Genome/org.html

Human genome, http://www.ornl.gov/TechResources/Human Genome/research.html

Human diseases which have been identified as being linked with particular genes. This site can be accessed at the NCBI home page, http://www3.ncbi.nlm.nih.gov/omim

OWL, http://www.biochem.ucl.ac.uk/bsm/dbbrowser/OWL/OWL.html

PCR primers, *xprimer*, http://alces.med.umn.edu/xprimerinfo.html and http://alces.med.umn.edu/webprimers.html

Positioning ORFs on 2D gels, http://expasy.hcuge.ch/ch2d

ProDom (protein domains), http://protein.toulouse.inra.fr/prodom.html

Protein Information Resource (PIR), http://pir.georgetown.edu

SWISS-PROT, http://www.expasy.ch/sprot/sprot-top.html

Sequence retrieval system (SRS), http://www.expasy.ch/srs5/man/srsman.html or http://www.embl-heidelberg.de/srs5/ c) http://www.expasy.ch/srs5/

Searching the *OMIM* Gene Map, http://www3.ncbi.nlm.nih.gov/Omim/searchmap.html

Searching the *OMIM* Morbid Map, http://www3.ncbi.nlm.nih.gov/Omim/searchmorbid.html

The address of the online version of this chapter, http://www.herts.ac.uk/natsci/Bio/informatics.htm

Washington University-Merck collaboration for EST sequencing, http://genome.wustl.edu/est/esthmpg.html

Washington University Medical School Genome Sequencing Center, http://genome. wustl.edu/gsc/index.shtml

4.1. Other URLs Concerned with Bioinformatics

1. University College London, *A Taste of Bioinformatics*, http://www.biochem. ucl.ac.uk/bsm/dbbrowser/jj. The aim of this tutorial is to provide a gentle introduction to sequence and structure function analysis.

2. DNA Learning Center Cold Spring Harbor Laboratory, *Online DNA Sequence Analysis and Comparison Tutorial*, http://vector.cshl.org/SequenceAnlaysis Exercise/index1.html.

   This tutorial uses reference human mitochondrial DNA sequences and online resources to:

   • Search for like DNA sequences in online databanks
   • Locate DNA sequences in genomes
   • Compare modern human DNA sequences
   • Compare modern human DNA sequences to Neanderthal
   • Compare modern human DNA sequences to other organisms

3. EMBnet, DNA analysis tutorial, http://www.ie.embnet.org/other/tut.html. This tutorial considers three popular sets of DNA and protein sequence analysis programs:

   • The *Staden Package*, from Rodger Staden et al., MRC Laboratory of Molecular Biology, Cambridge, UK
   • *The Wisconsin Package* (*GCG*), from the Genetics Computer Group, Inc., Madison, WI.
   • *EGCG* (*Extended GCG*) from a consortium of researchers mostly based in Europe at EMBnet Nodes.

4. University of Adelaide, *A tutorial on sequence analysis: From sequence to structure*, http://www.microbiology.adelaide.edu.au/learn/index.html

   An unusual tutorial by Harry Mangalam subtitled, "one person's cautionary tale of model building."

Index

A

ABI PRISM, 111, 122
Agarose, *see* Gel electrophoresis
Amplification refractory mutation
    system (ARMS), *see* ASO-PCR
Alkaline phosphatase, 18
Allele-specific oligo PCR (ASO-PCR),
    44, 47,
    cycling conditions, 49
    primer design, 49
Autoradiography, 20, 32, 69

B

Basic Local Alignment Search Tool
    (BLAST), 196, 201
Big Dye™ Terminators, 66, 71, 119
Bioinformatics, 193
BiomassPROBE, 184, 185
Biotin, 57
Bovine serum albumin, 13
Bromophenol blue, 4, 9

C

CA repeat, 30
CDP-*Star*™, 171
Capillary electrophoresis, 160
Capillary transfer, *see* Southern
    blotting
Chemical cleavage of mismatch
    (CCM), 109
Chemiluminescence detection, 174
Chromosome paint, 73
CINEMA, 203
Cleavase fragment length
    polymorphism (CFLP), 165

Clustal analysis, 120, 202
Competitive *in situ* suppression, 86
Complementary DNA (cDNA), 40
Conformation sensitive gel
    electrophoresis (CSGE), 137
Counter staining, 82
Cycle sequencing, *see* Sequencing
Cystic fibrosis, 152, 157, 158, 159

D

DAPI, 78
Databases, genetic, 194
Denaturing gradient gel electrophoresis
    (DGGE), 109, 125,
    preparation of, 130
    gel staining of, 132
Denaturing high performance liquid
    chromatography, 110
Dideoxy fingerprinting, 165
Digoxygenin, 19
Dimethylsulfoxide, 44
Dithiothreitol (DTT), 41
DNA,
    affinity capture, 60
    end-labeling, 17, 170
    extragenic, 29
    GC content, 44, 49, 62
    hairpin structures, 44
    internal labeling, 13
    intragenic, 29
    interphase, 73
    isolation, 115
    melting temparature (T_m), 44,
        126, 128
    metaphase, 73

FUTURE TALK

FUTURE TALK
The Changing Wireless Game

Ron Schneiderman

IEEE
PRESS

The Institute of Electrical and Electronics Engineers, Inc., New York

This book may be purchased at a discount from the publisher
when ordered in bulk quantities. For more information contact:

IEEE PRESS Marketing
Attn: Special Sales
PO Box 1331
445 Hoes Lane
Piscataway, NJ 08855–1331
Fax: (908) 981–9334

For more information about IEEE Press products, visit the IEEE Home Page at
http://www.ieee.org

Printed in the United States of America

10   9   8   7   6   5   4   3   2

ISBN 0–7803–3407–8

IEEE Order Number: PC5679

Library of Congress Cataloging-in-Publication Data

Schneiderman, Ron.
    Future talk: the changing wireless game / Ron Schneiderman.
        p.   cm.
    Includes bibliographical references (p.) and index.
    ISBN 0-7803-3407-8
    1. Personal communication service systems.   2. Telecommunication
equipment industry.   I. Title.
    TK5103.485.S56   1997
    621.3845—dc21   96-49304

For my wife, Susan, who bought me
my first cellular phone

Contents

Preface

We used to call it two-way radio. Today, "wireless" personal communications represents the fastest-growing segment of the global telecommunications market.

Two things we know: People love to talk, and they're doing a lot more of it wirelessly. Worldwide, 50 new cellular networks came on line in 1996. Hundreds of billions of dollars are being spent to expand and upgrade existing wireless networks throughout the world.

An industry-sponsored study of consumer attitudes toward wireless telephones calls them the "smoke alarm of the '90s." Twenty years ago, no one had smoke alarms; now they're everywhere. Today, 20¢ out of every dollar in revenue for telecommunications providers comes from wireless communications. By the year 2008, that figure is expected to increase to 80¢.

Signs of explosive industry growth are everywhere. More than half the homes in the U.S. now have cordless phones. More than 70 million people in the world subscribe to cellular telephone service—a number that could easily more than triple to 280 million by the year 2000. Nearly 10 million people are expected to be using wireless networks to transmit data by the year 2000.

The ubiquity of wireless communications even shows up in our popular culture. The Jim Dyer character on the TV sitcom "Murphy Brown" says, "Anyone who makes cellular calls from a stall in the men's room is capable of anything." A cartoon in the *Wall Street Journal* depicts an angry motorist on her mobile phone saying, "No, I'm *not* interested in aluminum siding."

The demand for pagers (about 70 million people worldwide now use them) continues and has led to a host of innovative new products and services. Specialized Mobile Radio (SMR) dispatch networks are positioning themselves to compete with cellular and other wireless services. The emergence of personal communications services (PCS) is rapidly broadening the market for highly mobile communications with cheaper and smaller portable phones.

Wireless data are going the way of the personal computer, developing as a platform for very specialized vertical applications and moving slowly into larger, more broad-based, horizontal markets. Today, portable personal computers—from laptops to smaller notebook and subnotebook models—account for nearly 30 percent of all PC sales as more people depend on communications-enabled portable PCs as their primary computing platform.

International markets have never been stronger and they continue to represent a huge opportunity as the economically powerful countries deregulate and enhance their telecom services, and underdeveloped and developing nations begin to expand their telecommunications industries.

Industry analysts believe that wireless communications could generate $100 billion in revenue by the year 2000. But not without some hurdles along the way. A 1996 study by the U.S. National Academy of Sciences agrees with just about everyone else that wireless communications is going to be a huge success, but it also quotes several industry sources who readily admit that they don't know how this growth is going to occur. Everything is moving too fast, they say—particularly the technology. Business plans are getting shorter, and they're being reviewed more often. Everyone is trying to figure out what's happening. And what's coming.

Maybe this book can help.

"They told Marconi, wireless was a phony . . ."

From the song, "They All Laughed" (1937);
words and music by Ira and George Gershwin.

1

Cellular Phones and Personal Communications Services

On April 1, 1995, the Associated Press reported that after nearly a century of monitoring telegraph distress calls, including the *Titanic*'s after her collision with an iceberg in 1912, the U.S. Coast Guard turned off its Morse code equipment.

The Coast Guard didn't make this decision lightly. After all, even the early U.S. manned space vehicles were equipped with Morse code systems in case other communications equipment aboard the spacecraft failed. But with technology advancing so rapidly, supporting Morse code hardware and training people to use it no longer made sense. The simple and slow dots and dashes have been replaced by satellites and a myriad of terrestrial communications choices.

The Morse code announcement sent a strong signal that the communications industry is changing, and not just the technology.

Long-distance phone companies are pushing into local phone services. Local phone companies are edging into long-distance and cable television markets. Cable TV companies are adding phone service and are connecting customers' home computers through cable modems and coaxial cable lines.

Broadcasters can hardly wait to install their new digital television system so they can begin to offer new revenue-generating data services. Utility companies are stringing fiber-optic cable in anticipation of offering new communication services. State and regional highway authorities are ready to start running wires along their roads to offer telecommunications services.

Meanwhile, wireless companies including cellular carriers are developing and enhancing their products and services, in some cases almost faster ı the market can absorb them. Wireless service has become the fastest growing segment of the world telecommunications market and the most dynamic.

Cellular service is growing at the rate of more than 50 percent annually worldwide. By the year 2001, if market projections are correct, there will be more than 280 million cellular subscribers in more than 120 countries. That's more than four times today's global subscriber level of about 65 million.

The U.S., with about half of the cellular subscribers in the world, added as many customers in 1995 as it did in cellular's first nine years of existence. With more than 90 operating networks, Europe is becoming one of the most dynamic cellular markets in the world. Currently only 2 out of 100 Europeans use a cellular phone, but one industry consulting group, MTA-EMCI, is projecting 78 million cellular subscribers across Europe by the year 2000.

The Global System for Mobile Communications (GSM), the European digital standard for cellular and personal communications networks (PCN), as personal communications services (PCS) are called in Europe, is ex-

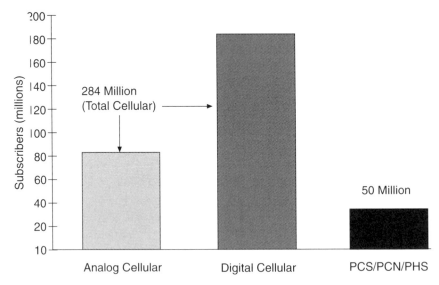

Distribution of world cellular service and PCS subscriber base by technology, 2001.
Note: PCS/PCN/PHS includes U.S., Canada, U.K., Germany, France, Japan, Singapore, Malaysia, Thailand, and Puerto Rico.

pected to account for 74 percent of the subscriber base, followed by analog cellular and DCS-1800, a GSM derivative system operating in the 1.8 GHz band. The U.K., France, Germany, and Italy will likely be the dominant markets, accounting for more than 60 percent of the total cellular subscriber base by 2000.

PCNs based on the DCS-1800 standard and currently licensed in the U.K., Germany, and France are expected to be introduced in the Nordic countries, Italy, and Spain by the end of 1998.

MTA-EMCI expects Eastern European countries to triple their share of aggregate European cellular subscribers, increasing from 2 percent market penetration in 1994 to 6 percent in 2000, reaching almost 5 million subscribers. Most of the demand for cellular service in Eastern Europe will be in Hungary, Poland, and Russia.

The numbers on the infrastructure side are no less staggering. Worldwide, 50 new cellular networks came on line in 1995, a pattern of growth

The World's Largest Cellular Operators

Company	Millions of Subscribers
AT&T (U.S.)	4.03
SBC Mobile Systems (U.S.)	2.98
GTE (U.S.)	2.34
Telecom Italia	2.24
BellSouth (U.S.)	2.16
Vodafone (U.K.)	1.82
Cellnet (U.K.)	1.73
Bell Atlantic (U.S.)	1.68
DeTeMobile (Germany)	1.60
DGT (China)	1.57
AirTouch (U.S.)	1.56
NTT DoCoMo (Japan)	1.32
Ameritech Cellular (U.S.)	1.30
Telia Mobitel (Sweden)	1.18
Sprint (U.S.)	1.04
US WEST (U.S.)	.97
Korean Mobile Telephone Co.	.96
NYNEX Mobile (U.S.)	.91
Mannesmann Mobilefunk (Germany)	.85
Cantel (Canada)	.79

Source: Data from International Telecommunications Union (ITU).

"Well, I know how to be at one with nature, and
***still* have my phone with me."**

From the *Wall Street Journal*, April 26, 1995; © 1995 by
Sidney Harris.

that will be easily matched in 1996 and 1997. In addition, billions of dollars will be spent to expand and upgrade existing systems.

Cellular carriers in the U.S. alone estimate they will need 15,000 new cell sites over the next 10 years to complete their coverage, upgrade their systems, and meet the growing demand for cellular service. Another 10,000 sites will be needed for new PCS providers, and thousands more are planned for new Enhanced Specialized Mobile Radio (ESMR) networks to expand and enhance commercial dispatch radio services.

There's more than enough business and competition to go around.

SOME BACKGROUND

In Europe, most of the excitement has developed from the digital Global System for Mobile Communications (GSM), which the European Telecommunications Standards Institute (ETSI) has adopted as the European digital standard for cellular and personal communications networks. Initially known as Groupe Speciale Mobile, GSM evolved out of the rapid growth of conventional analog cellular networks in Europe and the need for a system with much greater capacity than analog could offer. In 1982, the Conference of European Post and Telecommunications (CEPT) formed a special working group to develop GSM. By 1987, 18 European nations had committed to the technology by signing the GSM Memorandum of Understanding. In 1989, the effort to develop a GSM standard was shifted to the European Telecommunications Standards Institute (ETSI) and plans were put in place to begin GSM service in 1991.

Things didn't quite work out as planned; the entire effort was slowed by almost a year when GSM equipment wasn't widely available and technical problems began to emerge; some were software related, but the major obstacle was the incompatible equipment from different manufacturers. Roaming became an issue when GSM operators tried to work out agreements on how to charge subscribers who use their GSM phones outside their home territories. Most of these problems have been solved and GSM appears to be on its way to becoming the world's most widely used digital cellular standard—at least for a while.

GSM has emerged in two phases. Initially, GSM handled basic voice service and some emergency calling features. The second phase, which began in 1994, added call waiting, caller information services, and improvements in subscriber identity module (SIM) cards, which contain a microchip with information on the caller. By inserting a SIM card in a GSM phone, the caller can gain access to the GSM network and be billed for the call, even if it isn't his or her phone. SIM cards can also store abbreviated phone numbers for speed dialing.

From the users' point of view, the obvious difference between GSM and the cellular systems now emerging in the U.S. is that virtually all cellular phones in Europe operate only digitally—in the GSM mode. In the U.S., cellular carriers are offering dual-mode (analog/digital) phones and U.S. subscribers will be switched—transparently—between the two modes, depending on their location (in other words, what service is available in the calling area).

Another important development is the emergence of Digital European Cordless Telecommunications, or DECT. Originally intended for wireless private branch exchange (WPBX) service, DECT will eventually find a place in the residential and telepoint markets as well. DECT supports bo voice and data with encryption, and seamless handoff between cells, using spectrum already allocated throughout Europe in the 1,880–1,900 MHz band.

DECT has the support of ETSI, the European Commission (EC), and five major European telecommunications companies. Alcatel, Ericsson, Nokia, Philips, and Siemens have all announced plans to show DECT products.

Another new service emerging as Europe's PCN is Digital Communication Service, a GSM derivative operating at 1,800 MHz (DCS-1800). In fact, with some inherent capacity problems, GSM carriers will likely be forced to market multiband digital handsets to ensure that their subscribers can make or receive calls from virtually anywhere at anytime.

The U.K. represents the largest wireless market in Europe. It may also be the most confused market. The British government licensed four companies to provide CT-2 (cordless telephones–second generation) telepoint service shortly after the publication of the landmark *Phones on the Move* by the Department of Trade and Industry, but it couldn't handle the high operations costs, the competition, and, least of all, the lack of interest. All four soon disappeared. Today, the only company offering telepoint service in the U.K. is Hutchison Personal Communications Ltd., which wasn't even one of the original licensees.

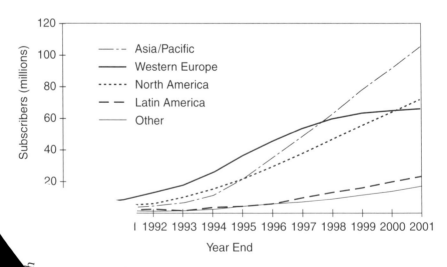

I cellular subscribers by region, 1989–2001.

Hutchison invested a reported $100 million to build more than 8,000 base stations in train stations and along major highways in the U.K. But the service, known as Rabbit, never really caught on. (The joke in the U.K. is that there are more telepoint base stations than subscribers and, in fact, only a few thousand people subscribed to the service.) Hutchison has tried cutting its losses by selling CT-2 phones directly rather than through retail outlets.

Shortly after it licensed the telepoint operators, the British government issued PCN licenses to three companies—Mercury Personal Communications, Unitel, and Microtel—which said they planned to offer nationwide service. Mercury was initially a joint venture between Mercury, a subsidiary of Cable & Wireless, and US WEST. Unitel decided to abandon its PCN plans shortly after receiving its license and merged with Microtel, which is now owned by Hutchison. (Microtel had been owned by Pacific Telesis and Millicom, both U.S. companies, and Matra, a French company, but they all dropped out of the venture.) The Microtel–Unitel partnership still owns a telepoint license, but it's not clear what they're going to do with it.

According to *British PCN Policy Pitfalls: Implications and Lessons for the U.S.*, a report commissioned by the Cellular Telecommunications Industry Association (CTIA) and written by Alan Pearce, president of Information Age Economics Inc., one of the problems in getting a wireless personal communications network off the ground in the U.K. was the high cost of building three separate PCN infrastructures—about $1 billion each. Another was the way in which the British government doled out spectrum for the PCNs. As Pearce pointed out in his study, the three PCN licenses were a huge spectrum giveaway. The government, he noted, assigned 150 MHz of spectrum at 1.8 GHz (1,800 MHz) and said that the licensees were each entitled to 50 MHz once their networks had been built. That caused an immediate outcry of unfair competition, since cellular carriers in the U.K. only have 30 MHz of spectrum.

Government licensing rules also call for universal PCN service, but licensees said they planned to launch limited service focusing only on major metropolitan areas. In other words, said Pearce, universal PCN would have to wait until PCN proves itself in the marketplace. The dilemma here, according to the Pearce report, was that the PCNs can't have the full 50 MHz of spectrum until they offer universal service, and they can't afford to build a universal infrastructure until they prove that PCN service is in demand in the marketplace. Meanwhile, the PCN licensees were already operating on 10 MHz of spectrum, temporarily quieting the two cellular carriers, Vodaphone and Cellnet.

The loudest response to the Pearce report came from US WEST, a founding member of CTIA and a partner with Cable & Wireless Plc in the U.K. PCN service called Mercury One-2-One, which was scheduled to begin service in the fall of 1993. US WEST publicly disassociated itself from the association-sponsored study. In a letter to FCC Interim Chairman James Quello, US WEST said CTIA was using the Pearce document to support its campaign in Congress to limit the potential of PCS. "The CTIA's recommendation that the public would best be served if the U.S. were to authorize five PCN licensees [per market] is not based upon the U.K. experience and represents no more than an attempt by the cellular industry to avoid viable competition," US WEST said in its letter to Quello.

US WEST also said it believes that the record of the U.K.'s PCN experience bears out the wisdom of certain policies, including awarding no more than three PCS licenses and allowing these licensees "latitude in meeting customer needs," using large regions for licensing PCS, and providing each licensee with access to a substantial block of spectrum to reduce the cost of providing service and to stimulate a mass market.

In response, CTIA called the US WEST assertions that it was using the report to gain influence in Congress "spurious," noting in its own letter to the Federal Communications Commission (FCC) that "CTIA and its members have been active, vocal advocates in the commission's PCS proceeding, urging a broad definition of PCS and open entry for as many competitors as possible."

The telecommunications industry is keeping a close eye on the Mercury One-2-One service, which, so far, is confined to London. Despite call charges that are up to 20 percent lower than cellular service in the U.K., service restrictions and reports of minor technical problems may make it difficult to wean cellular users over to the newer service.

With little acceptance of PCNs, the two major cellular carriers in the U.K.—Cellnet, a subsidiary of British Telecom, and the Vodafone Group, which is owned by Racal Electronics—have restructured their tarrifs to be more competitive with PCN services. Vodafone launched a regional GSM service in December 1991, but it met with little initial success. The service was essentially relaunched in early 1993, targeted at business users. Since then, Vodaphone, the larger of the two U.K. cellular operators, has signed roaming agreements with SIP, the Italian telecommunications concern, and

SFR, one of the French cellular system operators. Vodafone has also introduced a "short message" text transmission service using GSM protocols.

Originally, the German government took a monopolistic approach to
cellular communications, licensing the Deutsche Bundespost Telekom
(DBT), Germany's national public phone network, as the only analog cellular service operator. More recently, however, the German government
launched a major modernization program called Telekom 2000 to upgrade
the communications network in eastern Germany and link it with the rest
of the country. To do that, Germany issued two GSM licenses: one to the
DBT and the other to Mannesmann Mobilfunk GmbH, a privately held
consortium whose members include several European and U.S. telecom
organizations (PacTel International has a 26 percent interest). The difference is that Mobilfunk's license requires that the network be based exclusively on the GSM standard. BDT's license covers both analog and digital
cellular service.

Mobilfunk began its D2 Privat service in June 1992; the DBT system,
called D1 Privat, began operation in July 1992, and both services have
grown steadily. Typical of fast-growing cellular services, prices of vehicle-
installed GSM phones dropped from about $1,563 in June 1992 to $812 in
February 1993. Transportable models cost slightly more, and the first handheld portables, which arrived in late 1992, were priced at about $1,406, con-
siderably more than in today's competitive market.

In February 1993, Germany issued a third license to the German consortium E-PLUS for a new national service based on the DCS-1800 standard. E-PLUS began rolling out its service in Berlin and Leipzig, the
largest cities in eastern Germany, in 1994. Consortium members are led by
Thyssen AG and Veba AG, each with a 28 percent stake in the organization. Other investors include BellSouth Enterprises of the U.S.; Vodafone
Group Plc of the U.K.; and a number of small- and medium-sized companies from eastern Germany and the Caisse des Depot et Consignations
Group of France, which operates the Cofira telecom network in France
with BellSouth, Bau GmbH, Industriemontagen Leipzig GmbH, Minol
Mineralolhandel AG, and Part'Com S.A.

France's France Telecom operates the Radiocomm 2000 analog cellular
system throughout the country, and began its GSM service, called Itineris,
in 1992 in Paris and Lyon. The country's second analog carrier is Societe
Francaise de Radiotelephonie (SFR), a private firm.

France's version of the cordless telephone, called Bi-Bop, is another telepoint service based on CT-2 technology. Launched in April 1993, the phone allows calls to any number provided it is no more than about 200 meters from a base station. At last count, Bi-Bop only had about 2,000 subscribers, and most of them were participants in the initial trial of the system.

Ameritech International, a subsidiary of Chicago-based Ameritech, helped build and operate a GSM-based cellular network in Norway, one of Europe's leading per capita users of mobile phones because so many Norwegians own second homes, few of which have fixed telephone service. Ameritech International and Singapore Telecom have won government approval to acquire 49.9 percent interest in Netcom GSM, a Norwegian firm. The three companies have been working together to build and operate the cellular network in Norway.

Analog cellular service was introduced in Japan in 1979, almost four years ahead of the U.S. Nippon Telegraph & Telephone Co. (NTT), Japan's government-owned public corporation, had the cellular market to itself until 1985 when Japan enacted the Telecommunications Business Law, which essentially abolished the legal monopolies held by NTT, the Telegraph Public Corp., and Kokusai Denshin Denwa (KDD), and privatized the NTT Public Corp.

In 1986, the Ministry of Posts and Telecommunications (MPT), which regulates the cellular industry in Japan, licensed two new service providers, Nippon Ido Tsushin and Daini Denden, Inc., to compete in the cellular market with NTT. However, neither company received a national license similar to NTT's. Ido—whose backers include Toyota, Nippon Electric Corp. (NEC), the Japan Highway Authority, and Tokyo Electric Power—is licensed to operate only in the Tokyo–Nagoya area. DDI Corp., which is made up of eight affiliated Japanese cellular companies, can operate only in the remaining, mostly residential suburban, regions of the country.

In fact, it wasn't until April 1994 that Japanese citizens could actually buy a cellular phone; prior to that time they could only lease a phone. The results of deregulation and new competition in the cellular market in Japan have been dramatic: More subscribers signed up for cellular service in the first 14 months following deregulation in April 1994 than subscribed in the first 14 years of cellular service in Japan.

Japan's digital mobile communication system is called Japan Digital Cellular (JDC). It is a TDMA-based system in the 800 MHz and 1.5 GHz bands, and therefore similar to the American time division multiple access (TDMA) network, but with one major exception: It will not be dual mode (analog and digital). The digital market in Japan will be much more competitive than analog. Customers will be able to choose among four providers of digital cellular service. NTT and the new entrants will be licensed to offer digital cellular service nationally, although they are expected to begin operations on a regional basis. Ido and DDI also will be licensed to offer digital service, but only in their current regions.

NTT selected Motorola, AT&T, and Ericsson along with six Japanese manufacturers to develop its digital cellular network, and Fujitsu, Matsushita, Mitsubishi Electric, and Motorola are supplying NTT with digital phones. Motorola, NEC, and Ericsson will also supply DDI. Ido has selected AT&T, NEC, Fujitsu, and Nokia Mobile Phones as its key equipment suppliers. Ericsson and Toshiba of Japan have formed a joint venture to develop Ericsson's digital cellular equipment business in Japan, an agreement that called for Ericsson to supply Toshiba with digital cellular networks in Tokyo, Osaka, Kobe, and other Japanese cities.

Another emerging service in Japan is the two-way (send and receive) PCS type of Personal Handy Phone (PHP), which looks very much like most cordless phones. Operating in the 1.9 GHz band, PHP field trials began in the fall of 1993 and commercial service started in June 1995. PHP, now called the Personal HandyPhone System (PHS), is being offered through private networks and is not subject to the same foreign ownership restrictions applied to common carriers. As a result, the PHS market may be open to more competition than the cellular market. But with a range of only 100–200 meters, more base stations will be required than for the typical cellular system. Ultimately, DDI envisions PHSs operating building to building with base stations in homes, office buildings, and stores.

Japan has high hopes for PHS. The Ministry of Posts and Telecommunications (MPT) is predicting that the low cost of PHS could push sales to 40 million units by 2010. DDI is also trying to introduce PHS to Korea, Hong Kong, Singapore, Taiwan, and Thailand. DDI officials have met with several American companies, including the regional Bell companies, cable television companies, and MCI Communications, in an effort to promote PHS as an international standard.

Hong Kong is the largest per capita market for cellular phones in the world and the most competitive, and CT-2 is far more successful in Hong Kong than any other area. The government introduced cellular service in

1983 by granting Hutchison Telephone Limited, Pacific Link Communications Limited, and Hong Kong Telecom CSL Limited licenses to operate cellular networks. But three carriers were not enough to satisfy Hong Kong's "telephone fever." Capacity limitations created a slump in cellular handset sales in 1992, forcing the Hong Kong Telecommunications Authority to license a fourth cellular system. It also ordered the three existing cellular operators to switch their systems from analog to digital by mid-1995. To protect their existing customer base and allow continued roaming into the People's Republic of China (PRC), they most likely will switch to dual (analog/digital) systems. Building the fourth digital network and switching the other three to digital will cost an estimated $150 million for telecommunications equipment, not including handset sales. The Hong Kong Telecommunications Authority anticipates the need for a fifth cellular network operator in 1995.

The equipment for the Korean network will be supplied by four Electronics and Telecommunications Research Institute (ETRI)–designated Korean manufacturers: Goldstar Information & Communications, Ltd.; Hyundai Electronics Industries Co.; Maxon Electronics Co.; and Samsung Electronics Co. Under license from QUALCOMM, the developer of code division multiple access (CDMA) technology, Maxon will produce subscriber equipment only, while the other three companies will develop both subscriber and infrastructure equipment. In addition to providing equipment for the Korean market, these manufacturers will become alternate sources of CDMA equipment for networks in the U.S. and other countries implementing CDMA.

Eastern Europe and the former Soviet republics, once restricted from buying Western high-technology products, have become major customers of virtually every major telecommunications equipment manufacturer. Rather than spend years trying to bring their telecommunications systems up to the standards of the rest of the world by installing new telephone lines, most of these economically depressed areas have turned to wireless systems. In the reunited eastern and western states of Germany, for example, the Bundespost is implementing wireless local loop services to provide immediate, short-term telecommunications services. One of the key features of this system is that international calls are possible without having ̶ ᵢ through the fixed network.

ᵗ all eastern European and new Commonwealth of Independent (CIS) countries are equally developed in terms of phone service.

The CIS has the third largest telephone system in the world, with 27 million access lines, but has a very low teledensity of nine phones per 100 inhabitants. What was Czechoslovakia probably has the best telephone system in the region, despite a relatively low penetration of 25 phones per 100 people. Hungary's phone system is not nearly as good, but it is making rapid progress, partly as a result of trade between the East and West growing and becoming less complicated politically.

Cellular service is also now available in every region in Mexico. Nationwide roaming became available in 1992, and Mexico has signed international roaming agreements with the U.S. With a population estimated at close to 86 million, Mexico has plenty of room for cellular growth.

In Canada, cellular systems operate on the same frequencies as the U.S., but they are licensed differently. Block A, the 800 MHz service, is provided by a single nationwide carrier—Mobility Canada—while Block B, the 900 MHz service, is provided by provincial carriers. Canada's 1992 population was 27.2 million and the CTIA estimates that roughly 85 percent have access to cellular service. According to Canada's Department of Communications, approximately 1.02 million people subscribe to cellular service.

In December 1992, Canada awarded four consortia licenses for PCS while Mobility Canada won a national license to provide a Personal Cordless Telephone (PCT) service using CT-2 Plus. An enhanced version of CT-2 developed by Northern Telecom Ltd. of Canada, CT-2 Plus features an optional dedicated common signaling channel, cellular-like handoff, enhanced transmission speeds of 19.2 k/bits per second, subchanneling to allow capacity increases as the technology improves, improved security, and integrated voice, data, and imaging. Mobility Canada's PCT public zone service, launched in major Canadian city centers in 1994, covers stadiums, hotels, shopping malls, airports, factories, offices, and public buildings. PCT private zones would cover the area in and around a customer's home or office. The service should be available to approximately 18 million Canadians by 1998.

Owners and/or shareholders of Mobility Canada include AGT Cellular Limited, BCE Mobile Communications Inc., BC Tel, Edmonton Telephone Corp., Island Telephone Co., Manitoba Telephone System, Maritime Telegraph & Telephone Co., New Brunswick Telephone Co., Newfoundland Telephone Co., Quebec Telephone Co., Saskatchewan Telecommunications, and Thunder Bay Telephone.

The four licensed PCS companies operating in the 944–948.5 MHz band are Canada Popfone Corp., Mobility Personacom Canada Ltd., Rogers Cantel Mobile Inc., and Telezone Inc.

S NEW?

There are now more than 130 competing phone systems in the U.K. and they have to contend with some new and very large competitors, such as AT&T, MCI, and Sprint.

Finland has 52 independent telephone companies, many of them owned by their subscribers. Telecom Finland, the national phone service provider, is expanding its GSM mobile network with a new DCS-1800 system, and is installing Ericsson's MiniLinks, which are radio links operating in the microwave radio band for voice, data, and multimedia applications. Coverage of the MiniLinks is 50–60 km. Telecom Finland is also considering developing a wireless local loop service to compete with the local phone companies.

Belgium awarded its second cellular license in September 1995 to Mobistar, a consortium led by France Telecom. The group plans to invest $500 million in order to provide coverage to 97 percent of the country by early 1997.

Brazil passed a constitutional amendment to allow private operation of telecommunications services in late 1995. Cellular service was one of the first sectors liberalized through the licensing of systems to private companies that will compete with the regional government operators. The country will be divided into six or seven cellular regions.

In Chile, CTC Celular plans to spend $25 million in 1996 and 1997 to digitalize its cellular network.

Singapore issued a new cellular license to MobileOne, which will build both a CDMA and a GSM network. But the company cannot start operations until Singapore Telecom's monopoly runs out on April 1, 1997. Equipment contracts went to Motorola for CDMA, and Nokia for GSM.

Israel is expected to issue a third cellular license in 1997.

Taiwan passed new telecom legislation that will allow for competition in cellular service. It also opened the cellular and paging sectors to private participation with up to 20 percent foreign ownership.

In Russia, regional license winners continue to build their networks.

ASIA MAJOR

With more than half of the world's population, the Asia/Pacific region has the greatest potential for wireless personal communications growth in the world. MTA-EMCI projects the region will have 78 million cellular subscribers by the year 2000. Much of the market has been slowed by restrictive government regulations and incompatible systems, but that is all changing.

The 1995 MultiMedia Telecommunications Association's (MTA) *Review & Forecast* says the mature Asian/Pacific mobile communications markets (such as Japan, Australia, and Singapore) are growing faster than the U.S.—an average rate of 60 percent per year. The level of international telephone traffic confirms the increased importance of telecom trading links within the region. According to MTA, intraregion calling now accounts for nearly 55 percent of all the traffic originating in Asia and the Pacific—a percentage equivalent to that of the European Community. China's international calling traffic has been growing at more than 50 percent a year, pushing its global telecom revenues to $2.6 billion annually. A growing share of that traffic is wireless. To meet the demand for new mobile and fixed wireless installations, China (which has about two phone lines for every 100 people) is signing contracts for the construction of new networks valued in the hundreds of millions of dollars.

Cellular technology is also quickly spreading into the developing economies of Southeast Asia where virtually every country faces a severe shortage of telephone lines. Thailand, Indonesia, Malaysia, Bangladesh, Cambodia, India, Laos, and Vietnam are all strong wireless markets.

Japan is still a small wireless market by western standards, but Japan's MPT projects the number of radio terminals—mostly cellular and PCS-type services—will reach 104 million by 2010 and could go as high as 130 million.

Japan's MPT has already issued 1.9 GHz PHS licenses to 21 carriers, several of which began service in mid-1995. PHS operates as a cordless phone with no airtime charges in the home or office and as a cellular phone when used outside. The downside is that PHS will not work from a moving vehicle; the systems's architecture simply will not allow it to hand off calls quickly enough for mobile use. Nevertheless, projections for PHS indicate the market could easily reach 20 million subscribers by 2005 and 38 million by 2010. To compete, Japan's cellular operators have been cutting their registration fees, handset prices, and monthly charges to offset some of the gains made by PHS.

Wireless service continues to do well in other Asia/Pacific areas. Korea Mobile Telecom Corp. has doubled its analog Advanced Mobile Phone Service (AMPS) cellular subscribers each year since 1984. South Korea has selected CDMA as its national digital cellular standard, and several Korean electronics manufacturers plan to produce CDMA-based equipment for domestic use and to sell to other CDMA carriers around the world.

Singapore Telecommunications is adding new cellular subscribers at a rapid pace, but it will have new competition from four international organizations beginning in April 1997: MobileOne, Singapore Press Holdings, Hong Kong Telecommunications, and Cable & Wireless of Britain. Singapore Telecom's own projections call for a better than 25 percent market penetration rate for cellular phones by the year 2000.

Taiwan's Directorate General of Telecommunications (DGT), under pressure from consumers and legislators to meet the growing demand for cellular services, has expanded its AMPS network capacity to 590,000 subscribers, which is already oversubscribed. The waiting list for cellular phones at that time topped 120,000 and was growing at a rate of 10,000 a month.

India's addressable market for cellular subscribers is expected to grow from zero in mid-1995 to more than one million by the end of the decade. At one point, 33 consortia involving more than 60 companies were bidding for cellular licenses in India. That's not surprising since India's 920 million people have the fewest telephone lines of any major country in the world—about two lines per 100 inhabitants. New phone subscribers must wait two to three years for a phone line. However, the explosion of cellular systems in the country seems to be changing all that.

The Indian government has issued at least 39 licenses to fill out the country's wireless infrastructure. The eight cellular operators (Max/Hutchison, BPL/France Telecom, Bharti/Generale Mobile/SFR, Sterling/Cellular Communications, Modi/Telstra, Crompton Greaves/BellSouth/Millicom/DSS Communications, Usha Martin/Telekom Malaysia, and RPG/Vodafone) have launched cellular service in India's four biggest cities—Bombay, Delhi, Calcutta, and Madras.

Overall, EMCI expects digital subscribers in the Asia/Pacific region to surpass analog users by the end of 1997. By 2000, EMCI expects subscribers on GSM, Japan's Personal Digital Cellular (PDC), CDMA, and other digital systems to account for more than 70 percent of the total subscriber base in the region. "The driving forces behind this phenomenon," EMCI says in a study, "are the introduction of GSM in China and India, the two most populous countries in the world, and rapid subscriber growth of PDC systems in Japan and CDMA systems in South Korea."

EMCI expects China to replace Australia as the second largest cellular market by 2000, with almost 30 percent of the total subscribers in the region. (However, Australia will likely have the highest penetration rate by 2000 at 31.5 percent.) EMCI has also estimated that by the end of the decade, India will become the fifth largest market in the region with more than 6 million subscribers, all on GSM systems. Meanwhile, Japan is expected to account for about one-third of the region's cellular subscriber base, with much of that coming from PDC growth.

A BEEPING GOOD MARKET

Paging continues to grow at a record rate. Thirty-eight percent more subscribers signed up for the service in 1995 than the previous year. The last time that happened was in 1981. The 7.5 million subscriber increase in 1995 boosted the installed base in the U.S. to 27.3 million paging users, exceeding the previous year's gain by 3 million units. Worldwide, about 75 million people use pagers and the number is increasing steadily.

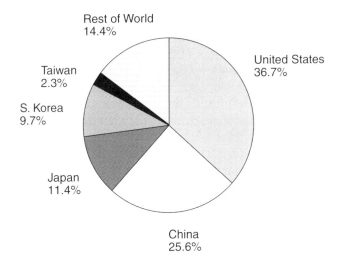

Distribution of paging subscribers by country, 1995.
Source: MTA-EMCI.

The pager market has also been marked by consolidation. At the end of 1994, 59 percent of the market was held by only 10 paging companies, according to the Yankee Group. A year and several service provider consolidations later, that number jumped to 72 percent. As a result, some companies have drastically changed their ranking in the market.

Research also indicates a kind of perverse pent-up demand for paging in that more people want nationwide paging coverage than actually subscribe to it. Most subscribers still use pagers on a local or regional basis.

Falling prices for equipment and service have been a key factor in paging's increasing popularity among business and consumer users. According to EMCI, revenue per pager continues to decline, although at a slower pace than in previous years. The industry's 1994 average revenue for digital display rental pagers was $13.10 per month, down from $14.20 in 1993. The reduction continues the trend of falling prices for paging services; in 1989, digital display rental revenues per pagers were $18.20 per month.

Digital display models continued their dominance as the most popular type of pager, accounting for about 87 percent of pagers in service in 1994. Tone/voice pagers accounted for 3 percent of the market in 1994, and tone-only units accounted for 2 percent. Alphanumeric pagers remained stable at 7 percent, but new models should help increase their market share.

Paul Kagen Associates estimates that paging revenues will more than triple to $11 billion and pagers in service will almost quadruple to nearly 100 million by the year 2006, a projection the telecommunications consulting firm bases on two facts: 90 percent of communications are one-way and 70 percent of e-mail and voice mail need no response. Some of the new, more highly featured pagers are already on the verge of creating a whole new category of wireless personal communications products.

Two-way paging has been slow to grow and could stay that way unless the service is improved and it is priced more competitively with other messaging services. Mtel launched SkyTel 2-Way, the first nationwide two-way paging and messaging service, and expects to offer two-way service in more than 300 U.S. cities.

Mtel sold 15,400 of its SkyTel 2-Way units by the end of 1995, outpacing initial sales of one-way nationwide pagers in 1987. The downside is that two-way paging is expensive—almost in the same league as cellular service. SkyTel 2-Way starts at $24.95 plus $15 for the lease of the pager. The price includes 100 messages plus the ability to receive and send nationwide messages for an additional 95¢ per message. Monthly nationwide SkyTel 2-Way service starts at $74.95 plus $15 to lease the pager. Subscribers can also buy, rather than lease, the two-way pager for a hefty $399. Of course, competition will push down the price of two-way paging. More significant is the fact that 75 percent of Mtel's two-way paging subscribers are new customers.

SkyTel has joined with Wireless Access to create and build a second-generation two-way messaging device that can, among other new features, initiate messages directly from the pager. SkyTel 2-Way subscribers can also send and receive messages with an HP 100LX or 200LX palmtop PC, or with an HP OmniGo 100 handheld organizer. In addition to SkyTel 2-Way, Mtel's other services include SkyPager, SkyWord, SkyTalk, SkyFax, SkyNews, SkyQuote, and SkyMail. Mtel has also acquired 10 nationwide Narrowband PCS licenses.

PAGING THE INTERNET

If recent market research holds up, information—in the form of e-mail and Internet services—will become more important to pager users than personal messaging. Motorola has teamed up with ESPN Enterprises, the all-sports cable TV network, to offer ESPNET To Go, a pager service that tunes into a network of satellites to pick up sports news, analysis, and commentary. Most of the sports data will be downloaded from SportsTicker, a sports news service jointly owned by ESPN and Dow Jones & Co.

Motorola also offers SportsTrax with PageNet. There are two SportsTrax models: One is dedicated to baseball and the other to basketball, and they provide updated information every day of the year.

Panasonic didn't enter the paging market until mid-1996, not so much because it liked the way paging was growing (which it did), but because of the high number of e-mail uses and the extraordinary growth of the Internet. Panasonic's extensive market research confirmed what its executives had long suspected—that pager users were willing to pay for information as well as person-to-person messaging.

Panasonic's three new alphanumeric models allow subscribers to access a broad array of basic services, including sports scores and headlines, weather forecasts, stock quotes, daily financial index and mutual fund updates, and local and national news. Subscribers can also contract for mailbox service to get extended PC-generated text messages. Total mailbox

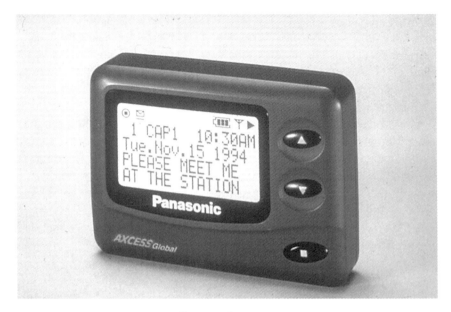

Panasonic pager.

capacity of the new Panasonic pagers allows users to scroll messages of up to 2,000 characters. The Panasonic pagers use the FM subcarrier radio broadcast data system (RBDS) to send and receive information, the system used by broadcasters as a data transmission channel on their existing FM services.

SEIKO Telecommunication Systems is also using a FM subcarrier network under the Receptor trademark to provide personal messaging, news, sports, weather, and traffic information using a numeric display wristwatch receiver. The system features time division multiplexing wherein time is subdivided into a system of master frames, subframes, and time slots. Each slot contains a packet of information. In multiple-station systems, each station's transmissions are synchronized to universal time code (UTC), ensuring synchronization between stations. Each receiver is assigned a subset of slots as times for monitoring transmissions. Multiple receivers may share time slots due to the random nature of expected communications. Each slot is numbered and each packet contains the slot number that permits rapid location of assigned time slots.

Alphanumeric pagers may also soon be used to transmit a patient's electrocardiogram to a doctor, and to send telemetry-type messages to computer printers and other devices. PageNet and Lexmark International are installing self-diagnostic wireless devices into their high-end computer printers; when technical problems are detected, a call is automatically sent to a technician's pager.

Upgraded paging systems and high-speed protocols, such as Motorola's FLEX, will allow paging carriers to increase their incremental revenues by entering previously unavailable niche markets. Enhanced 911 service is another potential application: Using two-way pager technology, onboard position location systems can be linked to airbags in cars and trucks to send out an emergency message in case of an accident.

Hewlett-Packard Co. and Apple Computer are collaborating with paging companies to make their computers work with the new two-way paging networks. Casio Computer Co. and Sharp Corp. have teamed with AirTouch Communication's paging operation to develop a pocket organizer that lets a user wirelessly retrieve or send updated information to a desktop or portable computer.

The availability of digital voice paging, which can receive voice messages, will be an important addition to consumers' product choices. Callers can dial an 800 number and leave a voice message, which is then transmitted to the pager. Voice pagers can store and play back up to four minutes of messages, but the network can store additional messages for a fee. New pager voice mail offerings have been introduced by PageNet (VoiceNow, developed with Motorola, will use a narrowband PCS network on frequencies licensed to PageNet), BellSouth Mobilcomm (ReadyTalk, developed with ReadyCom, uses the existing cellular nationwide phone network), and PageMart. Although somewhat larger than the other models, the ReadyTalk model lets the user talk into the pager to record a message, which is then forwarded to the number of the person who called. The voice pagers also operate on different communications networks.

The FCC's auction of nationwide and regional narrowband PCS licenses and the auction of smaller, metropolitan and basic trading area licenses will expand the paging market since most of the new Narrowband PCS licensees are expected to launch two-way paging networks. In June 1996, the commission granted 18 exclusive nationwide licenses to eight private paging carriers in the 929 MHz band. This is in addition to the 10 carriers that were granted exclusive nationwide licenses in May 1994.

The FCC unintentionally put at least a slight dent in the expansion of the paging market in early 1996 when it issued a three-month freeze on new paging licenses. In an effort to develop a new set of rules to auction geographically based paging licenses while preventing fraudulent application speculation, the commission stopped accepting applications from all paging carriers except those licensed to build nationwide networks. Under pressure from trade groups, Congress jumped into the fray, expressing concern that damage could be done to a healthy industry. On April 23, 1966, the FCC ruled that incumbent regional and local paging companies could expand by erecting new antenna sites within 40 miles of another operational site that was in operation on February 8, when the FCC issued its initial freeze order.

The 28 nationwide licenses are the result of new FCC rules announced in 1993 that allowed exclusive licensing on 35 private paging carrier channels. The ruling was followed by a flood of applications for these frequency bands. However, the game changed again in August 1996 when private carriers became common carriers under the Omnibus Budget Reconciliation Act of 1993. Most private carriers have been expanding for several years, from smaller market operations to virtual common carrier status. The law places the carriers on level regulatory ground.

In Canada, where market penetration for paging is only 3 percent, or about one million people, there are only two big players—Mobility Canada with an estimated 42.3 percent market share, and Rogers Cantel with 17.6 percent market share. Telezone has another 3.5 percent and the rest of the market (36.6 percent) is covered, according to The Yankee Group's research, by 250 providers who operate their paging service as secondary businesses. Two-way paging is scheduled to start in Canada in 1997.

Pagers are hugely popular in India, where a Motorola pager production facility in Bangalore, India, began running at full capacity when it opened in mid-1995, and in Pakistan, even though there are few telephones in that country to respond to a message. Initial orders when pagers were introduced into India were in the hundreds of thousands. South Korea's Samsung Electronics is working with Larsen & Toubro Ltd., a Bombay-based builder of industrial machinery, to produce pagers, competing directly with U.S. companies that have so far dominated the Indian pager market. The two companies plan to spend $44 million through 2001 to expand India's pager market.

Current market statistics would indicate that Short Messaging Service (SMS) could challenge paging. SMS is already a component of the GSM package, and is bound to be offered in some form in other cellular and PCS systems. An estimated 36 percent of cellular subscribers already use pagers, which could provide a major attraction for new digital phone users. In fact, most handset manufacturers plan to offer SMS capability in their new digital phones.

However, many analysts don't see SMS as a big threat to paging services. Their rationale is that basic paging continues to grow because it serves a specific requirement at a very low cost. In fact, the introduction of PCS with its own messaging features will very likely help increase the awareness of wireless service options and serve as a major boost to the entire market for advanced messaging, including basic paging services.

IN-BUILDING COMMUNICATIONS

If there is one wireless communications product that can claim a shot at ubiquity it is the cordless phone. Currently, there are several hundred million cordless telephones in use around the world, with 18 million sold each year in the U.S. In fact, about half the homes in the U.S. are now equipped with cordless phones.

Most of these products operate in the 46–49 MHz frequency band and exhibit several drawbacks, such as limited range and voice quality, and questionable privacy. The problem is that the number of callers using the cordless phones at any given time often exceeds channel capacity in many densely populated areas. Low transmission power can greatly limit the range of a cordless phone. Also, it is not unusual for someone living in an apartment house, for example, to be able to listen in to a conversation of someone living in the same building—perhaps one floor away.

Consumers can avoid these problems, however, by acquiring a cordless phone that operates in the 900 MHz range. These phones are available in the U.S. and Europe. They exhibit very little, if any, static or any other kind of interference and their operating range is up to several hundred yards. But they cost at least twice and sometimes three times as much as the 46–49 MHz models.

However, the market for 900 MHz cordless phones is growing, which should help reduce the price over time. It's also a strong replacement market; an estimated 12 percent of the available market switched to 900 MHz phones in 1995, which translates to 4 million units. Sales trends favor 900 MHz phones, with forecasts predicting that by the year 2000, 900 MHz phones will claim half of the cordless phone market.

In dollar terms, there's an even larger in-building market in wireless local-area and wide-area communications. Portable phones or personal information terminals can now operate with no airtime charge inside a home or in or around a business or other campuslike facility, and then transparently become a standard cellular or PCS device when the subscriber moves outside.

The typical in-building system ties into an existing wireline telephone system. This could be a private branch exchange (PBX), Centrex, or key system. Since they are connected to the building's wireline system, they operate just like a desk phone. However, some in-building systems operating in licensed frequencies will tie seamlessly into public wireless networks.

Bell Atlantic NYNEX Mobile employees in Connecticut were the first customers in the U.S. to use INReach, a new wireless in-building communications system developed by Motorola. INReach connects digitally with multiple PBX manufacturers' equipment. The system uses the SpectraLink Pocket Communications System at noncellular frequencies and requires no incremental airtime charges or monthly fees. Calls are charged to the regular landline system.

Initially, in-building wireless customers will purchase and use these systems to improve productivity and communications. The sell is that you will always be accessible by phone. It's a legitimate pitch: A study by Opinion Research Corp. estimates that people spend an average of 302 hours each year listening to voice mail and responding to pages, costing businesses more than $3 billion in lost time.

The market's growth has been slowed by a dearth of products, but that should change with the introduction of several new systems by AT&T Network Systems, Ericsson, and Nortel, as well as a number of innovative products from smaller equipment vendors.

A survey by Alexander Resources has found that at some point, as the price and performance of these systems reaches that of a wired telephone network, the purchase of an in-building communications system will be made more on cost. Alexander Resources expects the in-building wireless systems market to reach $1.3 billion by the end of the decade. InfoTech Consulting estimates that manufacturers shipped more than 88,000 handsets in 1995 for in-building wireless systems, and projects that number could jump to 1.6 million handsets in the year 2000.

Systems using the 900 MHz industrial-scientific-medical (ISM) band have accounted for most of the market, but InfoTech expects this market segment to plateau as vendors migrate to the new unlicensed PCS frequencies. Market research by MTA-EMCI indicates that many manufacturers will take advantage of FCC rulings, which dedicate 20 MHz of unlicensed spectrum for the development of wireless office telephone systems and wireless local-area networks (WLANs).

Both cellular and PCS carriers are expected to aggressively pursue in-building wireless service subscribers. In fact, several cellular and PCS handset manufacturers have established original equipment manufacturer (OEM) relationships with PBX manufacturers.

Interviews of U.S. telecommunications managers have revealed that awareness of wireless office telephone systems (WOTS) increased from 23 percent of telecom managers in 1991 to 31 percent in 1994. Awareness was strongest among manufacturing, financial services, and health-care managers. EMCI also found that 17 percent of the respondents had a "definite" or a "probable" need for wireless PBX/Centrex or key systems.

By 1999, EMCI expects WOTS usage in the U.S. to grow from a few thousand lines to 2.4 million lines. The widespread availability of dual-mode handsets, which can also access cellular or PCS networks, could generate even higher growth rates. As in-building use begins to take hold, carriers will have to develop new pricing schemes for indoor-outdoor service, and then refine those schemes as competition increases and new, enhanced services are introduced.

In Europe, MTA-EMCI expects that the development of DECT and the introduction of dual-mode handsets will be the primary drivers of market growth for in-building services. EMCI projects the installed base of wireless office telephone users in France, Germany, and the U.K. to grow from several thousand today to more than 2 million lines by 1999. Despite the use of what seems like ancient CT-2 (cordless telephones–second generation) technologies today in PCN and other networks, EMCI expects the majority of the European growth to occur in DECT-based office telephones.

The initial acceptance of WLANs has been slow. Data rates, distances covered by different technologies, and distribution systems also differ. And there are no formal or de facto standards for WLANs. At least not yet.

The Institute of Electrical and Electronics Engineers (IEEE), the largest technical society in the world with more than 300,000 members, has established a subcommittee—designated IEEE 802.11—to produce a 1–10 megabit/sec wireless LAN specification covering everything from wire augmentation and ad hoc networking to factory automation. The IEEE group has been meeting regularly for more than six years, but progress has been slow. Part of the problem is that IEEE 802.11 meetings are attended by representatives of up to 100 companies of all sizes from different countries and different cultures, representing a mix of computer, communications, and components suppliers, with each company and industry sector promoting its own agenda.

Wireless private automatic branch exchanges (WPABXs) are used primarily for voice communications, although some companies have been using them successfully as a low-data-rate WLAN system. WPABXs are being pitched as a way to reduce wiring costs and virtually eliminate unanswered calls by freeing workers to roam around the office or plant with a cordless phone that is linked to the company's WPABX. This may be viewed as more of a curse than a blessing, but studies have shown that some 60 percent of business calls do not reach the intended party on the first try.

Although an all-wireless PBX system is not expected for several years, wireless adjunct systems are now available. The wireless communications system adjuncts that have already been announced tie in to a PBX via analog connections, although future plans call for specialized circuits (called T1 links) that transmit digital signals over wired telephone networks between the adjunct and main PBX.

Wireless PBX adjuncts usually consist of a control server and strategically placed transmitters/receivers, or base units, including antennas. Pocket telephones with a range of about 100 meters communicate with the base station, which is usually hidden away in a closet. As the user roams the facility, calls are handed off among the base stations, just as they are in a cellular network.

The wireless PBX market is off to a slow start for many of the same reasons that WLANs are experiencing slow market acceptance—management information system and telecom managers are trying to figure out how to implement wireless technologies into their office and factory systems. They're concerned about spectrum allocation, standards, and usage fees. Clearly, PBX users would be reluctant to pay for airtime when using their own system. And they haven't yet seen enough WLAN or wireless PBX installations to give the technology the credibility to commit to a full-blown, in-house wireless network.

Outside the U.S., several companies are building 2.4 GHz spread spectrum systems because of the potential international market opportunities in Europe and Japan at that frequency.

WIRELESS LOCAL LOOP

Wireless local loop (WLL) has been described as the "last mile" of phone service. Essentially, WLL, or fixed wireless, is the wireless version of the physical connection between a telephone and the telephone company's central switching center.

With worldwide telephone penetration rates at 11 percent and at less than 5 percent in many developing countries, the international WLL market may represent the greatest opportunity for growth in the wireless infrastructure equipment market. Globally, the infrastructure market is growing at 40 percent a year. By the year 2000, ArrayComm, one of the key players in WLL, believes that WLL will account for more than half of this market.

WLL can be used in point-to-point or point-to-multipoint radio phone links in analog or digital fixed cellular systems or limited-range cordless phone networks. WLL also fits into automated teller machine (ATM), point-of-sale (POS) terminal, point-of-banking terminal and kiosk, and on-line gaming and lottery terminal applications.

WLL is one of the hottest areas in wireless communications because it encompasses several technologies and expands telephone service in underserved developing and underdeveloped countries where wired infrastructures are too expensive to install. WLL can also be installed much faster than wireline systems.

There are 5.5 billion people in the world and only a fraction of them—about 500 million—have phones. Telephone line penetration in the less developed regions of the world averages about 10 percent, and most of that is in the larger urban but telecommunications-poor areas.

In developed countries with high levels of telephone line penetration, WLL is expected to serve two primary purposes: to implement competition in the local loop and offer new fixed wireless services in an emerging range of vertical and horizontal telecommunications services.

Worldwide, MTA-EMCI projects 60 million WLL subscribers in developed and developing markets by the year 2000, almost evenly split between the two market sectors. However, from 2000 to 2005, the market research firm expects developing markets to significantly outpace developed markets. By 2005, developed markets are projected to have 54 million WLL subscribers, while developing markets will have nearly three times that amount—148 million—as they work to bring their telecommunications networks up to speed with the rest of the world.

The biggest opportunities for WLL are China, India, Indonesia, Brazil, and Russia. These countries alone could account for nearly three-quarters of the total global market demand by 2005. Bangladesh, for example, is installing a WLL network that could grow to 250,000 subscribers. The first of several planned WLL systems in Paraguay is a single-cell, five-channel Advanced Mobile Phone System (AMPS) that will serve about 100 rural subscribers in Paraguay's agricultural region. WLL systems have also been installed in Mexico, Taiwan, and the U.K., where Ionica began WLL telephone service in May 1996 for residential customers. Ionica subscribers will connect to the local exchange using a 12-inch-diameter flat antenna that operates in the 3.425 GHz and 3.490 GHz bands, using technology developed with Nortel of Canada.

WLL will work in the U.S. for some of the same reasons it works elsewhere: It is cheaper and faster to install than wire. To some extent, WLL in the U.S. has fallen into niche applications. In New Orleans, for example, a local cellular operator has installed a WLL system that links the paddlewheelers that ply the Mississippi River with New Orleans banks. Using WLL, gamblers on the riverboat casinos can access an ATM at any time.

But there's a much broader WLL market across the country. Approximately 7 percent of the nation's population is without telephone service, either because some people don't have access to copper or fiber-optic cable, or it's just too expensive to reach some remote areas.

Competition among WLL equipment providers will be tough. At least 30 companies are now touting various WLL solutions.

NOW, THE SMRs

So far, the cellular industry seems to have everything going for it, even with the introduction of PCS. What the cellular carriers did not plan on until fairly recently was having to compete with Specialized Mobile Radio (SMR) operators. Like cellular service, SMR is a two-way service, used mainly for dispatching taxis and delivery vehicles, and for public safety.

SMR uses mobile phones and base stations that communicate through the public telephone network.

Although most SMRs have been struggling to make the transition to digital networks—either they can't find the financing to expand their networks or seem to be stuck in mostly vertical applications—most market projections give them the benefit of the doubt. MTA-EMCI expects SMR/ESMR (Enhanced SMR) operators to increase their share of the mobile data market from 13 percent in 1995 to 21 percent by the year 2000. While most of this growth is expected to come from digital SMR subscribers, MTA-EMCI believes the analog SMR operators will also increase their penetration of the mobile data market, adding significantly to SMR revenues.

Churn could be a big problem for SMRs, however. Churn rates are not as high as the rates being reported for paging or cellular service, where getting people to keep their pagers and cellular phones (and to use them) has always been an industry problem, but churn is growing for SMRs—reaching an estimated 21 percent in 1995.

The American Mobile Telecommunications Association (AMTA) blames a soft economy. Most SMR users are small businesses, such as taxi operators and small construction companies, many of which have gone out of business or cut back on their communications when business gets tough. Some SMR operators also are losing customers to cellular carriers, mainly because they like the mobility offered by cellular service. SMR operators may also lose customers to new digital mobile networks. Still, EMCI is predicting that the industry will add about 4 million subscribers by the end of the decade, close to half of which will be using digital systems.

Even as its revenues have more than doubled (mostly reflecting acquisitions), Nextel Communications, the biggest of the SMRs, has suffered large losses in recent years as it builds its nationwide wireless network.

In February 1991, the FCC gave Fleet Call, Inc. (now Nextel), permission to convert its conventional 800 MHz SMR systems in six major markets to digital, cellular-like networks with at least 15 times their original capacity, handling both voice and data. Understandably, the cellular industry, led by CTIA, McCaw Cellular (now AT&T Wireless Services), and the regional Bell cellular operators, strongly opposed the FCC's decision, fearing that a new service so closely resembling theirs may force them to cut prices.

The FCC took a different view, noting that competition is good and that, at any rate, Nextel wouldn't be competitive with cellular for years, especially on a national scale. Indeed, SMRs will have to build a totally new infrastructure to handle the new service, including new subscriber handsets, which should make it more expensive than cellular service. Nextel says its phones will be priced competitively with digital cellular handsets, although the handsets may be a little bulkier than cellular models.

Nextel's new system, called Enhanced SMR (ESMR), was launched in the greater Los Angeles metro area in August 1993. It's a huge "footprint," stretching from Santa Barbara to San Diego. Three cellular metropolitan service areas (MSAs) cover the same area. Nextel also agreed to buy a large segment of mobile radio licenses from Motorola for $1.8 billion in stock, and announced in November 1993 that Nippon Telegraph and Telephone of Japan would invest $75 million in the company and would design a system that would enable Nextel to link all its local systems into a single network within three years. The Motorola licenses, along with those it already owns, would expand Nextel's network across 21 states, with the potential to serve 180 million people, making it considerably larger than the McCaw system. However, Nextel still has a long way to go to reach its operational goals, while cellular has easily exceeded its market expectations with more than 30 million subscribers in the U.S.

ENTER PCS

The entire competitive picture changes for wireless personal communications with the introduction of personal communications services (PCS). Billions of dollars are being spent to develop PCS in the U.S.

Accounts of preparations for competition in the Dallas market between AT&T Wireless, Sprint Spectrum, and PCS PrimeCo, for example, refer to a "war zone" mentality with sales teams taking on names like "82nd Airborne."

An analysis by the Deloitte & Touche Consulting Group gives PCS almost half (48 percent) of the current allocation of land mobile radio frequency spectrum by type of service. Cellular service only has 20 percent. Smaller allocations are held by public safety (9 percent), SMR (8 percent), and "Other" (15 percent).

Consumers in some areas may now choose between different technologies, different pricing plans, and distinct coverage areas. The challenge for both cellular and PCS carriers will be to differentiate themselves.

A rundown of MTA-EMCI's end-of-year 1995 PCS market analysis calls for critical milestones in the implementation of commercial PCS in the U.S. to include the first direct competition with cellular and WLL applications. EMCI believes that PCS revenues will be generated by both traditional mobile and new WLL services. The dominant service offering is assumed to be a mobile radio service directly competitive with cellular service, including portable and mobile voice, mobile data, mobile messaging, and hybrid voice and data services.

After accounting for the competitive environment, consumer preference, market rollout, and geographic rollout, EMCI projects a PCS subscriber base of 20 million in the year 2000 and 27 million in 2005. Net subscriber growth will peak between 1998 and 2000, when between 5 million and 7 million net subscribers will be added each year. Revenues for mobile services are projected to exceed $12 billion by 2001.

EMCI assumes that some carriers will offer differentiated PCS focused on providing basic telephone services in and around the home. There will be a significant overlap between mobile and local loop subscribers. Even where mobile and local loop services are used by the same subscriber, EMCI assumes that local loop services represent an additional revenue stream to the carrier.

In fact, the number of local loop subscribers could be significant, reaching 7 million by the year 2002. The PCS industry should generate approximately $1.7 billion in annual revenues from WLL services by 2003. Looking at both mobile and WLL income, PCS carriers are likely to generate about $10 billion in annual revenues by 2000.

Projections for PCS handset revenues call for the market to peak at almost $3 billion in 1999. EMCI projects the PCS handset market to exceed 500,000 units in 1996, climbing to 12 million handsets annually in 1999 and 2000.

PCS carriers are expected to dramatically change the competitive characteristics of the wireless market. Today, cellular carriers essentially dominate the mobile telephone market while ESMR represents a relatively insignificant number of subscribers. By the year 2000, however, PCS carriers could capture about 27 percent of the mobile communications market. By 2005, the mobile market will still likely be dominated by cellular service, with almost two-thirds of the market. At that point, PCS is expected to have a 28 percent share of a much bigger market, with ESMR pulling down 7 percent of the market.

In fact, the emergence of PCS is already showing signs of boosting cellular sales by helping to promote the entire concept of portable, anywhere, anytime, communications. That doesn't mean that cellular carriers won't have to make some changes in the way they promote their products and services. Cellular One in the Washington/Baltimore area was forced to make some adjustments when APC/Sprint Spectrum became the first PCS carrier in operation in the U.S. in the same area. Cellular One immediately introduced several new services to compete with PCS, including a TDMA digital service, paging, enhanced directory assistance, and a long-distance service. Cellular One also beefed up its customer retention effort and introduced preferred customer and upgrade programs. Clearly, lower-priced, smaller, and lighter phones with more features are in the future.

Why "Cellular" Communications?

In nontechnical terms, cellular service works by dividing a city or region into small geographic areas called cells, each served by its own set of low-power radio transmitters and receivers. Once a cellular call or data message reaches a transmitter/receiver tower, it is plugged into the regular land-line phone system. Each cell has multiple channels to provide service to many callers at one time. As a caller moves across town, the signal to or from the cellular telephone is automatically passed from one cell to the next, without interruption.

AT&T Bell Laboratories developed the concept in 1947, but the first tests weren't conducted until 1962 to explore commercial applications. It then took another eight years before the Federal Communications Commission (FCC) set aside new radio frequencies for land mobile communications. That same year (1970), AT&T proposed to build the first high-capacity cellular telephone system. It was called Advanced Mobile Phone Service, or AMPS.

The FCC decided to license cellular systems in the 306 largest metro areas first (called metropolitan service areas, or MSAs), then to the less populated 428 rural service areas (RSAs). With its rules in place, the FCC began accepting applications for the 60 largest cities during 1982. In early 1983, when 567 applications were filed just for the 30 markets ranked 6lst to 90th in size, the FCC knew it had a problem: Its traditional system of issuing licenses following comparative hearings would take forever. In May 1984, the commission amended its rules to allow lotteries to be used to select among competing applicants in all but the top 30 markets. On October 13, 1983, the first cellular system began operating in Chicago.

2

From POTS (Plain Old Telephone Service) to PANS (Pretty Advanced New Systems)

In the next few years, every sector of the communications, information, and entertainment industries will have some wireless communications component. Very few people would disagree with the assessment of Deloitte & Touche Consulting Group's Telecommunications & Electronics Services, which says, "The world is going wireless, and the business opportunities are huge." The trick for industry companies and hangers-on will be to move fast.

Product developers and service providers still test new services and technologies, although for shorter periods than they used to, and they are much more in tune with what their customers want in a portable communications product. Motorola's approach to testing wireless products is to get into the market as quickly as possible, even if that means they won't have all the bells and whistles in Motorola's technology arsenal. The industry is developing new tools almost faster than the market can absorb them.

Motorola's Air Web, an open-technology software product that works with RAM mobile data and cellular digital packet data (CDPD) wireless networks, lets laptop computer users surf the Internet over wireless networks. The most significant new feature offered by Air Web is that it allows users to access information from the Internet without downloading all the graphics.

Personal Productivity Tools' new EtherWeb supports Intranet desktop-to-wireless communications using any paging service to send information to an unlimited number of alphanumeric pagers. Pages can be generated manually simply and instantly from desktop computers using any 2.0 HTML-compliant Web browser.

Motorola StarTAC cellular phone.

SkyTel introduced its two-way pager shortly before Motorola announced its "wearable" 3.1 ounce StarTAC cellular phone; the phone is smaller than the pager and costs just about the same to use.

Phones and pagers are already being integrated into a single unit. Cellular phones with a global positioning system (GPS) functionality are also available, and will become more popular when the price plunges well below the early adapter level and people figure out why they need such a device. There are now more than 300 software vendors in the world that claim to offer products dedicated to wireless connectivity.

With cellular phone users facing the prospect of having to communicate with phones based on different standards (most likely, CDMA, TDMA, or GSM), manufacturers will have to produce dual-mode (analog and digital) and dual-band phones, at least for the U.S. market. This is particularly true for GSM carriers who can't build their networks in the U.S. fast enough to compete across the entire U.S. market. The GSM MoU Association admits that both GSM and Advanced Mobile Phone Systems (AMPS) will be required during the first three years of the PCS rollout to ensure access to mobile phone services when outside a GSM service area.

Channel capacity is already an issue for GSM in Europe, where few regions have reached the market penetration levels of the U.S. The concern is that frequencies in the 900 MHz range may soon be used up. Trifrequency phones aren't out of the question.

Several proposals have been suggested for dealing with the issue, such as producing a phone with both PCS (DCS-1800, the GSM-based network, which operates at 1.8 GHz) and digital cellular services (the GSM-900 MHz network).

Swiss Telecom's commercial GSM network, launched in October 1995, already operates a dual-band (GSM/DCS-1800) network and has tested a GSM/DCS-1800/Digital European Cordless Telecommunications (DECT) network. Adding DECT could improve indoor coverage.

With GSM penetration expected to exceed 10 percent in some European countries in the near future, and with a limited number of frequencies to go around, the rising demand for GSM mobile access could lead to capacity bottlenecks, mainly in large cities and business and industrial centers. In rural areas, GSM capacity should be sufficient for a long time.

Multiband/multimode networks open up new roaming options. At the end of 1995, GSM-900 provided roaming with more than 55 operators in 39 countries. At the same time, five operators offered automatic roaming with DCS-1800 in four countries: Germany (the E-Plus network), England (Orange, Mercury One-2-One), Malaysia (Sapura Digital), and Thailand (TAC). Subscriber identity module (SIM) cards would, for example, allow roaming from GSM to DCS-1800 in Switzerland and from DCS-1800 abroad into the Swiss GSM network.

The system would be transparent to the user. Multimode/multiband terminals would automatically select the preferred mobile communications network, usually the one with the lowest charges.

The completion in 1995 of a second-generation program by the GSM association will assure DCS-1800 compatibility. It also sets forth a plan for supplemental features more familiar to anyone using landline telephones, such as line identity, call waiting, information on call charges, and other business- and consumer-oriented services. Work is now underway on Phase 2+, the advanced portion of the second phase of the GSM development program, which covers more than 80 items, including wireless local loop, high-speed data, enhanced full-rate speech codec, and special Short Messaging Services (SMS).

Another possibility is the integration of GSM with Enhanced Specialized Mobile Radio (ESMR) into one phone. This would allow ESMR-equipped commercial fleet drivers to operate as independent route salespeople, making their own phone calls to customers from their delivery vehicles.

Why so much interest in GSM in the U.S.? One reason is that after spending hundreds of millions of dollars to win licenses, most PCS carriers decided to play it safe by going with a readily available, widely tested technology with a broad family of products from several manufacturers. Given GSM's reputation for capacity limitations, those decisions may be tested at some point in the future.

For the consumer, most of the news is good. Products and services are becoming more diversified. The handheld products keep getting smaller, and the batteries last longer. The next big push is to multimedia. Not only will you be able to communicate (voice and data), but you will be able to conduct transactions (banking, etc.) and be entertained (audio and video).

PCS ARRIVES

The development of PCS has been phenomenal. PCS grew out of a report published in Britain in 1989 called *Phones on the Move: Personal Communications in the 1990s.* Barely seven pages long, the report got the telecommunications community in the U.K. to thinking that it had the potential to become a "world leader" in mobile telecommunications. "More and more," the report stated, "U.K. business is coming to rely on mobile communications, and government has acted as an enabler, making sure they get the services they need." The British government didn't disappoint. It quickly licensed four companies to provide so-called telepoint services. Also known as CT-2 (cordless telephone–second generation), telepoint is essentially a cordless pay phone, allowing subscribers to originate, but not receive, short-range phone calls in public areas equipped with telepoint base stations, such as shopping centers, train stations, and airports.

Telepoint seemed like a good idea at the time; London has very few public pay phones. But high subscriber costs ($200 for a handset, plus a $60 service connection fee and a monthly service charge of $15) didn't play well in a weak British economy. Also, the systems licensed by the government were not compatible. Today, only one company is offering limited telepoint service in the U.K.

With all of its problems, the introduction of telepoint in the U.K. sent a wake-up call to the rest of the world. In the U.S., the FCC responded by issuing more than 200 experimental licenses for PCS trials. Cellular telephone carriers, cable television system operators, and independent telecommunications operators are spending hundreds of millions of dollars to develop new products and services based on PCS concepts.

Europe offers a unique opportunity for new telecom services because there are fewer entrenched players and spectrum use is lighter and less of an issue than in the U.S. While Japan's cellular telephone market has grown 80 percent annually in 1994, 1995, and 1996, its market penetration rate remains low—slightly more than 1 percent of Japan's population uses a cellular phone. Nippon Telegraph & Telephone (NTT), which accounts for about 550,000 of the country's 870,000 cellular subscribers, expects the market penetration in Japan to double by the year 2000. As a result, every cellular system operator and equipment supplier in the world is pursuing these markets with a vengeance.

DESIGNING TOMORROW'S TELEPHONES

Designing a personal product like a cellular or PCS phone has become a growing challenge for the industry. The phones keep getting smaller and lighter, the batteries are lasting longer, and new models are coming out with new features and more functionality. However, like cars, the new phones may have to say something about the user and his or her individual style.

Phone manufacturers are usually concerned about such things as basic features and cost, but this hasn't stopped them from developing the next generation of phones that are smaller and lighter than current models, and that offer features that are totally new in portable communications products.

Hewlett-Packard Co. and others are trying to fit an entire telephone into a credit card. BT Research Labs in the U.K. has been working on a wearable personal communicator that combines video telephony with data communications and a personal digital assistant (PDA). Sony has been working for years on a system that can translate languages in real time so that people (presumably in different countries) could have a normal conversation in different languages. In time, the goal is to get the phone down to a single chip or chip set with the computing power of a 486 microprocessor.

In fact, it is becoming increasingly difficult to tell the difference between the new notebook computers and some of the more recent handheld

PDAs. Even the newest "smart phones" offer many of the functions available in handheld computers and PDAs.

IBM Japan, for example, has introduced a $2,000 palm-size computer that doubles as a cellular phone, but the company says market studies indicate that it won't sell in the U.S.

AT&T Wireless Services has introduced a phone and service that enables users to send and receive data over existing cellular networks and display e-mail on the phone's three-line liquid crystal display (LCD) screen. Called AT&T Pocketnet and expected to be available commercially early in 1997, the service uses CDPD technology. AT&T Pocketnet can also browse the World Wide Web (within the limits of the phone's memory of 60,000 bytes) and provides interactive data services; for example, users can look up flight information and then use their cellular phone to make a reservation.

Motorola and Nokia have also introduced a cellular phone that can tap into the Internet, as well as send and receive faxes and e-mail.

Nortel has already introduced a GSM phone that combines digital voice and data services and also serves as a personal electronic organizer. The new Nokia 9000 Communicator can send and receive faxes, e-mail, and short messages, access Internet services and corporate and public databases, and function as a calendar, address book, notepad, and calculator. But at $2,000, it's going to be an item for the "early adopters" until enough buyers come along to shrink the price to something more manageable by many consumers.

Nokia also offers a GSM phone with an infrared (IR) port; pointing an IR-equipped notebook computer at the phone provides wireless access to the network (all notebook computers introduced in 1995 were IR equipped) and has teamed with Hewlett-Packard Co. to produce a new version of the HP 200LX pocket computer that can be linked to a GSM cellular phone.

Alcatel and Sharp Electronics have developed GSM terminals with graphic display screens, icons, and one-touch direct access keys. The two companies also plan to develop and distribute Alcatel-made "personal mobile communicators."

Because of its size (3.1 ounces, which is lighter than most alphanumeric pagers) and advanced features, Motorola's MicroTAC has received a tremendous amount of attention, as has its initial price ($1,100 as advertised by most retailers). This is the type of innovation that is expected to drive the industry in the next few years.

AT&T's new PocketNet phone lets users access Internet-based e-mail wirelessly. Access is controlled through a menu-based interface with the cellular phone's keypad.

Making phones smaller and lighter comes with some very significant technical challenges. Obviously, a certain amount of space is required for all the chips and connecting circuitry, antenna, and batteries. The human engineering or ergonomics part usually comes at the end of the design cycle and creates its own set of trade-offs for product designers.

TESTING THE WATERS

AT&T's highly miniaturized "Dick Tracy" wristwatch/phone is a real-life example of what's possible—and what's coming. The AT&T watch/phone has two bands—one keeps the watch on the wrist while the other contains the cellular phone. When the phone receives a call, an electronic bell sounds or the user feels a vibration in the wrist. Pressing down on the top of the

AT&T "Dick Tracy" wrist phone.

watch releases an outer speaker band, automatically placing the phone "off hook." The outer band pivots into the palm of the hand. The user listens by cupping the palm over the ear and speaking into the microphone in the wristband. In a quiet or private environment, the communicator can function as a speaker phone by keeping the speaker band connected to the watch. Like many of the futuristic cars at automobile shows, it's only a design concept at this point, but the prototype model actually works.

Not much bigger is Nippon Telegraph & Telephone Corp.'s prototype 2.5 ounce wrist phone. There is no keypad in this unit; users dial by speaking into it. The lithium-ion battery and antennas are embedded in the wristband. NTT believes that it can develop batteries powerful enough and small enough to introduce the unit in 1998. NTT has designed the wrist phone to work with Japan's Personal HandyPhone System (PHS).

Another company, Aura Communications Inc., in Wilmington, Massachusetts, has developed an even smaller phone by putting only the necessary electronics in an earpiece with a microphone extension; the rest of the hardware fits comfortably in the user's pocket. Rather than radio waves or infrared, the earpiece and pocket electronics communicate through a magnetic induction system patented by Aura, a technology Aura plans to sell to other phone manufacturers.

Sony very cleverly deals with some of the more critical design issues in its CM-RX100 phone. A tiny knurled "push and scroll" knob on the side of the phone allows the user to thumb through up to 99 preprogrammed names or phone numbers displayed on the phone's two-line, 11-character LCD screen. Once the name and number is found, the user simply presses the knob and the phone dials the number.

The Sony phone's microphone has a foldaway stem that pops out when needed. An optional earpiece allows the user to make calls with the phone in a pocket or with the phone clipped to a belt. There's also an infrared remote device that stores and dials programmed numbers.

Another Sony model enables users to interact intuitively with the phone. User-friendly icons indicate signal strength, battery status, calling mode, message waiting roaming status, and other key user information. It also offers speed dialing for up to 199 numbers, caller ID information, SMS for brief text messages, and notification when voice and e-mail messages are received.

Mindful that cordless phones have become so popular that it is sometimes difficult to get a dial tone in densely populated areas, Sony has also

Sony CM-RX100 phone.

introduced a 25-channel cordless phone that automatically scans all frequencies and selects the clearest possible channel at that moment.

VOICE SYSTEMS: STILL MOSTLY TALK

It's not likely that voice recognition will replace other data input technologies on a broad scale; however, features such as voice dialing ("Call

Mom") will become more important as wireless communications products become smaller and as phone manufacturers are under more competitive pressure to differentiate their products. The ability to use voice commands also reduces the training required for new wireless devices.

Work on speech recognition has been underway for more than two decades. The development of low-cost digital signal processing (DSP) chips that can perform digital voice recognition and recent progress in algorithm research will help drive mobile voice recognition into viable consumer-level applications. The trick so far has been accuracy; it is very difficult to design speech recognition systems that perform the way they are supposed to 100 percent of the time.

Displays are also important. The first question designers ask themselves is, What type of information do people need and want? Cellular phones with displays in a variety of languages are already available for world distribution. Help menus and a personal database are obvious applications; consumers and business users will want much more. Just as cellular phone users want the same sound quality they get on their wired phones, data users will want to be able to access the same data sources wirelessly as they do from the desktop and other computer platforms. Space and cost are important, but these issues will have to be solved if wireless products are going to reach their full market potential.

Reflection Technology has introduced one of the more innovative display products, the FaxView personal fax reader, which allows the user to view a full page of text or graphics in a handheld device. FaxView uses the same technology as Nintendo's 3-D immersion game system, Virtual Boy.

More special ease-of-use features are showing up in portable phones. These include:

- One-touch function keys that minimize the key strokes necessary to place calls for emergency assistance or to list the last five phone numbers dialed
- A red #9 key that is preprogrammed to dial 911 or an emergency service number
- A "keyguard" feature that protects the phone from accidental operation, which can be automatically activated when the END key is pressed and held
- Simplified programming to allow quick activation for retail distributors

FaxView.

- Enhanced 911 to allow 911 and other emergency operators to determine the location of anyone calling for help or assistance

By mid-1997, wireless phone companies will be able to provide the technology to report the nearest cell site receiving an emergency call. Within five years, the technology will be in place to report the exact location within one-tenth of a mile.

The FCC now requires that all wireless calls going to 911 services be transmitted even when callers are "roaming" outside their regular service area. This, however, is the existing industry position. The decision about whether or not to adopt the new location technology plan rests with local governments that operate 911 services. The FCC ruling clarifies that wireless phone companies will implement the new technology if required to do so by local governments, which also provide funding.

Power management becomes more important as users become more dependent on their mobile products. The next generation of devices will make it possible to monitor battery consumption, reduce or shut off power during idle periods of use, and display battery status. Standby time for cellular phone batteries is now comfortably in the 50 hour range. By mid-1997, using batteries of equal size, standby time will jump to 90–100 hours. Meanwhile, product designers are only half joking when they predict that some wireless devices will be powered by body heat.

Handheld mobile communications satellite terminals will have their own unique requirements. Most of them will be dual band (analog and digital) and dual mode (cellular, or some other terrestrial service, such as PCS or SMR). Additional features will likely include a satellite/cellular service indicator, a screen that will support short messaging services, e-mail status indicators, and a data/fax port.

Also coming are portable phones and advanced (nonvoice) messaging designed specifically for the hearing and visually impaired. Hitachi Ltd., for example, is developing a sign-language phone for the hearing impaired.

Other new functions, such as position location, directory information, and e-mail, will have to be shared by the phone and the network.

Increased competition in handsets, mainly from Japanese and Korean consumer electronics companies like Sony, Panasonic, Toshiba, and Samsung, will encourage the development of more highly featured but lower-cost handsets, challenging such telecommunications giants as Motorola, AT&T, Ericsson, and Nortel.

A growing number of nontraditional manufacturers will likely enter the market. Automobile manufacturers, particularly those with a strong capability in electronics (such as GM, which owns Delco Electronics and Hughes Aircraft Co.), will offer mobile communications and in-vehicle navigation products that operate as an integral part of a vehicle's stereo system, which already has many of the required components, like speakers and a display, as well as some of the circuitry. Small appliance companies such as Krups, with their own strong technology and design backgrounds—and high-profile brand names—are also potential players in the portable communications market. Countries such as Russia, China, India, and Turkey, which have vast experience in military electronics, will enter the market at some level with wireless products of their own designs.

Some wireless local loop (WLL) alternatives are already based on GSM in the Asia/Pacific region. Internet access from GSM via laptop computers is in use in some countries, although it is not very user-friendly.

Geoworks and Ericsson have formed an alliance to develop a range of next-generation Smart Phones based on Intel Corp.'s 386 microprocessor technology that may be able to tap into virtually any wireless or wired network in the world. The new Ericsson phones will support most leading communications protocols and will be able to track and organize a variety of messages (e-mail, fax, voice, and paging).

The new phones with all of their new features will require much more flexibility in networks than is currently possible. Increasingly, new technologies will be integrated into existing infrastructures.

CONVERGENCE WITH NETWORK SUPPORT

Clearly, the wireless evolution will follow the trend to merging mobile communications and portable computing and there is a great deal of work underway to develop new applications for both commercial users and consumers.

One of the more ambitious programs is the InfoPad project supported by IBM, Motorola, Hewlett-Packard, GE/Ericsson, and the Pentagon's Advanced Research Projects Agency (ARPA). InfoPad is an attempt to create a system that offers ubiquitous information access with a lightweight, low-power, wireless multimedia terminal that can operate just about anywhere, including indoors. The goal is to develop a terminal that will support up to 50 users with up to 2 Mb/s of bandwidth.

InfoPad can access the Internet, display text and graphics (including compressed video), record and play audio, and receive data from pen-input terminals, such as personal digital assistants (PDAs). It features a high-speed wireless network interface to connect to a wired network. It also recognizes voice commands that usually require using function keys.

InfoPad is the leading-edge project of three related programs at the University of California at Berkeley. A second program, called Daedalus, is focused on using the lessons learned in developing InfoPad to produce a system for more general-purpose, programmable, computer laptops and PDAs. Daedalus researchers are developing network management software to allow portable computers to dynamically adapt to any changes in their network connectivity. Daedalus is designed to operate in any wide-area network environment, from inside buildings to campus, regional, and metropolitan areas. The third research project, called Medley, is targeting multimedia transport issues for wireless systems.

Another program, the Seamless Wireless ATM Network (SWAN), is designed to allow PDA, laptop, and portable multimedia terminal users to seamlessly access multimedia data in a wired asynchronous transfer mode (ATM) backbone network.

Under development at AT&T Bell Laboratories, the goal of SWAN is to create an integrated network that will support audio and video on wireless communications devices, as well as paging, multimedia messaging, and video cameras. Bell Labs researchers have already built a prototype model of SWAN consisting of base stations connected by a wired ATM backbone network, with wireless ATM links to mobile ATM hosts.

Several wireless ATM research programs are underway in Europe and microwave ATM technology is already in use in Scottsdale, Arizona, where it links the Civic Center and the Via Linda Complex, an area that includes most of the city's government offices. Meanwhile, the ATM Forum is developing a set of specifications to facilitate the use of ATM technology over wireless networks. The forum hopes to work with wireless organizations worldwide to promote the use of wireless ATM.

DEDICATED HARDWARE

Special requirements call for special hardware. Law enforcement agencies can now access the National Crime Information Center (NCIC) and check out criminals, missing persons, and stolen property from FBI records. The next step, under a program called NCIC 2000, will allow these agencies—and even cops on the beat—to call up mug shots and fingerprints on a pager and notebook-size computers.

Several law enforcement agencies are already using laptop and notebook computers in the field, both as investigative tools and to simply speed up the process of issuing traffic tickets. What is needed, and what most of these products do not yet offer, is the ability to simply push a single button that will instantly call for help and accurately transmit the location of the officer in trouble.

The U.S. military is getting the wireless message as well. TRW and Litton Data Systems are leading a team of companies developing lightweight handheld computers under a U.S. Army program to digitize the army's battlefield information and communications operations. The program, called Force XXI—Applique by the Army, calls for the development of three different systems.

One is called the dismounted soldier system, which includes a helmet-mounted computer with a miniature video camera and eye-level monitor. The second system is a book-size computer tied into a wireless local area network for use by field officers. The third unit is a position/navigation device built into a handheld computer capable of receiving, processing, and displaying precise battlefield situational awareness information and geographic locations transmitted from global positioning system (GPS) satellites orbiting the earth.

NASA owns the rights to technology that could lead to a computerized system based on wristband radio transponders for real-time tracking of prisoners. A digital code would be assigned to each wristband to identify the wearer and to pinpoint his or her location. The wristbands would also be tamperproof: An embedded wire loop would cause the wristband to send an alarm if broken. Each of these units would resemble a hospital wristband and would cost only a few dollars.

How far can we take the technology? Obviously, pretty far, given our recent history. Microsoft's Bill Gates predicts that everyone will eventually carry wallet-size personal computers. How about portable phones powered by body heat? The technology is already available to do that with wristwatches.

Remember "decoder" rings that you could get with so many cereal box-tops? Dallas Semiconductor Corp. has demonstrated a ring with a 64,000-bit microchip implanted with a replica of a driver's license, credit-card numbers, and a digitized photograph of its owner. The ring's database can be accessed when part of the head of the ring comes in contact with a data reader. The ring can also be formatted so that it can only be used with a password. Dallas Semiconductor employees have replaced their company ID cards with the ring to access their offices and PCs. Another application for the ring that Dallas Semiconductor is pursuing is accessing automatic teller machines.

Philips Electronics N.V. of the Netherlands has been working on a video telephone that can be worn on the wrist and that would access the Internet. The device would include a video camera, radio, television receiver, and computer that connects users to the World Wide Web, and that can display graphics and video images. Philips says it hopes to have a working model by the year 2002, which may or may not be equipped with voice recognition technology that would eliminate the need for a keypad to control all of the device's functions.

An informal and unscientific survey by the *New York Times* turned up much more interesting and innovative products from consumers' wish lists, such as voice-activated universal translators for international travelers. The ultimate in wireless communications might be microchips that would connect very small personal computers and even machines to the brain to send commands directly to a computer or a car. Another idea: surgically implanted pagers.

"SMART" ANTENNAS

Complaints from cellular phone uses about coverage, dropped calls, and noisy connections, as well as the rapid growth of PCS, have also opened up a huge market for antenna manufacturers.

Cellular carriers in the U.S. alone estimate they will need 15,000 new cell sites by the year 2005 to complete their coverage, upgrade their systems, and meet the growing demand for cellular service. Another 100,000 sites will be needed for new PCS providers, while thousands more are planned for new ESMR networks.

Kenneth W. Taylor & Associates expects the world market for base-station antennas to climb from $64.2 million in 1994 to $325.5 million in the year 2000. But the antenna market is changing. Most traditional antenna makers have tended to focus on the 900 MHz market. Now, the trend is toward higher frequencies, lower power levels, and smaller, more compact designs. There are also several new players in the market, mainly defense/aerospace companies with access to advanced "smart" antenna technologies but with little experience in commercial markets.

Smart or "intelligent" antennas are designed to extend the service area for portable phones, reduce co-channel interference at the base station, improve overlapping coverage, and allow the construction of fewer base stations. The one significant problem with these antennas is that they are expensive and have not been widely tested in cellular or PCS systems.

In the past, antennas were fixed in a single direction, with a cellular subscriber passing through the "fat" part of its beam only at certain positions. One supplier, Engineering Research Associates, a subsidiary of E-Systems (which has a lot of experience designing communications equipment for the military and intelligence community), claims that with its Diversity line of smart antennas, the best part of the signal literally follows customers as they move through the system. The antenna works equally well on analog or digital systems.

If there is a downside to the antenna market it is in the public policy area. At the local level, many homeowners do not want an antenna tower in their neighborhood; they view it as an eyesore. As a result, several cities and towns have begun to regulate antenna installations and, in some cases, are charging a fee for each new installation. On a national level, it has become even more complicated.

While the Federal Communications Commission (FCC) is responsible for regulating the industry, the Federal Aviation Administration (FAA) must approve the construction of antenna towers if they are a certain height—usually 200 feet (61 meters) or higher. But the Pesonal Communications Industry Association claims that the FAA has slowed the installation of new cellular and PCS base stations and antennas because it cannot keep pace with new antenna tower applications. The problem is complicated further by attorneys for PCS licensees who are advising their clients to file with the FAA for antennas of any height, even down to 10 feet (3.05 meters), to reduce their liability in case a helicopter hits one of them.

The integration of computer technology with communications and solid-state electronics could be the basis for future multimedia systems with tremendous processing power.

This is the future of wireless communications. People are coming up with new ideas, new applications, and new products every week. By the turn of the century, virtually everyone will have access to wireless communications and will be sending e-mail, faxes, and video mail, most often on a personal, portable device.

3

First of all, you have to love the numbers. Today, more than 145 commercial communications satellites are operating high above the equator in geostationary (GEO) orbit. Another 155 GEOs are expected to be launched between 1996 and 2000, with hundreds more low-earth-orbit (LEO) and several regional communication satellites planned for launch over the next several years.

The earth station market is growing as well and by the year 2000 could generate several billion dollars in added revenue to the fast-growing satellite market.

The Federal Communications Commission (FCC) has already set the stage for the launch of hundreds of so-called "Big LEOs" over the next three to five years by licensing three systems—Motorola's Iridium, TRW's Odyssey, and Globalstar, a system developed jointly by Loral and QUAL-COMM—that, by design, will provide mobile telephone service anywhere in the world. Other Big LEOs, such as the 48-satellite Russian Signal system, are being developed by the KOSS Consortium to provide space-based voice, data, fax, and messaging services.

So-called Mega-LEOs, like Teledesic, an 840-satellite LEO system proposed by Microsoft's Bill Gates and Craig McCaw (who sold his McCaw Cellular Communications Co. to AT&T), which features intersatellite links using fast packet switching technology and gateway terminals on the ground to serve groups of users, and Alcatel Espace's proposed 64-satellite Satlvod, differ from the Big LEOs mainly in that they plan to provide video and interactive multimedia as well as voice, data, and advanced messaging services. Teledesic plans to market its fixed and mobile network to other service providers in the U.S., acting as a wholesaler of services rather than selling directly to end-users.

Market analysts are projecting sales of portable mobile satellite service (MSS) phones at 3.5 million units by the year 2004, but they continue to hedge their bets. Does the world really need all of these systems? Probably not. Will some of them fail? Most likely, they will all at least survive, although some of them may have to merge their services.

Certainly, the price of making a satellite phone call will have to drop before MSS becomes a mass market item.

After 16 years in operation, the London-based International Maritime Satellite Organization (Inmarsat) has more than 40,000 subscribers paying an average rate of $7.50 per minute. Cellular service, after 12 years, has well over 70 million subscribers paying an average service rate of less than 83¢ per minute (considering roaming charges).

Several "transportable" satellite systems are available, but a more recent innovation is COMSAT Mobile Communications' Planet 1, a portable, laptop computer–size terminal that offers a global, personal voice and data

COMSAT Planet 1 terminal in use.

communications service. With the $2,995 Planet 1, users can call anywhere in the world via Inmarsat-3 satellites for $3 per minute. But as an innovative new product, Planet 1's days may be numbered. The Big LEOs are coming. Why carry around a small box with a phone in it when you can just carry the phone?

HOW DO MOBILE SATELLITES WORK?

Most communications satellites maintain a geostationary orbit 22,300 miles above the earth's equator. The satellite moves at a velocity that makes it appear to stand still over a fixed point above the equator. In operation, each satellite produces a series of beams that divides its coverage into a pattern of overlapping cells, or "footprints," on the earth's surface. The total area visible to the satellite commonly includes one or more regions that are usually heavily populated. As a result, satellite antennas are designed to provide coverage to only a portion of the total area visible to the satellite. In some systems, the antennas can be steered to provide service to specific areas over extended periods of time.

But this high orbit causes some problems. Because of the satellite's distance from earth, it takes at least a half-second for the signal to bounce between ground stations and satellites. The time delay can be annoying during a normal conversation. Also, the distances at which geosynchronous satellites must operate require technical capabilities that make them very costly, ranging into the hundreds of millions of dollars.

Geostationary satellites also have certain advantages. Satellite communication carriers don't have to actually own a geostationary satellite. They normally lease channel capacity on a satellite and sell communications services, usually to vertical or niche markets, such as the long-haul trucking industry or maritime users. If they need more capacity, they simply buy it. As a result, they don't incur the high cost and risk required of launching and maintaining their own satellite system.

THE RACE INTO SPACE

MSS was initially conceived as a satellite link for cellular subscribers traveling to areas where little, if any, phone service exists today. Teledesic estimates that more than half the people in the world live more than two hours from a telephone. India has tens of thousands of villages without telephone service.

Motorola's vision is that Third World countries without a telephone infrastructure would license its Iridium system with subsidized, solar-powered, centrally located telephone "booths" in every town and village. In addition to serving these poor, underdeveloped countries, Iridium and similar systems would provide a premium communications service for international business travelers, government VIPs, and others.

Most of these very small satellite carriers expect to begin revenue-generating services by the end of 1998. By then, will some of the already established satellite services (such as COMSAT Mobile Communications) have a running start in MSS? And does the world really need all these satellites for personal communications?

Washington, D.C.–based MTA-EMCI has developed two possible themes based on the anticipated market demand, country and regional demand characteristics, technology, and the number of satellite operators planning service—a so-called base case and a high demand scenario. They differ primarily by their assumed service and equipment pricing levels.

In its base case, MTA-EMCI expects demand to reach 9 million subscribers worldwide by the year 2004, but in its high demand scenario, EMCI says there could be as many as 15 million subscribers by 2004. These projections are for mobile subscribers only and exclude the potential for fixed applications. Populations not yet covered by cellular services constitute the largest market opportunity, representing almost 4 million subscribers (or about 43 percent of the total global distribution) in the base scenario and almost 7 million in the high demand forecast. EMCI lists trucking, international travelers, areas not yet covered by mobile communications, air travel, and the maritime industries as the remaining market segments, but even narrower niche markets are likely and will add to the growth of MSS.

The Asia/Pacific region, with its vast populations and relatively low cellular usage in countries such as India and China, is expected to dominate global satellite demand by 1998. (Early in 1995, Hughes Telecommunications and Space Co. won a contract from Afro-Asian Satellite Communications Ltd. in Bombay to provide handheld satellite mobile phone service across Asia, the Middle East, and Africa. Using a geostationary satellite, Hughes will connect these calls to any other handheld or fixed phone in the world.)

Latin America, with its vast areas of limited terrestrial communications alternatives and growing demand for telephony services, has already been targeted by satellite providers as a major market opportunity. Brazil, in particular, represents a large future satellite market.

Cambridge, Massachusetts–based Pyramid Research projects close to 35 million MSS subscribers in developing countries in 2010, with the market for mobile communications equipment likely to grow to about $14 billion in the same period. Pyramid also expects annual service revenues to reach $24.7 billion in 2010.

Pyramid offers another scenario in which it assumes that technical feasibility and other issues will slow the uptake of MSS and delay the introduction of a Teledesic-type system until after the projection period. In this case, Pyramid expects MSS subscribers to reach 16.1 million, with the mobile communications equipment market growing to $6.5 billion in the year 2010 and service revenues exceeding $11.6 billion.

HERE COME THE LEOs

Rather than buy capacity on geostationary satellites, the LEOs plan to launch constellations of new satellites that are smaller and much less expensive than geostationary satellites, designed to operate in low-earth orbit. The altitude is critical. Too high an orbit would require a very large rocket and result in huge launch costs. The higher orbit would also require radiation-hardened components, driving up the cost of the satellites. Too low an orbit would add to fuel requirements needed to maintain orbit, increasing operational costs. Once they are in orbit, the LEO satellites will communicate with mobile or portable battery-powered telephones linked to the satellites through ground stations and the public switched telephone network (PSTN).

Whereas the Big LEOs operate above 1 GHz and will offer their subscribers voice, paging, facsimile, data, and radiodetermination satellite services (RDSS) for navigation and tracking applications, Little LEOs will operate below 1 GHz with a limited range of services—primarily position-location and two-way communication of very brief text messages, but not voice.

A study by the Commercial Space Transportation Division of the Federal Aviation Administration (FAA) says Little LEOs could provide data

Comparison of LEO, MEO, and GEO Systems[a]
(Handset or PCS Communications)[b]

	Iridium	*Globalstar*	*ICO-P*	*Odyssey*	*AMSC*
Altitude (km)	780	1414	10,355	10,380	35,786
Number of satellites	66	48	10	12	1 + 1
Space segment cost $B	3.8	1.9	2.3	2.0	0.87
Satellite mass (kg)	700	232	2300	1920	4000
Satellite power (watts)	1200	875	5000	3050	3600
Satellite capacity (Ckts)	4720	2800	4500	3000	7000
Access method	TDMA	CDMA	TDMA	CDMA	Either
Satellite antenna gain (dBi)	27	12	34	28	45
Number of satellite beams	48	16	160	37	19
Satellite antenna diameter (m)	1.6	0.3	3	2	10
Coverage	Global	Global	Global	Global	Nat'l
On-board processing	Yes	No	No	No	No
ISLsYes	Yes	No	No	No	No
Frequency band (GHz)	1.6[c]	1.6/2.5	2.0	1.6/2.5	2.0
Life (minimum, years)	5	5	10	10	10
Factor G/F^2*R^2 (dB)[d]	0	−15	−9	−13.1	−7.2

Notes: Different assumptions are made for amplifier efficiencies, noise temperatures, handset transmitter power, and margins. Time delay is significant only for GEO.

[a]GEO system has been filed but not authorized.
[b]Iridium has 15.7 dB margin with 0.6 watts subscriber transmitter power.
[c]Bidirectional operation.
[d]Factor is computed at 10 degrees elevation.

services to 15 million customers and generate $2 billion in construction, launch, and service revenues by the year 2000. But the second round of Little LEO applicants must first prove to the FCC that they have the financial strength to launch and operate at least two satellites and that they will not interfere with the first round of Little LEO licensees. And there's the possibility that the FCC may decide to auction Little LEO licenses.

The three first-round licensees—Orbital (Orbcomm) Communications Corp., Starsys Inc., and Volunteers in Technical Assistance (VITA), a non-profit organization that seeks to operate a two-satellite Little LEO network for Third World countries—have launched satellites.

By mid-1996, only Orbcomm had launched satellites with 2 in orbit and plans to launch 26 others. Orbcomm and Teleglobe Inc. of Canada plan to

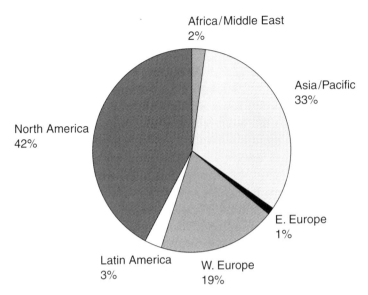

Distribution of Little LEO subscribers by region, 2005.
Source: MTA-EMCI.

jointly finance and operate the Orbcomm system. Teleglobe is expected to provide $80 million for the system, with Orbital providing another $55 million for the project. Under their agreement, Teleglobe would operate Orbcomm outside the U.S. under a new company, Orbcomm International Corp. Orbcomm has already won FCC approval to build, launch, and operate a network of up to 36 LEO satellites.

Starsys plans to transmit and receive very brief, nonvoice messages via a network of satellites and ground stations and small, low-power, low-priced portable terminals resembling pagers. Ground processing of Starsys signals will give subscribers the geographic location of these small terminals. Starsys's technology could be used to locate stolen vehicles and equipment, monitor fleet and cargo movements, transfer data, and for convenient and inexpensive two-way messaging. Starsys has already successfully tracked and processed data from an experimental satellite.

Second-round license applicants are GE Astro Space, Final Analysis, E-Sat Corp. (a subsidiary of Echostar Corp.), LEO One Panamerica, and CTA Corp. FCC analysts are not convinced there is enough spectrum allowed for the future expansion of the first-round licensees and to support all five of the second-round Little LEO applicants.

Motorola was the first to apply for a license to build and operate a Big LEO MSS in December 1990, nearly two years before the 1992 World Administrative Radio Conference (WARC '92) in Madrid, Spain, allocated global frequencies 1610–1626.5 MHz and 2484.5–2500 MHz for LEO satellite systems. At that point, the FCC awarded Motorola an experimental license to build five Iridium satellites.

Iridium was originally conceived as a 77-satellite system, named after the element whose atom has 77 orbiting electrons. (Motorola engineers have since figured out a way to cover the globe with 66 satellites, configuring them in six polar orbit planes of 11 satellites each; however, Motorola says it has no plans to change the name of the system to Dysprosium for atomic number 66.) Iridium satellites will orbit about 420 nautical miles above the earth's surface. They will be phased so that odd-numbered planes have satellites in corresponding locations, with satellites in the even-numbered planes staggered approximately midway between. The satellites will essentially travel in co-rotating planes, moving up one side of the earth, crossing over at the pole, and coming down the other side of the earth. Of course, the earth will continue to rotate beneath them. Each Iridium satellite will transmit 48 beams with each beam capable of handling about 230 calls.

Motorola has developed a gateway to interconnect the Iridium constellation with the PSTN. The gateway is designed to handle such functions as call setup, caller location, and billing. The mobile subscriber unit, when turned on, will link with the nearest Iridium satellite, which will then update the system's location register. Obviously, caller location is critical: An Iridium subscriber who lives in India could carry his or her phone to Moscow or Beijing and still make and receive calls through the Iridium network.

When Iridium was introduced in June 1990, Motorola estimated that it would cost $2.1 billion and would be in full operation in 1996. Since then, the cost has climbed to $3.7 billion and the program has yet to be fully funded. By the end of 1996, the Iridium program had raised most of the funding it needed to get its full system into service.

Motorola even set a deadline for itself: Fund Iridium by December 15, 1992, or seriously consider dropping the program. As the deadline passed, Motorola would only say publicly that it had received letters of intent from several potential investors and that, based on this level of interest, the company would proceed with the next phase of the initial round of financing.

Initially, Motorola declined to identify specific investors, citing nondisclosure agreements.

In late January 1993, representatives from 18 potential investors met for the first time as a group in Geneva, Switzerland, to review the status of technical, marketing, and financial plans for Iridium. They signed subscription agreements or letters of intent, which represented an investment level beyond the $800 million targeted for the initial round of financing. But the agreements were nonbinding, forcing Motorola to set a new schedule for launching the Iridium system.

By August 1993, Motorola had completed an initial $800 million equity placement for Iridium and identified the initial members of its Iridium consortium. Motorola would retain about a 34 percent equity interest in Iridium at a cost of $270 million. The other investors included Nippon Iridium Corp., an alliance of 18 Japanese companies, and two Saudi Arabian ventures led by the Mawarid Group. Companies with smaller shares in the program include BCE Mobile Communications/BCE Telecom International Inc. of Canada; Sprint Corp. of the U.S.; STET-Societa Finanziaria Telefonica per Azioni, an Italian telecommunications concern; United Communications of Thailand; and The Great Wall Industry Corp. of China. As part of its arrangement, BCE Mobile will operate and manage Iridium services in Canada.

Motorola also found an angel in, of all places, Russia. McDonnell Douglas Corp. will launch most of the Iridium satellites, but Motorola has also signed on the Russian Federation's Khrunichev Enterprise to launch a portion of Iridium satellites from the Plesetsk cosmodrome in northern Russia. In return, Russia will spend $40 million on the satellite project, which will be recouped from future hard currency income.

Then there's the cost of the satellites, which could come to as much as $30 million each. Sixteen of the satellites will have to be replaced annually, representing at least a $480 million annual investment by Motorola and its partners, and these costs could climb if Iridium runs into technical problems.

Motorola now plans to begin offering commercial service in September 1998. The business plan anticipates 75,000 subscribers by the year 2000; about 10 percent of those are expected to be Japanese. (At one point, Motorola estimated that it would need about 500,000 subscribers worldwide to break even financially, while 870,000 users would "represent a good business.")

At least three regional GEO systems are also in development. ASCOM, an Indian company, plans to provide handheld mobile phone service for India, Asia, Africa, and the Middle East. Another satellite group called APMT wants to serve the China and Singapore region.

In Japan, a consortium consisting of 25 companies has established Satellite Phone Japan Corp. of Tokyo to manage the Japanese portion of the proposed 10-satellite Inmarsat-P global mobile phone system. KDD Corp., Japan's leading international telephone carrier, heads the group with a more than 50 percent stake in the operation. Toyota Motor Corp. and NTT DoCoMo Corp. (Nippon Telegraph and Telephone Corp.'s mobile subsidiary), three trading houses—Sumitomo Corp., Marubeni Corp., and Nissho Iwai Corp.—have taken smaller stakes in Satellite Phone Japan.

Japan's Ministry of Post and Telecommunications (MPT) has also formed a group to study a technical proposal for another system that would likely use NTT's new N-Star satellite. Space Systems/Loral has already signed on with NTT to build two N-Star spacecraft to provide fixed and mobile communications in Japan. Japan's NEC Corp. is also developing small, low-cost satellites.

MORE GLOBAL TALK

AirTouch Communications and Sime Darby, a Malaysian multinational conglomerate, have formed a joint venture to act as the exclusive service provider for Globalstar in Malaysia.

In France, the Centre National d'Etudes Spatiales (CNES) is developing the Taos system, using five small satellites for messaging, position-location, and remote monitoring. France hopes to launch its first Taos satellites in 1997. The European Space Agency (ESA) also is considering the development of a high-earth-orbit (HEO) system to provide the best possible coverage for European users.

In Mexico, LeoSat Panamericana, SA de C.V. was formed in 1991 by Manuel Villalvazo, who owns Baja Celular-Mexico and is chairman of the board of the Latin American Cellular Association (Alacel), to develop an "affordable" LEO network offering messaging, paging, fax, electronic mail, and position-location services throughout Latin America. LeoSat has been using high-altitude balloon flights to perform frequency interference, communications protocol, and system tests. Following those tests, LeoSat plans to use two LEO satellites to validate its designs in an operational system. Another possible market entrant is Ariadne, a joint Ukrainian-Russian venture, which has proposed a Little LEO network.

Alcatel Espace is working with Space Systems/Loral and Mabuhay Philippines Satellite Corp., a Philippine-based private corporation created to provide transponder capacity for regional service.

Lockheed Martin Corp.'s Astro Space subsidiary has a $650 million contract to build a complete mobile telephone communications system for the Asia Cellular Satellite System, called ACeS. The consortium includes PT Pasifik Satelit Nusantara of Indonesia, Philippine Long Distance Telephone Co., and Jasimine International PLC of Thailand. ACeS will provide voice, fax, and pager services through new satellites to handheld mobile and fixed terminals throughout Southeast Asia, India, and China beginning in 1998. It will allow a person to use the same GSM phone number for both mobile and cellular telephones while traveling throughout Asia. Lockheed Martin is also working with state-owned China Orient Telecomm Satellite Co. Ltd. of Beijing to build the ChinaStar-1, a satellite for voice, data, and TV distribution services throughout the People's Republic of China, possibly as early as 1997.

MORE ABOUT LEOs

Will the LEO concept succeed? Some industry analysts are skeptical. LEOs do offer true global roaming, and they would help some terrestrial carriers fill in their service offerings with satellite coverage. But they have no certain market, and it's not yet clear that they can obtain service rights worldwide. There's also the prospect of FCC auctions looming over their heads. When the FCC raised more than $15 billion through its PCS and other license auctions, proposals were floated that called on the International Telecommunications Union (ITU) to auction global MSS frequencies.

Motorola liked the idea immediately, figuring that the auction price would more than offset the time-consuming and costly country-by-country approval process. But it didn't win much support from the rest of the communications satellite community that believes auctions would provide foreign governments with an excuse to block the entry of U.S. satellite service companies into their markets.

Several questions continue to stalk the MSS community. Can Iridium, Globalstar, Odyssey, and the others afford to put up several multi-billion-dollars worth of spacecraft and an earthbound infrastructure to support what appears to amount to a subset of a mature and well-entrenched global telecommunications service? How will international Post, Telephone, and Telecommunications authorities (PTTs), who fear being cut out of satellite service revenues, respond to competition from a well-financed international consortium? If Motorola wins the spectrum it needs for Iridium through an international auction process, will that get the Big LEOs off the hook, or will global telecom authorities ask for some form of tribute for similar consideration?

And why are there so many international investors? One possibility is that they're simply taking a defensive posture. By investing in an MSS, they're improving their strategic position by integrating satellite services into their terrestrial services. If they don't participate, one of their terrestrial competitors may get an edge.

NOT QUITE READY FOR INMARSAT

By mid-1996, the FCC had only given three companies—Iridium, Odyssey, and Globalstar—the go-ahead to build, launch, and operate global LEO mobile communications satellite systems. The FCC put LEO applications filed by Mobile Communications Holdings and Constellation Communications on hold because they could not meet the commission's financial requirements.

The tough one for the Big LEOs is the system proposed by an affiliate of the 79-nation International Maritime Satellite Organization (Inmarsat). Initially referred to as Project 21 (for the 21st century) or Inmarsat-P (for portable), ICO Global Communications (the ICO stands for intermediate circular orbit), is expected to begin service in 1999 and be fully operational by 2000.

The Inmarsat plan has been attacked on several counts. For one thing, Inmarsat represents the U.S. government in satellite treaties. In addition, Comsat, Inmarsat's U.S. signatory and largest shareholder (23 percent), is forbidden by U.S. law from offering a satellite-based telephone service. Inmarsat, which expects to retain a 15 percent interest in the affiliate company, says that ICO will operate as a private company with no tax or legal advantages over its rivals.

Inmarsat announced in September 1995 that it would invest up to $150 million in a new affiliate company to develop and operate a global telephone service. The new intermediate circular-orbit 12-satellite system will cost an estimated $2.56 billion and will provide all the services planned by Iridium and other mobile satellite systems.

Iridium and TRW, in particular, have been complaining loudly since 1994, when Inmarsat established ICO as a separate but affiliated company, about Inmarsat's "unfair advantage" over other mobile communication services.

The U.S. supported the formation of ICO on the condition that its structure include certain principles that favor competition. Inmarsat's member countries agreed to many of these principles. However, in July 1996 the U.S. General Accounting Office (GAO), the investigative arm of Congress, issued a report questioning the fairness, and even the legality, of several proposed restructuring plans developed by both INTELSAT and Inmarsat, including the formation by Inmarsat of ICO.

In its report, the GAO said that the treaty organizations, "as structured, may now be impeding the flowering of a private market and the benefits it can bring to consumers."

For one thing, satellite systems must gain permission from domestic licensing authorities for the right to use the necessary ground stations to receive satellite signals and to interconnect on the ground with the domestic telephone system. Because many of these licensing authorities or telecommunications companies are the signatories that own INTELSAT and Inmarsat, the GAO pointed out that they may have a financial incentive to favor the two organizations, or their affiliates, over other firms when determining who may do business in their countries. Analysis by the U.S. Department of Justice found that in certain areas INTELSAT currently dominates the market as a result of its large share of transoceanic satellite capacity and its signatories' ability to keep other competitors out of their domestic markets. The organization's treaty status also provides certain privileges and immunities, such as exemption from taxation and immunity from lawsuits. INTELSAT and Inmarsat also have easier access to orbital slots in space for their satellites and to spectrum.

The GAO says that Inmarsat and its signatories have both the incentives and the ability to provide ICO with market advantages over its potential competitors. For example, the U.S. Departments of State and Commerce, in a September 1995 letter to the FCC, expressed concerns about whether contractual arrangements between Inmarsat and ICO were conducted with sufficient independence to ensure that there was no cross-subsidization. In their letter to the FCC, State and Commerce concluded that ICO's organizing documents did not fully incorporate the conditions to which Inmarsat's members had agreed. State and Commerce asked the FCC to delay authorization of Comsat's share of Inmarsat's investment in ICO until it is clear that ICO is bound by the principles Inmarsat adopted. They also asked that Comsat state on the record that ICO is bound by the principles approved by Inmarsat and that supporting documentation be provided.

Comsat, which received a review copy of the GAO report before it was published, says that it had reported to the relevant U.S. government agencies in late May 1996 that at ICO's annual meeting on May 28, 1996, the shareholders approved an amendment to ICO's organizing documents that fully incorporates these principles. Comsat also assured the GAO that it would provide the supporting documentation "in the near future."

RESTRUCTURING PROPOSALS

Several options for restructuring INTELSAT and Inmarsat have been discussed. For example, some portion of the satellite facilities of each of the two treaty organizations could be privatized. Inmarsat's interest in reorganizing itself is focused on improving its ability to respond to more commercial opportunities than its original treaty structure allows. Under one proposal, Inmarsat would evolve into a privately owned international public corporation. An early version of that proposal would transfer all of Inmarsat's satellites to the new corporation, while a smaller intergovernmental organization with more limited responsibilities would be retained to ensure the ongoing provision of services related to safety and rescue at sea.

Inmarsat has already indicated that it would consider a future merger with ICO. However, Comsat does not support such a merger any time soon because the business plans of ICO and a restructured Inmarsat differ significantly.

The GAO believes that ownership ties between the ICO and a largely privatized Inmarsat could create a company with significant advantages in the market that would be free of any of the decision-making or operational burdens imposed by an intergovernmental structure.

Another proposal, developed by the U.S., would separate INTELSAT into two companies. From a competitive standpoint, this would reduce the size of the two entities, presumably putting them on a more level playing field with private competitors. Currently, INTELSAT has 24 satellites in space and seven empty orbital slots. The affiliate, with which others would most directly compete, would have about half of the INTELSAT satellites. By comparison, one private competitor, PanAmSat, plans to grow to an 8-satellite operation within a few years.

Rather than operate in geostationary orbit, ICO spacecraft will move slowly across the sky at an altitude of 10,000 km, with at least two satellites in view to subscribers at all times.

Will it work? Inmarsat has already spent close to $10 million for six separate studies to determine the size of the global MSS market and what type of satellites and networks are best suited to the MSS task. It assigned two teams to investigate its geostationary option—GE Astro Space/Matra Marconi Space and Hughes Space & Communications/British Aerospace/NEC Corp. Intermediate orbit studies were turned over to TRW and the GE Astro Space/Matra Marconi Space team.

Inmarsat then assigned a European alliance, including Aerospatiale and Alcatel of France, Alenia Spazio of Italy, and Deutsche Aerospace of Germany, to study the LEO option. It also hired Touche Ross, an international management consulting company, and Schema, a specialist in cellular market research, to study the MSS market. Schema and Touche Ross have prepared detailed country-by-country demographic and communications usage characteristic studies for eventual use by Inmarsat signatories. Inmarsat has also been working with Nokia Mobile Phones, Europe's leading producer of cellular phones, to develop satellite mobile telephone markets and technologies. (Inmarsat believes that with the development of ICO, the number of land-based switching stations compatible with its own system will climb from 50,000 to more than 400,000 by the turn of the century.)

The high stakes nature of the mobile satcom market began to emerge in July 1992 when Inmarsat first announced that it would offer a global satellite network for pocket-size phones by the end of the decade. Shortly thereafter, Inmarsat Director General Olof Lundberg suggested that Project 21 might be extended to other applications—"for instance, entertainment communications centers for your car, personal satellite navigation, position reporting and alerting services for your security." Lundberg believes the accumulated worldwide potential mobile satellite market will be large enough by the year 2000 to support an investment in new satellite systems of about $1 billion.

Despite all the homework, most Inmarsat members took their time deciding if they were willing to undertake the financial risk involved in the program. Sweden, which has a 10 percent stake in Inmarsat (mainly because of its maritime interests), was slow to support ICO. Comsat, on the other hand, with a 25 percent equity position in Inmarsat, may actually insist on a bigger stake in the program to help cover the potentially huge U.S. MSS market.

Most of the Big LEOs, led by Motorola, have worked very hard to kill off the Inmarsat program, but to no avail. Historically, the law has been on the side of the U.S.-based Big LEOs. Until recently, foreign investment in telecommunications companies has been limited to 25 percent and U.S. firms (like Comsat) have been forbidden from placing foreign officers on their boards of directors. However, new FCC rulings on foreign carriers and ownership are making it easier for foreign companies to enter the U.S. market, and even to purchase up to 100 percent interest in U.S-owned companies. The change is an attempt by the FCC to encourage foreign countries to open their telecom markets in exchange for a crack at the highly lucrative American market.

Initially, American Mobile Satellite Corp. sided with Inmarsat. On the day Motorola announced its plan to develop Iridium (June, 26 1990), AMSC signed a memoranda of understanding with Motorola, Inmarsat, and TMI Communications of Canada to jointly explore the potential of Iridium. That agreement has since been dissolved.

ICO has already signed Hughes Space and Communications International to build its satellites. Under the $1.3 billion agreement, Hughes becomes a strategic partner of the new affiliate company with a substantial investment share. Japan got into the act when a consortium of 25 Japanese companies formed Satellite Phone Japan Corp. to manage the Japanese portion of Inmarsat-P. KDD Corp., Japan's leading international telephone carrier, will head the Japan ICO organization with a 53.2 percent stake in Satellite Phone Japan. Other key investors include Toyota Motor Corp., NTT DoCoMo Corp., and NTT's mobile network subsidiary.

Even though its services will be competitive with the Big LEOs, ICO will not operate in the same way. As proposed, it will have 10 operational satellites in two orbital planes that cross at a 90-degree angle at 6,430 miles (or about 10,400 kilometers) above the earth. The full system requires 10 operational spacecraft, with two spares. The satellites will relay calls between the user and a satellite access node (SAN) within the satellite's view. SANs are interconnected using terrestrial facilities to form a network and are linked to the PSTN through gateways owned and operated by third parties.

AMSC ALREADY UP AND RUNNING

AMSC, whose shareholders include McCaw Cellular Communications, Hughes Communications Inc., Singapore Telecommunications, and Mobile Telecommunication Technologies (Mtel), began testing its AMSC-1 satellite in April 1995. Hughes Aircraft built the AMSC-1 and its Canadian twin (TMI-1). General Dynamics Commercial Launch Services and Westinghouse Electric Systems Group won the systems integration job for the ground network. Mitsubishi Electric and Westinghouse Electric are producing AMSC's cellular/satellite telephones.

In 1989, the FCC granted AMSC a license to provide the full range of mobile satellite services to the U.S. land mobile, maritime, aeronautical, and fixed site markets. Rather than wait for the launch of its own satellites, AMSC is already leasing capacity from Inmarsat, providing mobile data and position-location services to the transportation, maritime, rail, and remote monitoring industries.

In late 1995, AMSC introduced four satellite services:

- Skycell Satellite Roaming Service, a dual-band satellite/cellular phone service, will be distributed by more than 155 cellular carriers across the U.S.
- Skycell Satellite Telephone Service, featuring five separate mobile phone configurations targeted at different vertical markets—boaters and shipping operators, general aviation, a fixed site service to rural homes with no land line service, a vehicle-mounted satellite phone for transportation and other companies, and a transportable satellite phone for corporations and public safety agencies
- AMSC Fleet Communications Products, which includes the company's mobile messaging service available to transportation companies on leased satellite capacity since 1992
- AMSC Private Network Capacity, which covers satellite capacity, switched service, and private network management services for companies that create specialized mobile communications services to serve niche markets

AMSC says it has identified 14.5 million mobile cellular customers that the company considers "high probability" candidates for its services.

TMI Communications of Canada, a BCE subsidiary, will use the AMSC-1 until it can launch its own MSAT Network. Voice services for both mobile and fixed applications will be introduced first with data services to follow in 1996.

The MSAT-1 is identical to the AMSC-1. The two systems will back each other up, providing a seamless MSS for all of North America. Together, the AMSC and MSAT systems will be able to cover all of North America, including Hawaii, the Caribbean, parts of Central America, and 400 kilometers offshore. However, unlike AMSC, which is marketing directly to users and cellular carriers, TMI is working with service providers (Bell Mobility, a subsidiary of BCE Mobile Communications, Inc., is the first to sign on), allowing them to package their own services based on the TMI system. Airtime on MSAT is estimated to cost $2.50 per minute, with the hardware priced at around $5,000–$6,000.

But AMSC is having its own financial problems. According to its annual Form 10-K filed with the Securities and Exchange Commission (SEC), AMSC needed an additional $150 million to pay its debts, run day-to-day operations, and for working capital in 1996, and at least $200 million in additional funding for 1997. In its filings with the SEC, the company says it "does not believe it can obtain, on commercially reasonable terms, adequate liquidity to support its operations in the near term without substantial credit support from its principal stockholders"— namely, AT&T Wireless Services, Singapore Telecom, and Hughes Communications. Despite its financial problems, AMSC was expected to build its AMSC-2 satellite, which is scheduled to be launched in 1999.

ODYSSEY AND GLOBALSTAR

TRW's Odyssey and Loral/QUALCOMM's Globalstar are no less serious than Motorola's Iridium; they simply haven't received as much attention.

The TRW Space & Electronic Group's 12-satellite Odyssey system is actually a medium-earth-orbit (MEO) system designed to orbit the earth at an altitude of about 10,354 km (5,591 nautical miles). TRW believes the MEO approach is more economical than LEOs because it allows satellites to stay in orbit longer—up to 15 years on-orbit, compared to only 5 years for Iridium satellites. The 12 satellites will be orbited in three planes, each inclined at 50 degrees, to provide seamless coverage of the globe. Two Odyssey satellites will be visible almost anywhere in the world at all times, which means the system is always at high line-of-sight angles, thereby minimizing obstructions by terrain, trees, and buildings.

The system employs spread spectrum code division multiple access (CDMA) architecture on both the uplinks and the downlinks. The ground network of earth stations is connected via leased lines. Calls to and from the Odyssey handsets will be routed globally, using the Odyssey network and existing PSTNs. Odyssey traffic will be regulated via local "gateways" connected to the Odyssey system.

Odyssey is targeting sparsely populated regions that may never receive cellular service because there are not enough subscribers to justify building the infrastructure, and institutional users, including government agencies and business travelers. TRW plans to wholesale its services to other operators, rather than providing services to end-users.

Loral Qualcomm Satellite Services, Inc., a joint venture of Loral Corp. and QUALCOMM, Inc., will operate Globalstar, a network of 48 satellites designed to orbit in eight 750-nautical-mile orbital planes. Like Iridium, Globalstar is running behind schedule. Its management anticipated FCC authorization by the end of 1992 with construction beginning in January 1993. It missed those milestones, but still expects its consortium partners to come through with the funding it needs to fully develop the system.

Loral, Alcatel, Alenia, and Space Systems/Alliance (SSA), the international satellite partnership backed by Aerospaciale, will design, develop, and produce the Globalstar satellites and ground stations.

Constellation Communications, Inc.'s Aries and Ellipsat Corp.'s 12-satellite Ellipso system are the other Big LEOs. In August 1993, Ellipsat and Constellation Communications received authorization from the FCC to launch four experimental satellites to validate their LEOs. The Ellipso system was designed to serve the Northern Hemisphere, but also plans to offer complementary services in the Southern Hemisphere through an additional constellation. Fairchild Space and Defense Corp. has agreed to be Ellipsat's prime hardware contractor, responsible for in-orbit delivery of its satellites.

HUGHES PLANS ITS OWN
SATCOM SYSTEM

Hughes is taking a different approach, engineering a nine-satellite GEO system to provide an interactive "bandwidth-on-demand" service for fixed telephony, high-speed data exchange, and high-resolution interactive video. Called Spaceway, the Hughes system will use the high-frequency Ka-band (around 28 GHz) and tightly focused spot beams to enable transmission to or from 26-inch antennas. Hughes plans to price these satellite dishes under $1,000, or only slightly more initially than currently available 18-inch DBS (direct broadcast satellite) television antennas. Hughes says that two-way video transmission will cost less than an international phone call does today, and an international phone call will cost less than current domestic phone calls.

Hughes sees two roles for Spaceway: One is similar to the Big LEO concept, providing basic telephony in underserved areas of the world; the other will offer wideband communications services to the global marketplace. Spaceway will feature true, full-motion desktop videoconferencing; CAD/CAM workgroup computing for computer-aided design and manufacturing; technical and medical teleimaging, with X-rays transmitted in under 8 seconds, versus 21 minutes via today's phone lines; and high-speed, low-cost access to the next generation of on-line multimedia databases, at rates from 16 Kbps to 1.5 Mbps.

Loral plans a system similar to Spaceway, called CyberStar, with high-speed voice, data, and video. As planned, it would begin operation in 1998, covering all 50 states with a single satellite.

MORE THAN A PAY PHONE

Motorola says it expects to charge $3 a minute for an Iridium phone call. Inmarsat believes it can provide virtually the same service for less than $1 a minute, and Ellipsat plans to charge 50¢ a minute. By comparison, a cellular phone call costs from 20¢ to about 95¢ a minute. Market research by TRW indicates that at $3 per minute, only 300,000 subscribers are likely to sign up for service. A lower rate—for example, 30¢ per minute—would generate 12 million subscribers, but none of the proposed systems will have that kind of capacity. TRW expects to make a substantial return on its investment with 2 million subscribers. But those projections could be upset if, as some industry observers have suggested, market conditions force Motorola and other LEOs to rent, as well as sell, their phones to their service subscribers.

The FCC adopted proposed operating rules for Little LEOs in early 1993. With their position-location capabilities, Little LEOs such as Starsys would like to market their services to the long-haul trucking and other commercial vehicle operators. That means they would have to compete with QUAL-COMM's OmniTRACS, a two-way text-messaging and position-location satellite-based communications system already used by more than 40,000 trucks in the U.S., as well as trucking fleets in Europe and Japan. Omni-TRACS leases channels on GTE Spacenets' GStar I satellite; Loran C receivers are used in the trucks.

In Europe, OmniTRACS is called EurtelTRACS because it operates from the Eurtelsat satellite network. The system is managed by Alcatel under a joint venture agreement with QUALCOMM and is projected to grow at a fairly good rate as the European trucking industry works to improve their efficiency under the rules and regulations of the single market European Community. In Japan, QUALCOMM has formed OmniTRACS KK, a partnership with C. Itoh & Co, one of the world's largest trading companies, Nippon Steel Corp., Clarion, a major consumer electronics company, and Maspro, a supplier of DBS receivers. OmniTRACS's primary competition in the U.S. at the moment is from Motorola's Cover-agePlus, which operates through Motorola's existing network of trunked Specialized Mobile Radio (SMR) systems.

Increasingly, OmniTRACS is being used by companies other than truckers. Delta Airlines has installed OmniTRACS on vehicles used to transport jet engines. The U.S. Navy and U.S. Coast Guard are using OmniTRACS to monitor certain types of cargo. And with 24,000 locomotives in the U.S. alone, QUALCOMM also sees the railroads as a major market opportunity for OmniTRACS.

GUESS WHERE I'M CALLING FROM?

One segment of the satellite market that is already up and operating and growing fast is satellite-based phone services for commercial airlines.

In 1993, Skyphone, a London-based consortium owned by Singapore Telecom, Norwegian Telecom, and BT (formerly known as British Telecom), predicted that 1,000 planes would be equipped to handle satellite calls by 1998. In fact, by the end of 1995, GTE Airfone alone had equipped more than 1,200 aircraft with its digital phones.

The FCC has also licensed In-Flight Phone Corp. (formed by Jack Goeken, who also founded MCI, the FTD Mercury Network used by florists, and Airfone); Claircom Communications, a joint venture of Hughes Network Systems and McCaw Cellular Communications Inc.; Mobile Telecommunications Technologies Corp. (Mtel), and American Skycell Corp., a West Atlantic City, New Jersey–based startup.

Inmarsat provides its own international air-to-ground phone service, but only for private and general aviation aircraft. The international satellite carrier is forbidden by the FCC from offering mobile satellite services in the U.S., a fact that isn't expected to affect its proposed Inmarsat-P system because that service is targeted at the international market rather than the fast-growing U.S. cellular system.

Skyphone, along with Comsat, and ARINC, a U.S. airlines satellite communications system, is part of another consortium, Globalink, which offers satellite telephone service in the U.S. Another group, the Paris-based Satellite Aircom, provides service satellite communications services to Air France and Lufthansa, although Air France is considering testing a ground-based service during 1993. Satellite Aircom's owners include IDB Aeronautical of the U.S., OTC Australia, Teleglobe Canada, and the Societe Internationale de Telecommunications Aeronautiques, an international airline telecommunications organization.

Market dynamics will force some changes. The result will be a new industry with a highly competitive, fully integrated combination of satellite and terrestrial telecommunications products and services.

GPS GOES COMMERCIAL

Remember the Persian Gulf War when troops, tank commanders, and helicopter pilots had a big problem finding their way around the desert (no landmarks) until they were equipped with handheld terminals that provided their precise location? They used the global positioning system (GPS), a constellation of 24 satellites operated by the U.S. Department of Defense (DOD).

GPS satellites orbiting 11,000 miles above the earth can tell you precisely where you are or where you want to go in terms of longitude or latitude, as well as how fast you're traveling and how long it will take you to reach your destination.

DOD designed the system not only to help U.S. troops find their way in the deserts and jungles of the world but also for more accurate bomb and missile delivery. It is still used for those purposes. Today, however, portable GPS receivers that used to cost several thousand dollars are now available to anyone for less than $400. They have become so popular, especially among recreational users such as small boat owners and hikers, that more than 500,000 civilians now use GPS in the U.S. alone.

GPS has been integrated into cellular phones and other communications devices. Trimble Navigation's GPS/Cellular Messenger is one example. Several manufacturers and service providers, such as AT&T Wireless Services, plan to offer real-time GPS data as part of an expanded package of data network services.

GPS is being tested for use by commercial airlines as a primary navigation system. Railroads and trucking companies use GPS to keep track of their rolling stock. Avis and Hertz have combined GPS with mapping software in thousands of their rental cars to help guide drivers in unfamiliar areas.

The success of GPS and efforts to further enhance the usefulness of GPS have created problems with the Pentagon. Civilian GPS users have had to work around a GPS security feature known as selective availability (S/A). Imposed by the DOD, S/A deliberately degrades the navigation and timing signals of the system for most nonmilitary users. The U.S. military relies on S/A and antispoofing techniques to deny full GPS accuracy to the enemy while maintaining the use of a highly accurate spoof-resistant signal. Antijam antennas and antenna electronics are often deployed on many weapons systems to provide increased jam resistance, while integrated GPS/inertial navigation systems make it possible to find a target, even with successful jamming.

When S/A is "on," the signals received from GPS satellites can only define a position to within 100–330 feet. With S/A "off," the accuracy increases to 16–66 feet.

Now, with a directive signed by President Clinton, the rules are changing. The directive accomplishes several things: It establishes guidelines for dual-use (military and civilian) GPS, it eliminates S/A, and it requires the State Department to negotiate international agreements aimed at establishing GPS as the global navigation standard for civil aviation. It also gives the Pentagon, which initially did not support the directive, 4–10 years to develop the means to ensure that U.S. and allied military services will continue to have an advantage during a war or other hostile action. The directive ensures the continued growth of GPS in the commercial and consumer electronics markets. Indeed, the U.S. Global Positioning System Industry Council expects the market to grow from $2 billion in 1996 to $8 billion in annual sales by the year 2000.

The Federal Aviation Administration (FAA) plans to improve the accuracy and integrity of GPS for flight operations by using a concept known as the Wide-Area Augmentation System (WAAS) for landing approaches when the pilot's visibility is no less than 0.5 miles and the ceiling is no lower than 200 feet.

WAAS will consist of 24 ground-based, wide-area reference stations, two wide-area master stations, and two satellite uplink sites. Differential corrections (which involve the precise measurement of the relative positions of two receivers tracking the same GPS signals) and integrity data derived from the ground-based network, as well as additional ranging information, will be broadcast to users from the geostationary satellites using a GPS L-band signal at a frequency of 1574.42 MHz. The FAA expects the WAAS to be in place by the end of 1997, and several countries are interested in participating in WAAS.

The U.S. Coast Guard has established a GPS network that will meet, for the first time, the extremely accurate navigation requirements of commercial and recreational mariners in U.S. harbors. Now fully operational, the system is expected to reduce the number of navigation-related collisions and grounding incidents by 50 percent over existing navigation methods.

Internationally, the Russian Global Navigation System (GLONASS), which—like the U.S. GPS network—uses 24 satellites, but employs only unencrypted navigational signals, is being integrated into GPS systems to create even more accurate satellite-based navigation systems that can now utilize 48 satellites rather than the 24 available with only one system.

4

Wireless Data

Despite all the hype, most of it from the industry itself, the wireless data market is just beginning to develop. The mobile data industry served just over a million customers in the U.S. by the end of 1995 and most of them were involved in vertical or niche applications, such as transportation, warehousing, public safety, and health care. So far, consumers have expressed relatively little interest in wireless data.

Most independent market studies of wireless data have been well off the mark, predicting fairly large gains year to year. Equipment manufacturers have been no less positive about the market and several have launched initiatives to develop and promote Internet-based data communications. One of them, Nokia, believes that data will account for half of all cellular traffic by the year 2000. Carriers also are getting into the act: Cellular One in central California has offered network-based Internet access since the beginning of 1996.

But no one argues that there is considerable room for growth of wireless data services—even within companies that already employ wireless data solutions. One out of every five companies interviewed by MTA-EMCI said that at least one of their employees was already using a wireless data service. For most companies, less than 10 percent of their employees were using wireless data, although a few trucking companies had outfitted their entire fleet with mobile data terminals.

On average, industry decision makers estimate the number of workers who could benefit from a wireless data service at 40 percent of the entire work force. At the time of the survey, in mid-1995, cellular was the predominant network used for wireless data, serving more than 70 percent of current wireless data users. Close to 12 percent of all corporate managers said they would be very likely and 22 percent said they would be somewhat likely to purchase a wireless data application by the end of 1997. Out of the 125 million employed in the U.S., 40 million are mobile workers. That number is expected to surge by as much as 50 percent of the working population by the year 2001. This enormous pool of people-on-the-move has already benefitted from wireless voice. Most of the growth is expected to come not just in vertical applications but also for electronic mail, faxes, file transfers, portable computing, database access, and on-line services.

Why has wireless data service been so slow to grow? The problem isn't so much that wireless data are growing more slowly than expected as that projections for the market are too often gauged against cellular voice, which has been hugely successful. The success of wireless communications today, many consultants believe, is viewed as a jumping-off point for wireless data. In other words, wireless data should be feeding off the success of cellular and, eventually, PCS voice services.

The problem in comparing the data and voice markets is that there is a demand for cellular voice—people love to talk; they need to talk. There is far less demand for data and relatively few products have been developed to date to feed that demand. Worse, some of the products and services currently available have not been well promoted. Essentially, wireless data service providers aren't creating a demand for wireless data.

Another reason that wireless data service has been slow to develop is that most of the portable terminals used in business applications are clumsy or too costly. Service is another issue. Some wireless carriers that are trying to get a foothold in the wireless data market are not used to providing the end-to-end solutions and support often required from a data network. Cellular carriers, for example, usually sell phones and then essentially forget about them. Wireless data carriers will have to develop application-specific software and hardware, and train users.

It's just a matter of time before all of these elements fall into place. The introduction of digital cellular phones and other wireless devices that can access personal and public databases and send and receive faxes and e-mail is expected to create a huge market for wireless data services. Improved reliability, new entrants, and the inevitable price wars will also help sell the concept into more horizontal and consumer markets.

KEY VERTICAL APPLICATIONS

United Parcel Service (UPS) is a major user of cellular data, mainly for package tracking. FedEx uses a solar-powered wireless data system to determine if its drop boxes are full enough to warrant an immediate pickup by one of its delivery trucks. Ameritech Cellular Services has equipped American Family Insurance with a system that allows the property/casualty insurer to process claims and write checks to its policyholders at disaster sites more quickly than it could in the past. The Virginia Beach, VA, Department of Transportation employs wireless data to remotely change expressway traffic signs. Duke Power Co. uses the technology to collect, store, and transmit data from commercial residential sites during off-peak hours.

Some applications are more creative. The San Jose *Mercury News*, for example, utilizes cellular phones to transmit color photos of news events from the field to the home office. The newspaper's photographers use Nikon/Kodak and Canon/Kodak digital cameras, which record images onto PCMCIA cards. They simply remove the card from the camera and plug it into an Apple PowerBook, then edit their shots, write captions, and dump the pictures into the PowerBook. Then, they plug the cellular phone into the laptop computer, dial the modem for the art department's computer server, and transmit the images.

A photographer for the Seattle *P-I* did something similar during the inaugural flight of the Boeing 777. He took several pictures during the flight using a digital camera. He then loaded the images into a laptop computer, cropped the pictures he selected, wrote captions, and then downloaded the pictures and text to his office in downtown Seattle using the aircraft's inflight GTE Airfone system. The pictures were on the press before the plane landed.

Impressive stuff, but these are dedicated, niche applications. To seriously grow the wireless data market, the industry needs at least one so-called "killer app"—on par with what the spreadsheet did for the personal computer market. That could easily be the Internet, whose features are already available in some cellular phones and pagers. Leveraging the ability of these phones to send and receive text messages, customers will have access to information services, such as news, sports, and stocks; notification of voice mail, fax mail, and e-mail; and interconnection to existing paging networks. Callers wishing to contact a personal communications service (PCS) subscriber can send messages from a PC or be connected to a cellular service operator who will transcribe messages and forward them to the PCS phone.

It is possible, for example, with certain new portable phones, to check on a movie you have been wanting to see at a distant location. By keying in the zip code and a few additional key codes, the phone's display will indicate where the movie is playing within the zip code area and start times.

Basically, there are two technologies that enable consumers to transmit data over existing analog wireless networks.

Circuit-switched data transmission uses a dedicated circuit or channel to establish a link between two modems during a data call, just as a dedicated circuit or channel is used to establish a link between two wireless phones during a voice call. This technology is best suited for larger data transmissions—for example, for sending faxes, downloading files, and graphics or images. It is also used for wireless voice communications.

Cellular digital packet data (CDPD) is an overlay system to the existing circuit-switched system. A fully open specification, CDPD breaks the data into small "packets" of digitized information and routes the data to their destination in idle channels. Because a dedicated, end-to-end circuit is required, CDPD is a "connectionless" technology and lends itself to short bursts of data—for example, for credit-card authorizations, database queries, short e-mail messages, and utility meter readings.

As the worldwide cellular market continues to grow, there will be increased demand for wireless data services. In fact, CDPD is being deployed in Canada, Mexico, and South America and is being promoted in other areas of the world.

The rollout of CDPD didn't begin in earnest until mid-1995. By the end of the year, the CDPD Forum could identify 78 different CDPD service offerings, or "deployments," of the technology. About 45 percent of these offered full CDPD coverage. The rest provided either core or initial coverage.

Internet protocols are built into the CDPD network. IBM's ARTour Wireless Web Express software, for example, is designed for mobile employees using a browser to access corporate data. It works with leading Web browsers so users receive only the data in the Web site that has changed.

A typical application for Wireless Web Express might involve a mortgage lender whose field sales force accesses its corporate Intranet for the latest information on terms and rates for its offerings.

CDPD-enabled automatic teller machine (ATM) technology is expected to simplify the task of using an ATM in an area for a short duration of time;

for example, at outdoor festivals and sporting events. Wireless ATMs have been a primary target market for CDPD since its inception, but the technology to make this application real has not been available until recently. San Diego–based PCSI has developed a personal access link to check stock quotes, check and send e-mail (with a stored selection of prepared messages or the ability to use the keypad to send messages), or check weather conditions at the airport. It's a phone with a serial port, which allows users to connect a PC directly to either the circuit-switched Advanced Mobile Phone System (AMPS) cellular network or CDPD. Other possible uses for the phone would allow the main office to communicate instantly with a field sales force or delivery driver, aid health-care providers in verifying and updating patient information, and access listing information for real estate agents.

Dataquest estimates there were about 13,000 CDPD subscribers at the end of 1995—not a very impressive number. But CDPD is expected to grow rapidly as more systems integrators become more knowledgeable about the technology. NBI is projecting that CDPD users will account for 27 percent of mobile data subscribers by 1999, and that about 68 percent of the revenues from wide-area wireless data services will go to CDPD carriers by the end of 1999.

NBI expects mobile data revenues to grow to almost $1.5 billion, the Specialized Mobile Radio/Enhanced Specialized Mobile Radio (SMR/ESMR) industry to $933 million, and mobile voice to $25.7 billion by year-end 1999.

MTA-EMCI's research indicates that while cellular carriers are experiencing moderate success in attracting circuit-switched users, CDPD coverage and equipment options will remain limited for at least the next few years. Out of a market potential of nearly 5 million wireless data subscribers in the U.S., EMCI believes that cellular will serve 1.4 million by the year 2000.

Four carriers—Ameritech Cellular Services, AT&T Wireless Services, Bell Atlantic NYNEX Mobile, and GTE Mobilnet—led the way in marketing CDPD, but others are coming on board. SNET Cellular, Comcast Cellular, and 360 Communications (formerly Sprint Cellular) have deployed CDPD in several markets.

A relatively new feature in the development of CDPD is interconnectivity agreements that allow CDPD customers to essentially roam into another CDPD network when outside their home area. GTE, for instance, interconnects with Ameritech and Bell Atlantic NYNEX Mobile. Another good sign for wireless data is that the list of CDPD modem equipment manufacturers is growing. (Some modems can now handle both CDPD and circuit-switched cellular data.)

Most of the initial CDPD applications are in vertical markets, such as transportation, field service, financial transactions, and telemetry. That state of affairs should change as more users take advantage of CDPD's Internet protocol (IP) architecture, which works with a variety of horizontal market programs. CDPD should also improve its position with the introduction of more Smart Phones with integrated modems and PCMCIA card slots.

A hybrid system, called CS-CDPD, allows CDPD users to transmit packet data over existing nationwide circuit-switched networks for nationwide wireless data coverage. Developed by the CDPD Forum, CS-CDPD uses modem pools that connect to special routers known as mobile data intermediate systems.

If a CS-CDPD device is used in an area without CDPD coverage, it notifies the network via a circuit-switched call to the modem pool, which then converts the transmitted packets to the client or server application. Some carriers have packaged the two services to expand and improve their service offerings. CS-CDPD also offers carriers an opportunity to provide national coverage long before it would be available in a stand-alone CDPD format.

While many wireless carriers are incorporating these two technologies into their existing analog networks to support both wireless voice and data communications, two companies, ARDIS and RAM Mobile Data, have been providing data-only services over their wireless networks, mainly for commercial customers.

The RAM Mobile system is based on Mobitex, a nonproprietary protocol developed by Swedish Telecom and Ericsson Radio Systems in 1983 and maintained by the Mobitex Operators Association (MOA). (Anyone can produce subscriber equipment, but only Ericsson can supply infrastructure hardware.) At the end of 1995, RAM Mobile provided wireless data services to more than 40,000 mobile workers for 400 organizations, a 33 percent increase in the number of users since 1994.

There are currently 17 other Mobitex networks up and running throughout the world with more networks planned near-term in 14 other countries. Besides the U.S., Mobitex systems are operational in Canada, the U.K., Finland, Sweden, and Norway. Licenses for two Mobitex networks have recently been awarded to France. The MOA is also working with network operators in other European countries, Latin America, and the Pacific Rim to launch Mobitex services in those areas.

The BellSouth and RAM Mobile partnership, announced in October 1991, includes RAM's mobile data operations in the U.K., as well as cellular, paging, and other holdings in the U.S. BellSouth has contributed complementary paging properties to the joint venture, and provided more than $300 million equity funding to develop RAM's network in the U.S. and the U.K., and to pursue similar opportunities worldwide. RAM Mobile's other U.K. partners are France Telecom, Swedish Telecom, and Bouygues, a large French construction company.

Field service dispatch is a typical application for RAM, but wireless credit authorization, remote meter reading, and energy management by utility companies are being added to the RAM Mobile wireless network. Minneapolis-based National Computer Systems was using paging to call its field engineers on service calls, but now subscribes to RAM, allowing the information management company to access centralized databases and contact other NCS employees directly.

Several manufacturers have introduced or announced Mobitex-compatible portable and palmtop computers, radio modems, and other devices. Ericsson GE introduced a handheld, battery-powered wireless modem. Motorola and AT&T have also announced Mobitex-compatible modems. Mobitex radio modems will be available on a credit card-size PCMCIA module that fits into a slot in portable PCs. (The Personal Computer Memory Card International Association, or PCMCIA, was formed to develop and promote standards for memory cards and connecting slots for mobile computer products.) Motorola, RAM, and several software specialists are also developing application program interfaces (APIs) to make it easier for software packagers and systems integrators to develop new wireless data applications.

More recently, RAM Mobile has sought ways to cover less populated regions by teaming with other wireless network service providers, including satellite services. RAM Mobile has also been able to expand its geographic coverage by acquiring 83 licenses during the FCC's 900 MHz SMR auction. The new licenses expanded RAM Mobile's network to Alaska and Puerto Rico and will allow the data carrier to enhance its overall capacity and support new applications in its existing markets.

The data service provider has also signed reciprocal carrier agreements with U.K. operator GFD and TDR of France to extend the Mobitex wireless packet data network roaming area through the European Union. The agreement means that the U.K., Germany, France, Belgium, and the Netherlands will allow cross-border wireless access to Mobitex services.

ARDIS (Advanced Radio Data Information Service), the other two-way packet-switched data network, was inaugurated in early 1990 by combining the Motorola-developed private mobile data network used by IBM's National Service Division with Motorola's own shared-use radio data network. ARDIS is a proprietary system; manufacturers must license the technology from Motorola. (ARDIS started out as a 50/50 venture of IBM and Motorola, but it is now owned by Motorola.)

ARDIS currently holds FCC licenses for single-channel operation in most of the more than 400 metropolitan service areas (MSAs) nationwide. ARDIS transmitters are networked via dedicated land-based lines. The network transmits at 810–837 MHz and receives at 855–837 MHz.

Motorola designed, produced, and installed IBM's radio data network in 1983. When ARDIS was launched, the network was already being used by 16,000 IBM and 2,000 ROLM service personnel. So far, ARDIS has focused on vertical markets such as the field service industry. Regional trucking services, like moving companies, use the ARDIS wireless data network to control their delivery and pick-up operations. Household Finance Corp., a consumer financial services firm, uses the ARDIS network and GRiD Systems' pen-based GRiDPAD for remote real-time processing of consumer credit applications. Other users include Sears, UPS, and Avis Rent-A-Car.

SMRs ENTER THE MOBILE PICTURE

SMR and ESMR operators are expected to steadily increase their market share of mobile data subscribers from 13 percent of the 1995 market to 21 percent of an even larger market by the year 2000. While a majority of this growth is projected to come from digital SMR subscribers, MTA-EMCI believes that analog SMR operators will also increase their mobile data penetration of current subscribers. Mobile data services will become an increasingly important share of SMR operator revenues over the next five years.

Mobile satellite operators, led by QUALCOMM's OmniTRACS system, which mainly serves the long-haul trucking industry, covered an estimated 12 percent of all mobile data subscribers in service in 1995, but accounted for more than 35 percent of the total service revenues. Other mobile satellite communications operators will begin to emerge in the next several years to expand the market and increase competition.

With the right services and pricing, two-way paging operators could also be strong players in the mobile data market.

PDAs: GIZMOS ON THE GO

Many of the vertical data applications use either notebook computers or personal digital assistants (PDAs), or a dedicated handheld terminal that has been designed specifically for the task.

PDAs got off to a bad start with the introduction of the Apple Newton in 1993. The Newton had all kinds of problems. For one thing, its handwriting recognition system didn't work very well. Worse, the Newton couldn't communicate. But trying to develop products that are small enough and light enough to fit in a pocket, with enough computer processing power to perform many of the tasks available in a laptop, and that can communicate wirelessly, is an obvious challenge.

Even the category is confusing. Although the HP OmniGo looks like something between a subnotebook computer and an electronic organizer, it is usually listed as a PDA. This is probably because, like the Sharp Electronic Zaurus, it features a pen-based liquid crystal display (LCD) screen to enter appointments and for other to-do tasks. The Apple Newton MessagePad and Sony Magic Link fall more easily into the PDA form and function category.

Some PDAs, on the other hand, more closely resemble Smart Phones, like Motorola's Envoy and Marco, the Nokia 9000 Communicator, or the IBM/BellSouth Cellular-developed Simon, which combined a cellular phone with paging, fax, and e-mail features, with a calculator and organizer. The Simon featured a touch-screen cellular phone keypad with a separate data entry keypad and a pen-based scratchpad. But it was too big for even the most ardent early adapter and, at $900, too expensive.

The Pilot, developed by Palm Computing (which is now part of U.S. Robotics, the giant modem maker), may be the best PDA on the market. It actually fits in a shirt pocket and it can do anything a PDA is supposed to do and more, such as exchanging data with desktop computers. But the Pilot doesn't communicate wirelessly.

Newton message pad.

The more highly featured PDAs offer pen-to-text conversion, a PC Card slot, built-in modems, direct connection to a printer, fax and e-mail applications, and on-screen virtual keyboards. For lesser models, these are optional features. However, the one function that consistently sets PDAs apart from other data devices—its pen-based input—may be the biggest drawback in the future success of this product category as a computer/communications tool for consumers. Although PDAs are great for short, simple messages, most computer-centric users require a keyboard. Send a lengthy or somewhat complex message to a PDA user and you may have to wait a while for a response. Peripheral keyboards are available that can be attached to some PDAs (the Apple Newton MessagePad 120 offers one) for heavy data users. They may work nicely in the office, but not when portability is important.

Even with market projections that suggest PDA sales will continue to climb at a reasonably healthy rate through at least 1999, the industry has not been able to generate anywhere near the same excitement with PDAs that it has with cellular phones and pagers.

Despite some reservations, the market research firm Frost & Sullivan projects a 218 percent compound annual growth for PDAs, which is faster than the growth of the portable computer market. An F&S study indicates that PDAs "will continue to have success in niche markets but improvements will have to be made before they become the trillion dollar industry prophesied at its inception. Communication has been dubbed the killer application for PDAs but the market has yet to meet expectations."

Extended battery life is a key issue. Most users want their PDAs to operate for at least several days—a few weeks is better—without changing batteries. Newer models will undoubtedly improve their computing power, adding performance and expanding application opportunities.

PDA terminals are doing well in warehouses, for example, where remote, real-time inventory is important to many companies. Fleet drivers and delivery services like UPS have also found dedicated PDA-type products to fit their needs very well, particularly when the PDA becomes a productivity tool. The most successful models for these applications have been Motorola's Forte CommPad, Norand Corp.'s Pen*Key 6100, and Telxon Corp.'s PTC. All of these units operate wirelessly.

One of the better examples of how PDAs can be used effectively is the Indiana State legislature, whose members' pen-based NEC Ultralite Versa terminals help them keep up with what is going on in the statehouse. Legislators and staff can access information on any legislation from anywhere in the statehouse, and from as far away as a hotel across the street where many legislators stay. They can even remotely change proposed legislation, eliminating the delays and cost of additional printing for updated copies.

Several companies now offer PDAs that work with Japan's Personal HandyPhone System (PHS), providing e-mail, Internet browsers, and schedule organizers. Matsushita Electric's Pinocchio is a prototype PDA with PHS functionality. But without digital data transmission, the model relies on a modem to handle e-mail and faxes. Sanyo has also demonstrated a prototype of a handheld unit with PHS and personal computer functions, along with an LCD touchpad. Sharp Electronic's newest Zaurus PDA connects to wired and digital cellular phones, but no PHS interface. Toshiba's Windows-based palmtop Libretto (priced in Japan at $1,600) connects to any telephone system, including PHS, with the help of a PCM-CIA modem card. NEC's entry into the market is Mobile Gear, a $700 unit designed mainly to handle e-mail.

Engineers at Fujitsu Laboratories have proposed a form of distributed personal agent that would provide PDA users with personalized communications services using different communication media. Called a distributed user assistant for easy telecommunications (DUET), the Fujitsu plan would allow PDA users to transparently select and access the most suitable network available—fixed telephones, desktop personal computers, or mobile phones.

At the same time, new products like the Infopad under development at the University of California could quickly expand the PDA concept into many more applications, including some with highly specialized requirements.

The industry's current position is that PDAs are in an embryonic state, similar to where cellular service was in the mid-1980s. Most wireless data equipment manufacturers will focus on wide-area packet data networking and put other R&D programs on the back burner.

Laptop computers offer similar opportunities to operate wirelessly. One of the better examples is Ricochet, the system created by Metricom, Inc. Backed by Microsoft cofounder Paul Allen, Metricom has linked the entire San Francisco Bay area and is also hooking up Seattle, Washington, D.C., and several U.S. universities. By the year 2000, Metricom plans to cover up to 50 metropolitan areas with wireless access for laptop computer users across the country.

Increasingly, Ricochet subscribers are using their laptops instead of their desktop computers for data access, often to check their e-mail and to access the Internet. The service, which costs about $30 a month for a regular hookup ($20 for students), is also taking the place of a second phone line, freeing the Ricochet subscriber's main telephone line for simply talking.

In actual use, Ricochet users plug pocket-size wireless modems into a laptop computer. The modem sends signals to a small base station, usually hung on a nearby utility pole, which in turn relays signals to leased telephone lines. The signals are routed to Houston and are then linked to the Internet.

Intel Corp. and several personal computer makers and television broad-casters are ready to turn PCs into TV receivers that deliver computer data along with TV programs. Intercast, as Intel calls it, allows broadcasters to transmit data in a portion of the television signal known as the vertical blanking interval at speeds close to four times faster than the fastest modems now used to send PC data. This technology would bring a whole new set of features to television, including text, still images, and graphics. Interactive catalogs could be broadcast during commercials. Hidden clues could be displayed during quiz shows.

Portable PC/TV products can't be far behind.

5

Desperately Seeking Spectrum

Spectrum was never more precious than when Congress passed the Omnibus Budget Reconciliation Act of 1993, giving the Federal Communications Commission (FCC) the authority to auction off licenses for personal communications service (PCS).

In March 1995, the FCC's first auction of "A" and "B" block broadband PCS licenses had been completed with more than two-thirds of the licenses going to three bidders: Sprint Telecom Ventures (then known as Wireless Co., a partnership of Sprint, Tele-Communications Inc. [TCI], Comcast, and Cox Communications) with 29 licenses, AT&T Wireless Services with 21 licenses, and PCS PrimeCo (a joint venture of AirTouch, Bell Atlantic, NYNEX, and US WEST) with 11 licenses.

The very first round of PCS auctions pulled in nearly $8 billion. It was the largest sale of property in history (you can look it up in the *Guiness Book of World Records*), surprising even the most seasoned industry analysts and prompting Washington insiders to suggest that the FCC change its name to Federal Cash Cow.

Meanwhile, the so-called entrepreneurial "C" block auctions, reserved for small businesses, minority- and women-owned companies, and rural telephone companies were postponed by several court actions. Some of the less-well-financed bidders also challenged a number of early "C" block leaders whose winning bids were heavily financed by foreign interests, mainly Japanese and South Korean conglomerates.

The FCC also allowed unlicensed PCS carriers to use the 1,910–1,930 MHz band and, in May 1995, approved a plan for frequency coordination and the migration of existing fixed microwave licensees (mainly public utilities, railroads, petroleum companies, and law enforcement agencies) from the 2 GHz band to make room for the new PCS licensees.

Unlicensed, FCC-certified wireless equipment, such as cordless phones, may be deployed in accordance with FCC rules. To do that, the FCC has issued a set of technical rules, known in FCC parlance as Part 15D, to allow a variety of devices and technologies to coexist in the spectrum with minimal interference.

Originally, the FCC's PCS plan called for a minimum of 40 MHz for unlicensed technologies, split into two equal 20 MHz bands for asynchronous (largely data) and isochronous (largely voice) devices. But after more closely considering licensed PCS and the unique requirements of operating international satellite services, the commission decided to allocate 120 MHz to PCS, creating up to six new competitors in each market and tripling the amount of spectrum allocated to terrestrial commercial mobile telephone services.

Combined with more efficient digital technologies for cellular, PCS, Specialized Mobile Radio (SMR), and other services, the newly assigned spectrum should be sufficient to meet the demand for several years. However, with the emergence of new data or video applications and wireless Internet services, additional spectrum will have to be found.

In most countries, wireless licenses are assigned on a nationwide basis. The key cellular and PCS/PCN networks use the following frequencies:

NMT 450	450 MHz
AMPS/D-AMPS	800 MHz
TACS/NMT-900/GSM	900 MHz
PDC	1,500 MHz
PCN/DCS-1800	1,800 MHz
PCS-1900/CDMA	1,900 MHz

For cellular service, the FCC divided the nation into 734 separate market areas—306 metropolitan statistical areas (MSAs) and 428 rural services areas (RSAs). PCS licenses, on the other hand, are being allocated on the basis of basic trading areas (BTAs), or 51 major trading areas (MTAs). Two of the 30 MHz PCS licenses will be allocated on the basis of MTAs, while the other PCS licenses will be assigned to BTAs. A single MTA may contain several dozen BTAs in the same area. It's confusing, but it works. At least technically.

Under the Omnibus budget bill, cellular, PCS, and most SMR and mobile satellite services now operate under the same rules as Commercial Mobile Radio Service (CMRS) carriers. This came about because of disparities in the regulation of cellular and SMR carriers, which would become increasingly significant as the deployment of new technologies allowed SMR carriers to compete with cellular carriers.

To ensure that carriers don't sit on their licenses, the FCC required broadband PCS licensees to build systems that provide coverage to one-third of the population of their service area within five years after they're licensed, and two-thirds of the population in their market within 10 years. Obviously, the carriers' incentive is to get on the air as quickly as possible to begin generating a return on their investment, although financing that level of expansion has been challenging for many new PCS licensees, as well as time-consuming.

The FCC has also sought to maintain competition by limiting concentration of ownership. For example, the commission has not permitted cellular carriers to obtain more than 10 MHz of PCS spectrum in markets where they already own cellular licenses. Similarly, local exchange carriers (LECs) have been restricted to bidding on 10 MHz blocks of spectrum in their service areas, not the larger 30 MHz blocks. Also, carriers may not have more than 45 MHz of cellular, PCS, and SMR spectrum in any given market.

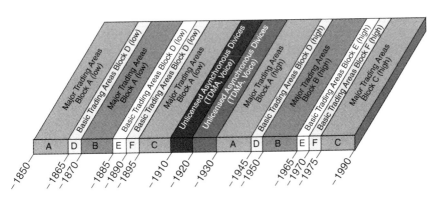

PCS band, 1,850–1,990 MHz.
Source: MTA-EMCI.

NEW FCC APPROACH

This is a major shift in the way the FCC has historically assigned spectrum. The first cellular licenses were allocated by comparative hearings; the commission selected licensees from applicants based on very detailed and highly technical proposals. But this proved to be a very demanding and time-consuming process for the commission, which regulates virtually all telecommunications in the U. S., including the broadcasting and cable television industries. With the passage of the Omnibus Budget Act, Congress authorized the FCC to use auctions primarily to reduce speculation on licenses by selling them at a huge profit, but also to help reduce the national debt. Money generated from the auctions went straight into the U.S. Treasury.

But this is where it gets a little complicated. The PCS spectrum allocated to unlicensed devices is, like all other PCS spectrum, already crowded with existing microwave licenses. By assigning PCS to the 2 GHz band, the FCC is forcing these point-to-point microwave services, such as public utilities, railroads, petroleum companies, and law enforcement agencies, to move their communications operations to higher frequencies. An organization called UTAM, Inc. (Unlicensed Transition and Management for Microwave Relocation), is charged with relocating the microwave incumbents by administering a system of clearing fees paid by manufacturers using the band. UTAM estimates that the cost to manufacturers to move the microwave users will run at least $67 million.

UTAM's plan calls for a fee on manufacturers based on the number of units they sell. These "commissions" are placed in a fund to relocate the existing microwave links into new spectrum. It could take years before the band is fully cleared and total coverage can be assured in the unlicensed band. Therefore, UTAM has been working to provide for the coexistence between the existing microwave systems and the new PCS systems. While not providing ubiquitous coverage, the UTAM plan at least allows for the deployment of unlicensed PCS systems in a shorter period of time.

This has not been a smooth process. Some of the incumbents, particularly law enforcement and other public safety organizations, simply don't want to move. Other microwave incumbents have delayed the process by demanding more money to move to higher frequencies, drawing threats from the FCC because they could easily slow the deployment of PCS, which needs to get "on the air" as quickly as possible to begin recouping the billions of dollars already spent on licenses.

THE NEED FOR NEW SPECTRUM

Studies by the National Telecommunications and Information Administration (NTIA), which oversees the federal government's use of radio spectrum and serves as spectrum policy adviser to the White House, indicate that the U.S. could fall about 204 MHz short of its spectrum needs for land mobile services over the next 10 years. According to the NTIA, "Commercial demand for access to the radio spectrum has already outstripped spectrum availability in many major U.S. markets."

The NTIA's study, which relied heavily on independent market research and equipment manufacturers and carriers, suggests that the demand for new spectrum will be led by cellular, mobile satellite, and in-vehicle navigation systems. But it also points to a need for new spectrum for transaction/decision processing, facsimile, slow- and full-motion video, and remote file-access applications.

Based on independent market projections and interviews, the NTIA has estimated that it would take four times the 50 MHz currently available for cellular service in the U.S. (or another 150 MHz) to meet the country's cellular service demands to the year 2006. But new spectrum-efficient digital systems will more than likely be able to provide at least three times the channel capacity of current analog systems. As a result, only an additional 33 MHz may actually be required to satisfy cellular needs over the next 10 years.

The need for spectrum may be the greatest in the unlicensed bands, where cordless telephones, garage door openers, on-site wireless data collection and point-of-sale devices, office private branch exchange (PBX) systems, and wireless computer networks operate. All of these devices require relatively low power and none of the trappings of wired systems such as the construction of expensive and complex infrastructures. But they are in high demand. Cordless phones now account for two out of every three new phone installations.

In its original petition for an unlicensed PCS allocation filed with the FCC in 1989, the Wireless Information Networks Forum (WINForum), an industry technical forum set up to promote wireless information networking, pointed out that at least 40–80 MHz of spectrum was immediately required for voice and data products.

WINForum says that auctions will not work for unlicensed services. Unlicensed allocations, according to WINForum, provide no right to exclusive use by any entity, or for their profit. WINForum also believes there is no practical way to evaluate comparative value for entities sharing the use of these frequencies because there is no means for metering airtime usage in an ad hoc network environment. "There is no equitable means to assess interested parties for contributions in advance of allocating such spectrum and, by definition, no further licenses (that could be auctioned) are needed," WINForum wrote to the FCC.

Meanwhile, Congress has been considering spectrum reform legislation that would loosen up some government- and privately held frequencies for commercial use. The Pentagon and broadcasters are obvious targets. New legislation may also lead to the consolidation of management of federal government spectrum (currently the responsibility of the NTIA) and a requirement for increased frequency coordination by industry.

NOW, THE NII/SUPERNet

With IEEE 802.11 not expected to be formally accepted as a wireless local-area network (WLAN) standard until early 1997 and the High-Performance Radio Local-Area Network (HIPERLAN) not yet approved as the European local-area network protocol, the FCC has said that it will support the allocation of spectrum for a new category of unlicensed equipment for the next generation of digital wireless products and services.

The FCC has said that it will consider making available 350 MHz of spectrum at 5.15–5.35 GHz and 5.725–5.875 GHz for NII/SUPERNet (National Information Infrastructure/Shared Unlicensed Personal Radio Network)—a short-range, high-speed wireless digital network that could support multimedia applications and eventually lead to the creation of new WLANs.

The commission plans to regulate NII/SUPERNet devices under Part 15 of its rules. Unlicensed Part 15 status would facilitate spectrum reuse and provide protection to incumbent and proposed primary operations.

Leading the charge for NII/SUPERNet are Apple Computer, WINForum, and UTAM. Formerly known as the Unlicensed PCS Ad-Hoc Committee for 2 GHz Microwave Transition and Management, UTAM is charged by the FCC with managing and financing the relocation of fixed microwave users in the 2 GHz band to make room for PCS.

Apple's proposal for NII/SUPERNet is similar to WINForum's in the types of applications they envision. It even embodies some of the HIPER-

LAN developments in Europe. But the Apple concept focuses on "community networking" with heavy emphasis on high-speed data communications for schools, libraries, businesses, and other local or regional community organizations.

But the commission is also considering a separate low-power wireless LAN for more commercial uses, which could get the jump on the SUPER-Net proposal.

WINForum wants NII/SUPERNet to offer asynchronous transfer mode (ATM) and broadband integrated services digital network (ISDN) type functionality, but in a wireless architecture. The key to the NII/SUPERNet is that it features a broad range of applications. Medical staffs could, for example, improve their efficiency with on-the-spot, real-time access to patient data, including X-ray and MRI images, video recordings, medical charts, and other records.

WINForum's original request contemplated use of the 5.10–5.35 GHz band for SUPERNet, frequencies the Federal Aviation Administration (FAA) had hoped at one time to use for microwave landing systems. WINForum eventually revised its original spectrum proposal to take into account the requirements of high-speed information, imagery, and video and emerging resource-intensive applications using multimedia.

Although NII/SUPERNet seems to complement HIPERLAN developments in Europe, the FCC has been urged to act quickly on the new network proposal to remain competitive with the faster HIPERLAN and other emerging communications concepts. While the IEEE 802.11 standard for WLANs in the 2.4 GHz industrial-scientific-medical (ISM) band was still pending in mid-1996, industry groups in Europe and other regions view HIPERLAN as a newer, faster technology. The Conference of European Post and Telecommunications (CEPT), the European organization of PTTs established to provide common, industry-wide standards for telecommunications, recently allocated 100 MHz of spectrum in the 5.1–5.25 GHz band exclusively for HIPERLAN and tentative protocols for sharing are already approved.

INCREASED EFFICIENCY

Excluding PCS-related networks, the NTIA expects the total number of point-to-point microwave systems to remain constant or decrease slowly under competitive pressure from optical fiber. At the same time, fixed networks used to support military test and training ranges will slowly begin to be converted to fiber—partly to reduce the cost of maintaining old analog microwave links and partly to increase bandwidth. The replaced microwave links may continue to be used for backup and for low-priority applications.

Most of the increased efficiency for satellite services will come from video compression techniques, but that could change over time as the demand for video programming increases.

The NTIA has already relinquished the federal government's primary status on 125 channels in the 220–222 MHz band, channels that were allocated equally to the federal government and to private users. The NTIA has recommended that the FCC auction these bands for commercial mobile services.

Over the long term, telecommunications policy planners will have to address other spectrum requirements, such as the five proposed U.S. low-earth-orbit (LEO) mobile satellite systems that are required to share 33 MHz in the 1,610–1,626.5 MHz and 2,483.5–2,500 MHz frequency bands. That won't be enough spectrum if the demand for this service catches up with market projections. The NTIA has estimated that an additional 60 MHz of spectrum may be required for mobile satellite services over the next decade. But with mobile communications satellite service not expected to be operational on a broad scale until at least late in 1998, regulatory agencies like the FCC and Geneva-based International Telecommunications Union (ITU) don't seem to be in a hurry to make any changes.

Organizations such as the Association of Public Safety Communications Officials (APCO) and the FBI worry about the availability of spectrum for their needs and have been developing plans for their spectrum requirements through the year 2010. Most of their attention is on new wideband applications. Transmitting fingerprints, mug shots, building diagrams, slow- and full-motion video, transaction/decision processing, and other high-speed data will require wider bandwidths than are currently available with conventional mobile communications systems.

The reality check for public safety agencies came in 1993, when bombs killed several people and seriously injured hundreds of others in New York's World Trade Center. Firefighters could not communicate with other rescue workers one floor away—even though they had portable phones. The Oklahoma City bombing was even worse. Local police, fire, and ambulance services operated on different radio frequencies, as did the U.S. Secret Service and U.S. Marshals Service. They could only talk to each other through a central dispatch command center, seriously slowing their search and rescue mission and the investigation of the bombing itself. The New York and Oklahoma City incidents led to the formation of a Public Safety Wireless Advisory Committee (PSWAC), which has been charged with identifying the operational needs of the public safety community, including basic voice, data, and Enhanced 911 services, as well as new wide-area, broadband telecommunication services.

Spectrum options for land mobile are generally limited to bands below 3 GHz because of propagation considerations. There aren't a lot of options, but one might be to transfer more bands from government use to commercial or public safety applications over the next several years. The NTIA has already shifted 95 MHz of spectrum to the FCC for assignment to the private sector and may release a total of 235 MHz for commercial use.

BROADCASTERS GET THEIR WAY

If there's a loose cannon here, it is broadcasters. Today's analog National Television Standards Committee (NTSC) television standard is considered to be very inefficient in its use of UHF spectrum. Each UHF channel allocation places restrictions on the use of 16 other UHF channels in the surrounding geographic area. These restricted channels represent wasted spectrum; they're not used because of the potential for interference to existing television service. The FCC has been working with the industry groups to use the restricted channels for a new digital television service—advanced television (ATV)—which would fit into an existing 6 MHz channel format. By loaning these 6 MHz "taboo" channels to existing broadcasters, each broadcaster can upgrade its service to full digital high-definition television (HDTV).

Several approaches have been considered. One would be to loan the 6 MHz channels to broadcasters, allowing each of them to upgrade their service to full digital HDTV. Another is for the FCC to auction new digital channel assignments to broadcasters. At some point the broadcasters would return their "old" analog channels to the FCC, which might then reassign that spectrum to other uses.

The broadcasters don't like either plan and have lobbied heavily against any plan that would require them to pay for the digital channels—even though they will use some of those channels to develop new advanced digital services in what by then will be a well-defined and highly developed market. According to Deloitte & Touche Consulting Group's Telecommunications & Electronic Services, "when the broadcasters give back their existing channels early in the next decade, the FCC will undoubtedly put the channels on the market, opening up vast new supplies of spectrum for wireless services." When that happens, broadcasters may have to compete with an entirely new set of players in data communications services.

There's also a tremendous opportunity in broadcasters' FM radio subcarrier networks for carrying wireless data and paging services. Widely used by European paging networks, FM subcarriers will likely get more attention from Panasonic's use of the technology in its paging system, and from agreements between Data Broadcasting Corp. (which provides sports news and on-line stock market information to more than 30,000 subscribers nationwide) and Mikros System Corp. (which will supply proprietary bandwidth optimization technology to increase DBC's data rate—currently at 9,600 bps—by a factor of 5 or more). The additional speed will allow DBC to pump more information through its network. DBC and Mikros also plan to codevelop high-speed wireless data equipment.

WORKING THE WRC

Globally, the U.S. has focused on advancing new international mobile communication satellite systems with most of the important work taking place at the ITU-sponsored World Radiocommunications Conference, usually held in Geneva.

The satellite industry tends to have a somewhat different view of the world than other carriers. For example, it opposes the broad allocation of spectrum, believing that the process would likely be biased against satellite services while favoring terrestrial services. The satellite industry is also concerned that auctions to license satellite systems might encourage satellite operators to look to foreign administrations for sponsorship, which could lead to the U.S. ceding regulatory leadership to other regulatory and multinational organizations. Since U.S. satellite manufacturers hold more than 70 percent of the world market for commercial communications satellites, their concerns are not taken lightly.

In the past, when there was less demand for spectrum, particularly in the higher frequencies, the FCC was able to plan ahead and set aside certain frequencies for future satellite development. This is how portions of the L band were allocated for mobile satellite service (MSS) in the 1970s and the Ku and Ka bands were allocated for satellite services well in advance of any specific proposed use. This process also worked well in the planning and technology development by the industry. Competition and the need for new services—and new spectrum—has changed all that.

Since the U.S. is the largest market, the U.S. State and Commerce Departments, the FCC, and even the Pentagon, which has a significant stake in how spectrum is assigned around the world, believe the U.S. should take the lead on new spectrum allocations. This would allow American manufacturers to bring products to market sooner and ultimately enjoy a head start in global markets. The rest of the world would have to follow the U.S.

The ITU's biennial World Radiocommuniction Conferences have become critical to the international spectrum allocation process. At the 1992 WRC, for example, the U.S. successfully proposed the basic allocation for the Big LEO systems and for other, new MSS systems.

In some circumstances, satellites can share spectrum with other services. In the C band, for instance, the Fixed Satellite Service (FSS) shares frequencies with microwave licensees by coordinating the siting of their facilities. The Big LEO systems will share frequencies with radio astronomers by controlling emissions from mobile equipment when in the vicinity of certain observatories that monitor frequencies in the Big LEO uplink band. Little LEO systems, which plan to offer a variety of data services, but not voice, will likely share their uplink frequencies with land mobile and other terrestrial services.

Unfortunately, the Little LEOs came up short at WRC-95. The working group of the Conference Preparatory Meeting, which develops proposals in advance of the WRC meetings, has estimated that Little LEOs would need 7–10 MHz below 1 GHz by the year 2000. (WRC-92 allocated 3.45 MHz of spectrum to mobile satellite networks operating below 1 GHz.) A number of studies are planned to determine what frequencies might be assigned to Little LEOs, but plans to launch new systems have effectively been delayed.

Many satellite services depend on intraservice sharing of spectrum. Two Little LEOs will share the same frequencies by using a dynamic band-scanning technique designed to prevent their systems from operating at the same time on the same frequencies. Satellite operators also use different polarizations to share spectrum.

As of mid-1996, only one Little LEO—Orbcomm—has been launched and is operating in a band already allocated for Little LEOs. Others may be licensed by the FCC, but they will not have the spectrum to operate. Applicants include CTA Commercial Systems, which has proposed a 38-satellite constellation of its GEMnet LEO satellites to offer tracking, monitoring, and messaging services; E-Sat, Inc., a partnership between Echostar Communications Corp. and DBSI, which plans to launch and operate a 6-satellite LEO network; Final Analysis Communications Services, Inc., which has developed a 26-satellite system; and GE American Communications, which hopes to launch and operate a 24-satellite Little LEO network.

If the Little LEOs are ever going to get off the ground and become a viable, competitive service, they will have to receive a lot more attention at the next ITU-sponsored WRC conference in Geneva in 1997.

The ambitious Big LEOs, such as Motorola's Iridium system, which will operate above 1 GHz and offers voice, data, and advanced messaging services, have been much more successful in winning spectrum assignments. The FCC has already licensed three nongeostationary mobile communications satellite systems to operate in the 1.6 GHz and 2.4 GHz bands: Iridium, TRW's Odyssey, and the Loral/QUALCOMM Globalstar system. Both bands were approved for Big LEOs at the WRC-92 meeting.

All three were assigned the feeder-link frequencies they requested—300 MHz. They were also virtually promised another 100 MHz in 1997 for their worldwide mobile satellite systems. The allocation, coupled with the user-link frequencies allocated in 1992, completes the spectrum allocation requirements for the entire Big LEO system. The decision enables the Big LEOs to complete their systems as originally designed and allows them to seek licenses in countries around the world.

Meanwhile, radio astronomers fear that the Big LEOs are a threat to their work. The problem is that the frequency assigned to the LEOs by the ITU—1,610–1,626.5 MHz for earth-to-space transmissions—is very close to the 1,610.6–1,613.8 MHz band used by radio astronomers to map galaxies and detect interstellar clouds. Emissions from handheld phones are small, but radio astronomers worry that they may still interfere with the highly sensitive radio telescopes used in their research. Global positioning system (GPS) satellites also concern the astronomers because they operate at a much higher orbit than the mobile communications satellites, and the number of earth-bound GPS receivers is growing at a rate of 30 percent a year.

Help may be on the way. Motorola and the American Mobile Satellite Co. have agreed to work with the U.S. National Radio Astronomy Observatory and its European counterparts to try to resolve the interference issue.

Although the WRC-92 conference allocated service-link spectrum for mobile communications satellites, it did not assign feeder-link spectrum, which is necessary before these systems can be introduced. Feeder-link spectrum for nongeostationary communications satellites have been proposed in the frequency bands allocated to, or proposed to be allocated to, the fixed satellite service.

WRC-95-designated frequency bands for feeder-link networks will operate on an equal basis with geostationary satellites and other radio communications services. The requirements identified for nongeostationary mobile satellite feeder-link networks prior to WRC-95 were 200–400 MHz in the 4–8 GHz and 8–16 GHz range and 200–500 MHz in the 16–30 GHz range. Unfortunately, the ITU's own projections indicate that spectrum currently available for mobile communications satellites will not support anticipated requirements, which the ITU estimates will range from 150 to 300 MHz by the year 2005.

Teledesic Corp., the satellite venture headed by Microsoft's Bill Gates and McCaw Cellular Communications' (now AT&T Wireless Services) Craig McCaw, won a rule change at the WRC in Geneva that gave LEO satellites primary status in 400 MHz of bandwidth at 28 GHz (the Ka band), with the promise of another 100 MHz in 1997. Rather than operate its own satellite service, Teledesic plans to sell all of its 500 MHz for a broadband digital fixed-data service using a network of 840 satellites, beginning in 1999.

Because the 28 GHz band is already in use, the WRC has recommended spectrum-sharing studies for these frequencies. The band is being used by fixed satellite services and NASA's Advanced Communications Satellite (ACTS) program. Launched in 1993, ACTS has focused on studying antennas with high-powered electronically hopping multiple beams, on-board processing and switching, and Ka-band receivers and transmitters.

Teledesic intends to operate in the 28 GHz band, which the FCC has already approved for Local Multipoint Distribution Service (LMDS), a new "wireless cable" service being developed for multimedia and video distribution, as well as telephony and data services. The FCC was expected to auction licenses for LMDS in the 27.5–30 GHz band by the end of 1996.

Internationally, the 27.5–29.5 GHz band continues to be allocated on a primary basis for FCC earth-to-space links, but under the FCC's LMDS plan, emerging global satellite communications services will share the 28 GHz band with the wireless cable television industry.

The rush to jump into the Ka-band market began late in October 1995 when AT&T asked the FCC for permission to build and launch a multi-billion-dollar global satellite network that would allow computer users to bypass local telephone networks and connect directly to the Internet using 2-foot-diameter satellite dishes. AT&T has been using satellites to supplement its fiber-optic, computer-wire, and undersea cable networks for years. But AT&T's proposed satellite system, which it calls Voicespan, would both enhance AT&T's global market share and allow the company to compete with the regional Bell telephone operating companies.

Similar proposals have been filed by at least 15 organizations, including AT&T, Lockheed Martin, Hughes, Loral, and TRW, that want to use the high-frequency Ka band to offer new satellite communications services, whereas others, like Hewlett-Packard, Harris Farinon, Alcatel, NEC, Stanford Telecom, M/A-COM, and Rockwell International, plan to build and sell LMDS hardware.

WIRELESS HITS THE ROAD

Intelligent transportation systems is another important market segment for wireless communications. Some of the newer electronic toll tag systems have been moved up to the 2,450 MHz band where certain European Electronic Toll and Traffic Management systems operate. Because of competing requirements at these frequencies, the 5,850–5,925 MHz band may have to be used to cover future growth and international operability.

The FCC also is making available 1.2 GHz of spectrum above 40 GHz for the exclusive use of vehicle radar systems. As a result, the FCC will relocate existing amateur radio service users.

This is the second time the FCC has dedicated a part of the spectrum to so-called intelligent transportation systems (ITS)–related applications. The first time occurred in 1995 and covered vehicle location and short-range communications.

In the recent action, the FCC is making available a total of 6.2 GHz of spectrum in the 46.7–46.9 and 76–77 GHz bands for vehicle radar systems in conjunction with ITS, and 59–64 GHz band for other, general unlicensed devices. The commission rejected the use of the 60–61 GHz band for vehicle radar systems that was under consideration by the Japanese Ministry of Posts and Telecommunications. The FCC is promoting the 59–64 GHz band as the only contiguous 5 GHz of bandwidth available, or ever likely to be available, for short-range broadband communications. In other words, the FCC does not want vehicle radar systems to share spectrum with other services (presumably because of the potential for interference). The 60–61 GHz band is not under consideration for ITS in Europe.

WHO'S WATCHING THE STORE?

As a major user and regulator of wireless telecommunications services (defense, public safety, emergency preparedness, and space communications) the federal government continues to play an important role in just about anything that happens in the field. Recently, however, government agencies have shown more interest in adopting commercially available systems and services to meet their own growing requirements. Several special-interest groups have been formed to stay on top of emerging technologies, and to coordinate and represent the federal government's interests in telecommunications.

In 1992, for example, several government agencies formed the Federal Wireless Users Forum (FWUF) to sponsor workshops on wireless services in order to enhance technical understanding and define the needs of the agencies for wireless systems and services.

Other federal government groups include:

- The Federal Law Enforcement Wireless User Group (FLEWUG) is composed of more than 60 representatives from several federal agencies and is open to state, country, and local agencies.
- The Joint Federal Wireless Review Office (JFWRO) was formed in January 1995 to consolidate federal wireless programs by eliminating duplication and incompatible systems, and to promote the use of more spectrum-efficient systems by government agencies. Although its duties seem fairly well defined, JFWRO has created jurisdictional concerns at the NTIA and the Independent Radio Advisory Committee (IRAC), which is made up of 20 federal agencies that use wireless communications. Historically, the NTIA has drawn on the spectrum management expertise of IRAC and the

Spectrum Planning and Policy Advisory Committee, a private-sector group with federal government members.

The executive branch has also established two committees to address specific wireless communications issues. The Federal Wireless Policy Committee, established in 1993, focuses on wireless policy, and the Untethered Networking Working Group examines the impact of wireless technologies on the National Information Infrastructure and the Global Information Infrastructure.

FUTURE APPLICATIONS

Wireless networks are also being implemented at some of the higher frequencies, with the FCC already approving three new bands. Two of them, at 47 and 76 GHz, are for vehicle crash-avoidance radar. The third, between 59 and 64 GHz, will most likely be adopted by computer manufacturers for wireless local-area networks, which need the capacity of the higher bandwidths.

Higher frequencies will allow new services and applications to be brought to the mass market more rapidly and at lower cost, partly because the higher channels require less power and therefore less infrastructure. The inevitable reduction in the cost of digital signal processing (DSP) and other integrated circuit technologies will also play an important role in advancing more spectrum-efficient systems.

Anticipating continued technological innovations, the FCC can be expected to do what it can to reserve additional spectrum for future applications. Near term, the increased integration of data services with voice communication and emerging multimedia services will create a demand for new frequency allocations. But forecasting the demand for new and emerging applications will not be easy.

Over time, telecom manufacturers believe that video, high-quality graphics, and data services will become as important as voice communication is today. The need for additional communications channels has become a major challenge for equipment manufacturers and regulators. The development of faster semiconductor technology is making it possible to build communications systems with wider bandwidths that operate at higher frequencies. At the same time, digital technologies like TDMA,

CDMA, and GSM are capable of increasing the data-handling capacity of a channel by sending more bits per second in a fixed amount of spectrum.

The FCC has been wrestling with a number of important policy issues. Among them: How should spectrum be allocated? What technological issues will improve current systems? What regulatory approaches would speed the development of interoperable equipment? What new regulations need to be developed to increase the efficiency of public safety communications?

The commission may consider establishing standards to protect U.S. interests. One concern is the increasing global dominance of the European GSM system, which has spread to the U.S. and is already competing with disparate and incompatible U.S. standards.

Of course, the U.S. Congress will continue to be involved as it develops spectrum reform legislation that attempts to deregulate much of the nation's airwaves. However, any changes will be hotly debated in industry and government and little serious action is expected until well into 1997, possibly later.

Any future expansion in mobile services will require technological advances that permit the economic use of higher-frequency bands for the reallocation of spectrum for new services or from other services.

A few things to keep in mind: The higher the frequency, the less distance can be covered. Also, as a general rule, the higher the frequency, the better the coverage in buildings. Another essential fact is that the higher the frequency, the more expensive the technology. It's also a fair bet that the introduction of new applications and services will outpace the FCC's ability to assign frequencies to these new services. In time, technology and market factors may reduce the significance of spectrum allocation debates. That day, however, may be well into the future.

6

Standards and Interoperability

There is competition and there are wars. The ongoing battle between digital cellular and personal communications service (PCS) standards is probably somewhere in between. Federal Communications Commission (FCC) Chairman Reed Hundt may not be far off when he called it "a war of religious dimension."

At least seven standards have been approved in the U.S. for emerging digital cellular and PCS communications services. Europe and much of the Asian/Pacific region are working with their own standards. The stakes are high. By taking an open architecture approach, cellular system operators will be able to purchase standardized components from different vendors. But which architecture?

At first glance, Europe seems to have done a better job than the U.S., picking a set of digital standards for specific applications and enforcing those standards. Japan has its own digital transmission and interoperability schemes. The FCC could have selected a single standard for digital cellular and PCS in the U.S., but Hundt says, "If you're for competition, you have to trust that markets will settle even fights about standards." This remains to be seen.

The key digital technologies competing in the U.S. are time division multiple access (TDMA), code division multiple access (CDMA), and the TDMA-based Global System for Mobile Communications (GSM).

TDMA works by dividing time at a channel frequency into parts and assigning different phone conversations to each part. For a time, TDMA, which claims three times the transmission capacity of present-day analog cellular systems, had virtually total industry support, mainly because it was a fairly well understood technology and was the only game in town.

The Cellular Telecommunications Industry Association (CTIA) even adopted TDMA as its digital transmission standard in 1990 and continues to support the system. But being the "older" technology in today's fast-moving wireless personal communications market has actually slowed its acceptance while other more advanced digital cellular systems have won support from cellular carriers and equipment suppliers.

There is IS-54, the dual-mode North American (analog and digital) standard, and the newer IS-136 version of TDMA, which is more flexible and offers more features than IS-54. Essentially, IS-136 is backward compatible with IS-54, but it supports several enhanced features. Along with an outdoor macro-cellular service, IS-136 offers a range of digital in-building (including fixed wireless) services, seamless networking with Advanced Mobile Phone System (AMPS)–based analog systems, dual-mode analog and digital service at 800 MHz, as well as dual-band service (digital at 800 and 1,900 MHz). IS-136 also features a sleep mode to boost battery life, and alphanumeric paging.

Several TDMA equipment manufacturers have organized under an umbrella group called Universal Wireless Communications (UWC) to promote TDMA IS-136. UWC's member companies include AT&T Wireless Services, Ericsson Radio Systems AB, Northern Telecom, Rogers Cantel, Hughes Network Systems, Tandem Computers, Celcore, PCSI, Motorola, Octel Communications, LanSer Wireless, and Lucent Technologies.

Initially, these companies plan to leverage their existing national 800 MHz coverage and get to market as quickly as possible with IS-136, by targeting high-end users, promoting new services, and adding new subscribers by marketing a "digital" brand. The UWC marketing strategy is to convert current AMPS subscribers to TDMA. The UWC's goal is to create a unified wireless network platform, giving subscribers seamless access so they can use their phone as an extension to the office phone system when they are at work, as a conventional mobile phone when they are out, and as their main home phone when they are at home.

This is the plan. But can it be done? UWC members demonstrated their technology at a conference in Singapore in mid-1996, which they believe indicated that their concept is possible using existing TDMA technologies. The demonstration highlighted three features of the IS-136 service: One is location identification; as people walked between the various cells, from one office to another, from an office into the public environment, or vice versa, a text display on the phone constantly reminded them to which service they were connected. Another is text messaging, which can be received via the phone display, and can be sent from any fixed or wireless phone. The third IS-136 feature is call delivery in which incoming calls to people's business numbers will reach them at whichever company site they happen to be working.

In the Singapore demonstration, the service was configured so a person's wireless phone rings first. The phone rings whether the person is on the company site or in the public access area. If the phone is not answered within three rings, the call is diverted to the person's fixed extension in the office. If that is unanswered within three rings, the call goes to the person's voice mail.

AT&T Wireless Services launched its IS-136 service at the end of 1995, installing its first system at Perot Systems in Texas. The Perot "network" includes two Ericsson microcells with a custom-designed, in-building radio antenna distribution system. The network is connected to the Perot private branch exchange (PBX) so that calls made to an employee's desk phone also ring his or her Nokia cellular phone and vice versa. Perot employees can also receive text messages over their cellular phones.

Ericsson has designed its IS-136 network to allow simple system integration by upgrading the software of the existing IS-54–based D-AMPS networks to IS-136, providing an evolutionary path for AMPS operators to migrate to digital cellular service and then to PCS. D-AMPS has been specified in the U.S. as a dual-band PCS technology for the 800 and 1,900 MHz bands. It enables dual-band terminals to support 800 and 1,900 MHz with national and international roaming.

Ericsson, which is totally committed to TDMA, expects the technology to grow worldwide from almost 2.5 million subscribers in mid-1996 to 7.5 million at the end of 1997.

Hughes Network Systems has developed and introduced its own version of TDMA—extended TDMA—which is fully compatible with TDMA, but employs digital half-rate voice coding and a digital speech interpolation (DSI) technique to increase subscriber channel capacity. DSI takes advantage of the quiet times that occur naturally in speech and, in real time, assigns the active speech from conversations to an inactive channel, doubling the call-handling capacity of each channel. According to Hughes, the combination of half-rate voice coding and DSI results in 15 times the analog capacity when trunking efficiencies are taken into account. Hughes claims E-TDMA improves transmission quality over analog because with DSI a person's speech "hops" between different radio paths, actually re-

Standards Table

	Analog Cellular		Digital Cellular					Analog Cordless	Digital Cordless/PCS/PCN				
	AMPS	ETACS	IS-54	GSM	IS-95	PDC	R-CDMA	CT	CT2	UD-PCS	DECT	PHP	DCS-1800
Frequency Rx Band MHz Tx	869–894 824–849	916–949 871–904	869–894 824–849	935–960 890–915	869–894 824–849	940–956 810–826 (& 1,477–1,501 1,429–1,453)	869–894 824–849	Varies by country	864/868		1,880–1,990	1,895–1,907	1,805–1,880 1,710–1,785
Radio access method	FDMA	FDMA	TDMA/ FDMA	TDMA/ FDMA	CDMA/ FDMA	TDMA/ FDMA	CDMA	FDMA	TDMA/ FDMA	TDMA/ FDMA	TDMA/ FDMA	TDMA/ FDMA	TDMA/ FDMA
RF channel	30 KHz	25 KHz	30 KHz	200 KHz	1.25 MHz	25 KHz	40 MHz	20 KHz	100 KHz	700 KHz	1.728 MHz	300 KHz	200 KHz
Modulation	FM	FM	π/4 DQPSK	GMSK 0.3	BPSK/ OQPSK	π/4 DQPSK		FM	GFSK		GFSK	π/4 DQPSK	GMSK 0.3
Channel rate	—	—	48 Kb/s	270.8 Kb/s	9.6 Kb/s	42 Kb/s	20 or 40 Kb/s	—	72 Kb/s	514 Kb/s	1.1 Mb/s	384 Kb/s	270.8 Kb/s
Number of RF channels	832	1000	832	124	10	1600	126	10, 12, 15 or 20	40	10	10	300	750
Voice channel per RF channel	1	1	3	8	20–60 per 3 sector	3		1	1	10	12	4	16
Duplex voice channel size	60 KHz	50 kHz	20 KHz	50 KHz	—	20 KHz	—	40 KHz	100 KHz	70 KHz	144 KHz	50 KHz	
Voice bit rate	—	—	8 Kb/s	13 Kb/s	8 Kb/s	8 Kb/s	16 Kb/s	—	32 Kb/s	32 Kb/s	32 Kbp/s	13 Kb/s	
Phone xmit pwr max/avg mW	600/600	600/600	3,000/200	2,000/125	200/6	3,000/200	100/1	510	10/5	100/10	250/10	250/10	250/10
Max cell	>32 km	>32 km	>32 km	32 km	>32 km	>32 km	450 m	100 m	100 m	500 m	500 m	500 m	500 m

Source: Analog Devices.

Notes: **AMPS** is the current North American analog system. **TACS** is the alternative analog system (as used in Britain, China, and others). **IS-54** is the digital system providing service within the currently allocated AMPS channels. **GSM** is the European digital cellular telephony standard. **QCDMA** is the Qualcomm CDMA. **PDC** is the Japanese personal digital cellular (formerly JDC). **RCDMA** is an alternative (wideband) scheme proposed by Rockwell. **CT** is the first-generation cordless phone standard using narrowband FM with 10 channels. **CT2** is a second-generation digital cordless system that supports telepoint service (i.e., a handset can initiate but cannot receive calls). **Universal Digital PC (UD-PCS)** is an alternative system that has been proposed by Bell Communications Research (Bellcore) for in-building cordless phone service. **DECT** is the European digital cordless telephone standard. **PHP** is the Japanese Personal HandyPhone System. **DCS-1800** is a variant of GSM, with lower power and higher carrier frequency, intended for PCS applications.

ducing the effects of interference or fading. Hughes is producing its own E-TDMA handset.

TDMA's strongest competitor is code division multiple access (CDMA), a proprietary system developed by San Diego–based QUALCOMM, Inc. CDMA uses a spread-spectrum technique originally used by the military to scatter signals across a wide frequency band, making transmissions difficult to intercept or jam. Other than its secure communications qualities, its most important technological claim to fame is that it offers at least 10 times the channel capacity of present analog cellular systems. CDMA also has an inherent "soft handoff" capability, which reduces the number of dropped calls when passing from one cell to another.

There's also a Broadband CDMA. In 1992, InterDigital acquired SCS Mobilcom, Inc., and SCS Telecom, Inc. With the acquisition, InterDigital picked up the SCS companies' broadband CDMA technology and its portfolio of 30 patents. Essentially, B-CDMA works by sharing the spectrum with the existing cellular telephone bands. Like QUALCOMM's CDMA, InterDigital claims B-CDMA provides additional network capacity and improved voice quality. InterDigital hasn't signed on nearly as many strategic partners as TDMA or CDMA, but it does have agreements with Siemens AG and Samsung Electronics that call for the three companies to jointly develop B-CDMA telephone handsets and then sell them under each company's own brand name.

Omnipoint Corp. has its own digital technology—IS-661 Composite CDMA/TDMA. Optimized for PCS, IS-661 is designed to take advantage of the best features of CDMA and TDMA as well as frequency division multiple access (FDMA), in which each conversation gets its own radio channel. Omnipoint's technical approach allows for in-building operation in the unlicensed ISM frequency bands as well as in the licensed PCS frequencies.

Nortel and Ericsson are the key suppliers of Omnipoint IS-661 as well as PCS-1900 network equipment and mobile phones. The Omnipoint system will be introduced initially in the New York City metro area. Ericsson has also agreed to a multi-million-dollar license to market Omnipoint's IS-661 technology.

The CDMA battle began to heat up in April 1993 when InterDigital filed a lawsuit against QUALCOMM, Oki Electric Industry Co., and its subsidiary, Oki America Inc., charging them with patent infringement. Oki was included because it had announced plans to make CDMA-based

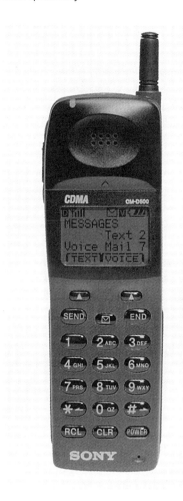

Sony CDMA cellular phone.

phones. The suit, which focused on a part of the technology that digitally passes a cellular signal from one calling area to another, came at a time when QUALCOMM was making significant progress in convincing the cellular industry that its proprietary narrowband CDMA technique was superior to TDMA. After years of work on CDMA, QUALCOMM was looking forward to collecting hefty licensing fees as the cellular industry began making the transition from analog to digital service.

But the suit only slowed the formal acceptance of CDMA by the industry's standards-setting body, the Telecommunications Industry Association (TIA). Rather than wait out the court, the TIA decided to publish QUALCOMM's CDMA technology as an interim standard, designated IS-95. The TIA believed that this would at least allow a more rapid updating of the standard as manufacturers and users gain practical experience with CDMA technology. But because of InterDigital's patent infringement charge, the TIA placed a cautionary legend on IS-95.

Like QUALCOMM, InterDigital submitted its B-CDMA technology to the TIA for consideration as a digital cellular standard, but not until the association indicated it was seriously considering the adoption of QUALCOMM's technology as a North American digital cellular standard. The TIA formed a volunteer working group to study B-CDMA a month before it approved IS-95, but no one from the industry attended the group's first meeting, and the TIA couldn't find anyone to serve as chairperson of the group. (When QUALCOMM approached the TIA two years earlier with its CDMA proposal, it had the formal support of five major cellular carriers.) TIA officials told InterDigital that it needed market support if it hoped to promote B-CDMA as an industry standard. Eventually, the two companies reached a settlement, the details of which have never been disclosed.

Carriers and equipment manufacturers—PacTel, US WEST, Motorola, AT&T, NYNEX, Ameritech, Bell Atlantic Mobile, Bell Mobility, and GTE Mobilnet—performed CDMA tests on QUALCOMM'S San Diego development system. Ameritech installed several hundred TDMA base stations and conducted field trials in suburban Chicago, but switched to CDMA when most of the people involved in the Ameritech trial said they thought the existing AMPS service was as good or even better in voice quality than TDMA.

CDMA is also attracting attention internationally. The first commercial installation of CDMA is in Hong Kong; however, contracts have been signed for additional installations in China, Africa, and Peru. The Chinese contracts are particularly significant because of their potential size.

Even though the China United Telecommunications Corp. (Unicom) has installed four GSM networks, China has conducted several CDMA trials. Also, China's Ministry of Post and Telecommunications, along with China Posts and Telecommunications Industry Corp. and the Hangzhou Communications Equipment Factory, have formed a joint venture with Motorola to manufacture, distribute, and service CDMA infrastructure products in China. Motorola is also supplying Xi'an Datang Telephone with CDMA-based cellular and wireless local loop equipment.

One possible approach calls for using GSM in China's urban areas (because it is a readily available system), while adopting CDMA (which is viewed in China as a technically superior system and less costly) in rural areas. Unlike GSM, which operates at 900 MHz, CDMA will most likely be allocated in the 800, 1,800, or 1,900 MHz bands in China.

In Europe, QUALCOMM has been promoting its technology as a possible alternative to GSM. CDMA has been tested in Switzerland and three cellular carriers in Australia are studying CDMA in anticipation of heavy cellular traffic in that country's largest cities. Ericsson, the largest cellular equipment manufacturer in Europe, conducted CDMA field trials, but has decided to stay with TDMA.

Toronto-based Bell Mobility Inc. will use CDMA in its PCS operation in Canada. The carrier is scheduled to roll out its 2 GHz PCS in early 1997 in selected metro areas. In addition, BC-Tel, another member of the Mobility Canada consortium, has chosen CDMA for use in its networks as part of a plan to make it easier for PCS users to roam across Canada as well as into the U.S.

The CDMA Development Group (CDG) has completed specifications for several advanced features including over-the-air service provisioning, enhanced roaming, and high-speed data.

Over-the-air service provisioning (OTAO) allows wireless carriers to activate and program their customers' phones quickly using the CDMA radio network. This capability enables separation of phone purchase from service purchase, allowing customers to use phones right out of the box, without waiting for dealer programming.

The CDG has defined specifications for both dual-mode CDMA/analog (IS-95) and CDMA PCS (J-STD-008) air interfaces. The enhanced roaming (ER) function allows a roaming subscriber to select a carrier regardless of frequency bands (cellular A/B, PCS A/B/C/D/E/F) and air interfaces (analog/CDMA). The subscriber is able to make this choice by providing preferences to the home carrier.

The CDG has also developed CDMA circuit-switched and packet-switched data specifications and approaches for integration with cellular digital packet data (CDPD) networks, and has begun work on further enhancements of CDMA to provide for high-speed data services. Today, the CDMA network supports 9.6 and 14.4 Kbps transmission. The CDG is developing data architecture and protocols to support applications that require 64–500 Kbps data speeds. This should open up several new applications for CDMA, such as high-speed mobile Internet access and high-speed mobile video.

By mid-1996, QUALCOMM said that it had shipped more than 100,000 CDMA handsets to customers in the U.S., Hong Kong, and Korea.

THE EUROPEAN ENTRIES

Initially called Groupe Special Mobile, GSM is a second-generation digital system. It was adopted as the European standard in the mid-1980s and introduced into commercial service in 1992. GSM is an open standard, which means that anyone can use the technology.

Interstandard roaming between the widely used AMPS networks in the U.S. and GSM networks came into operation in 1995, and subscriber identity module (SIM)–based roaming between Japanese PDC networks and GSM networks will be in operation in 1997. Roaming with all major mobile satellite services (Iridium, Globalstar, and Odyssey) is expected to be available in 1998. The GSM camp also expects to make a huge dent in the wireless local loop market.

With so much money to be made, competition among equipment manufacturers has been tough and aggressive, both in the U.S. and internationally. What isn't known is whether or not the outcome of the FCC auctions was influenced by the high-stake fight over the technical standards for digital cellular service and PCS. Some analysts think so. A case in point: South Korean companies with an interest in the success of one standard—CDMA—offered to help bankroll bidders who plan to deploy the CDMA-based hardware. The Koreans hope to become major suppliers of the technology in Asia and, eventually, elsewhere.

DECT THE HALLS

Another European system, Digital European Cordless Telecommunications, or DECT, operates in the 1,800–1,900 MHz band. DECT is designed to be used as a cordless phone and wireless office phone system in Europe. It also fits nicely into wireless local loop systems. By combining DECT

VTECH DECT phone.

and point-to-point microwave technologies, the system can provide the normal dial tones and call signaling as the public switched telephone network, completely transparent to the user, with voice, fax, and data services.

DECT phones transmit and receive on the same frequency; DECT transmitters must be turned off when receiving and their receivers must be turned off when transmitting.

In the United Kingdom, PCS is called PCN (personal communications network), so named because the government has some control over telecom networks—less so with systems and services. PCN operates in the 1.7–1.88 GHz band.

Europe also has its own SMR-type system. TETRA (Trans-European Trunked Radio Access) is used by dispatch mobile radio networks for truck, bus, and other fleet operators. TETRA can handle both voice and data communications and has a special band reserved for emergency services.

Japan's equivalent to PCS is the Personal HandyPhone System (PHS). Operating in the 1,895–1918 MHz band, PHS is a low-power, low-cost variation of cellular service. Priced lower than cellular service, PHS is competing more with public pay phones and pagers than with cellular phones, which continue to sell well in Japan.

Originally, PHS was called PHP, but that was changed when Japan's Ministry of Post and Telecommunications (MPT) realized that PHP is the name of a popular Japanese magazine published by Panasonic. Not wanting anyone to think that PHP was a Panasonic creation, the MPT changed the name of the system to PHS.

In fact, PHS is based largely on the Personal Access Communications System (PACS) developed by Bellcore, the research arm of the seven regional Bell telephone companies in the U.S. PACS is a low-tier/low-power radio system operating in the 1,850–1,900 MHz licensed PCS band. It is fully compatible with the local exchange telephone network and interoperable with existing cellular systems. The system is designed to support high-density mobile and fixed voice and data applications with low installation and operating costs. Bellcore, Hughes Network Systems, NEC America, Panasonic, PCSI, GCI, Motorola, ArrayComm, and Siemens-Stromberg Carlson have all been involved in making components or subsystems for PHS. At least 33 Japanese companies are making PHS equipment.

HOW BIG IS GSM?

MTA-EMCI, the Washington, D.C.–based market research and management consulting organization, is forecasting 200 million digital cellular subscribers in the world by 2001. A huge chunk of that will be Global System for Mobile Communications (GSM).

In fact, the global investment in GSM already exceeds $50 billion and it could easily top $80 billion by the end of 1999, fed mainly by new U.S. PCS carriers building their systems, as well as expansion and upgrades of networks in Europe and other areas of the world.

By mid-1996, 191 companies and government administrative agencies from 98 countries, including eight North American networks, had joined the GSM MoU Association, the umbrella organization for GSM service providers. At that time, more than 120 GSM networks were "on the air" in 70 countries, with about 13 million subscribers—a number that was expected to more than double by the end of 1996. About three-quarters of a million people subscribe to GSM service every month—more than 30,000 a day.

GSM supporters expect the standard to dominate the digital cellular and PCS/PCN markets as countries in Asia, Africa, Eastern Europe, and the Middle East expand existing GSM networks and launch new systems. By 2001, MTA-EMCI says GSM subscribers will account for 60 percent of the world's digital cellular subscribers, followed by the Japanese digital cellular standard, PDC, with 16 percent of the market. That leaves

Selected List of New GSM and DCS Licenses Expected in 1996 and 1997

Continent	*Country*	*Technology*	*No. of Licenses*
Europe	Austria	GSM	1
	Denmark	DCS	4
	Finland	DCS	3
	France	DCS	1
	Hungary	GSM	1
	Italy	DCS	1
	Norway	DCS	2
	Poland	GSM	2
	Romania	GSM	1
	Sweden	DCS	3
Asia/Pacific	India	GSM	?
	New Zealand	GSM	1
	Philippines	DCS	1
	Singapore	GSM	1
	Sri Lanka	GSM	2
	Taiwan	GSM	1
Africa and the Middle East	Botswana	GSM	1
	Israel	DCS	1
	South Africa	GSM	1
	Syria	GSM	1
	Zaire	GSM	1

Source: Goldman Sachs.

TDMA with 11 percent and CDMA with 13 percent of the global digital cellular market.

With TDMA-based installations well underway, and the emergence of CDMA, the GSM camp is working hard to maintain its momentum. Its goal was to achieve 25 million subscribers worldwide by the end of 1996.

The election of Gretel Hoffman Holcomb, director of technology at BellSouth Personal Communications, in 1996 as the first U.S.-based chairperson of the GSM MoU Association, and the selection of GSM-based systems for 7 out of the initial 18 PCS licensees in the U.S., has helped dispel any notion that GSM is a purely European technology. APC/Sprint Spectrum, the first PCS network to go on the air in the U.S. in the Washington, D.C./Baltimore area in November 1995, uses the GSM-derived PCS-1900 system.

With about 80 percent of its members now from non-European countries, the GSM association has developed a closer working relationship with the European Telecommunications Standards Institute to help ETSI retain its ownership and control of GSM specifications. The GSM group has also encouraged ETSI to work with other national standards bodies, such as the American National Standards Institute (ANSI), and its U.S. telecom branch, the Telecommunications Industry Association (TIA). There is serious concern in the GSM association and at ETSI that some component and equipment manufacturers may design products that do not follow ETSI standards to the letter, creating problems not only for ETSI but also for other GSM manufacturers and the six equipment-type approval organizations currently operating in Europe.

GSM is also moving into space. ICO, the Inmarsat spinoff, and Iridium, the Motorola-developed low-earth-orbit mobile satellite system scheduled for launch in 1998, have joined the GSM association. And COMSAT Mobile Communications is using a GSM phone in its new laptop computer-size Planet 1 personal satellite system. Comsat will bill GSM customers directly for airtime on calls made or received on the Planet 1 terminal.

Alcatel has already developed a GSM pay phone for countries with limited telecommunications infrastructures, and Ericsson is working on dual-mode (GSM/satellite and AMPS/satellite) handheld phones for the Asian Cellular Satellite (ACeS) system being developed by Lockheed Martin Corp. Car stereo manufacturers have begun to design GSM modules into their products. Others are developing GSM-based point-of-sale (POS) terminals.

FROM ANALOG TO DIGITAL

CDMA, TDMA, and GSM get most of the attention, but there is still plenty of life left in analog systems. More than half of the world's cellular subscribers continue to use the AT&T-developed analog AMPS system. In addition to the U.S., the list of AMPS countries includes Anguilla, Antigua, Argentina, Australia, Bahamas, Barbuda, Bermuda, Bolivia, Brazil, British Virgin Islands, Brunel, Canada, Cayman Islands, Chile, Colombia, Costa Rica, Curacao, Dominican Republic, Egypt, El Salvador, Grenada, Grenadines, Guadeloupe, Guatemala, Hong Kong, Indonesia, Israel, Jamaica, Kenya, Laos, Malaysia, Martinique, Mexico, Montserrat, Netherlands Antilles, Nevis, New Zealand, Pakistan, Panama, Paraguay, Peru, Philippines, Puerto Rico, Samoa, Singapore, South Korea, St. Kits, St. Lucia, St. Maarten, St. Vincent, Taiwan, Thailand, Trinidad and Tobago, Uruguay, Venezuela, and Zaire.

Narrowband AMPS (NAMPS), an analog cellular transmission protocol developed by Motorola, has a three-times capacity advantage over current

U.S. Cellular Technology Selections

Company	Market Size POPS (millions)	Technology
AT&T Wireless	78.757	TDMA
GTE	52.867	CDMA
BellSouth	40.401	TDMA
SBC Communications	38.634	TDMA
AirTouch	38.031	CDMA
Bell Atlantic	33.693	CDMA
United States Cellular	24.408	CDMA/TDMA
Ameritech	22.033	CDMA
360 Communications	21.578	CDMA
US WEST	20.346	CDMA
NYNEX	20.345	CDMA
ALLTEL	8.260	CDMA
Comcast	8.184	CDMA
Cellular Communications	8.132	CDMA/TDMA

Source: Data from Cellular Telecommunications Industry Association (CTIA).
Note: POPS = population or market available to a carrier.

AMPS systems. NAMPS works by reducing the amount of bandwidth required by each channel from 30 to 10 kHz.

Even though the Cellular Telecommunications Industry Association (CTIA) spent more than five years trying to rally the U.S. cellular industry behind TDMA when it was pretty much the only digital game in town, the CTIA and the Personal Communications Industry Association (PCIA) have been technology agnostics in their approach to the emerging digital cellular and PCS standards. Both organizations tend to focus on regulatory and legislative-type issues.

Now, cellular carriers and equipment manufacturers are upgrading their systems from analog to digital technology. The idea is to offer a higher-quality, higher-capacity, and more feature-rich service. However, it is not yet clear which of a number of proposed digital standards will serve the U.S. cellular market. The first phase of cellular digital carrier commitments breaks down as shown in the table on the facing page.

A GLOBAL VIEW

The Inter-American Telecommunications Commission (CITEL) of the Organization of American States has recommended the use of PCS throughout the Western Hemisphere. This would help ease international roaming and lower wireless costs for operators and subscribers using PCS frequencies (1,850–1,990 MHz).

CITEL has been in existence for more than 100 years, but was established in its current form in 1993. Its members include all the independent states in the Western Hemisphere and about 70 associate members representing 10 different countries and several private-sector organizations. CITEL's objective is to advance the development of telecommunications in the Americas.

Britain has been one of the busier areas in the world, opening at least three new mobile communications services, and licensing six telephone companies to provide wireless local loop services beginning in 1996. The U.K.'s Radiocommunications Agency has advised the country's two national cellular phone operators that they will have to switch off their analog networks by 2005, which means they must now convert at least 4.5 million analog phone users to digital phone service—hopefully, without losing any subscribers.

U.S. PCS Technology Selections

Company	Market Size POPS (millions)	Technology
Sprint Spectrum	163.305	CDMA
AT&T Wireless	111.431	TDMA (IS-136)
PCS PrimeCo	60.538	CDMA
Pacific Bell Mobile	33.209	PCS-1900
American Mobile Telecom	27.938	PCS-1900
Omnipoint	26.803	Omnipoint (IS-661)/PCS-1900
GTE Mobilnet	20.877	NAMPS/CDMA
Western Wireless	14.695	PCS-1900
BellSouth Wireless	12.163	PCS-1900
Powertel PCS (InterCel, Inc.)	8.984	PCS-1900
Ameritech	8.194	CDMA
American Personal Comm.	8.171	PCS-1900
Southwestern Bell Mobile	6.882	TDMA (IS-136)
Centennial Cellular	3.765	CDMA

Source: Data from Cellular Telecommunications Industry Association (CTIA).
Note: POPS = population or market available to a carrier.

Some of the analog spectrum will also be needed for a new, private GSM network, GSM-R, planned by 11 European railroad companies. Long-distance truckers and bus operators also expect to be assigned their own GSM frequencies in the U.K.

Will there be dual-mode/dual-band phones? Most GSM manufacturers believe the market will require them to make trifrequency phones, either GSM-900/DCS-1800/AMPS or GSM-900/PCS-1900/AMPS. But knowledgeable subscribers may opt for simpler solutions. Some cellular handset manufacturers, for example, may produce very small and cheap PCS-1900-only phones for PCS carriers that want to differentiate themselves from other cellular and PCS services.

The GSM association's North American Interest Group is focusing on the development of a GSM/AMPS dual-mode phone, and for good reason. Its members can't build networks fast enough to blanket the U.S. overnight with GSM alone. During the first three years of GSM's rollout, there's a share of the U.S. market—the association believes it's about 15 percent—

that will require both GSM and AMPS to ensure access to mobile phone service when outside a GSM service area.

Much of the new work in GSM is centered around setting specifications for higher-speed data, both packet-switched and circuit-switched applications. Interoperability specifications are also being developed to combine DECT with GSM for cordless in-building use.

NOT TOO LATE FOR WLANs

After six years of hammering out an agreement on a wireless local-area network (WLAN) standard, the IEEE 802.11 subcommittee responsible for producing a WLAN protocol was still at it into the fall of 1996, holding meetings and sending out round after round of ballots for approval of the latest version of the standard. But the group wasn't expected to finalize the WLAN standard until well into 1997.

Several WLAN hardware and service supplies decided not to wait and have begun to publish their own open interface specification enabling independent parties to develop compatible products. The Wireless LAN Interoperability Forum (WLI Forum) has also established a certification process for WLAN product interoperability.

WLI Forum member companies include ALPS Electric, USA (formerly Kalidor), AMP, Comtron, Data General, Kansai Electric, LXE, Matsushita Inter-Techno, Norand, NTT-IT, Proxim, Raytheon Electronics Seiko Epson, WiSE Medical Systems, and Zenith Data Systems.

The specification, which will document the RF interface used by the forum members' products, is based on the RangeLAN2 interface developed by Proxim and used in many products by forum member companies and other WLAN vendors. Several manufacturers, in addition to Proxim, are expected to supply specification-compatible WLAN adapters and design-in modules.

Forum members believe their specification will prompt more competitive pricing and ensure competitors that they're investing in a technology that will be supported and enhanced. Still, even the highly touted office market will take a while to penetrate in significant numbers.

Two issues that have slowed the development of the WLAN market are price and awareness. Currently a full-featured WLAN adapter card is priced at about $600; most WLAN vendors believe they must get that down to $300 if they hope to approach any critical mass. More to the point is the fact that no one wants to be told that they have to buy a WLAN product from a single supplier.

To help speed the availability of WLI Forum–compatible networking products from alternative suppliers, Proxim will license manufacturing rights for RangeLAN2 products to selected manufacturing companies. The WLI Forum has also implemented a program to test interoperability between suppliers' products. The certification process includes providing test suites and sponsoring the actual testing. The test suites are defined around a complete set of functionality, including basic communication, roaming, and power management.

To help fend off the perception that the WLI Forum interface offers a competitive alternative to the long-anticipated IEEE 802.11 WLAN standard, forum members have decided not to conflict with the 802.11 objective of developing new interface specifications. Once 802.11 becomes a formal standard, the WLI Forum says it will begin to incorporate it into its charter and will develop a transition path to 802.11 from the current RangeLAN2-based interface.

The situation is not much different in Europe where the European Telecommunications Standards Institute (ETSI) is developing the High-Performance LAN (HIPERLAN), a 20 Mb/s network operating in the 5,150–5,300 MHz band. Japan is adopting two different standards, one using 2 Mb/s at 2.4 GHz, and a much faster network at 10 Mb/s, and higher, at around 18 GHz.

NEW PAGING STANDARDS EMERGE

With paging continuing to grow, there is a lot more activity in paging protocols. Motorola has developed a next-generation paging protocol, called FLEX, which addresses the issue of limited channel capacity. FLEX is a fully synchronous multispeed signaling code that is optimized for throughput. FLEX can handle messages at speeds up to 6,400 bps. Operators currently using 1,200 bps technology can add FLEX to their system by upgrading their paging terminals while continuing to service existing pager units with relatively little investment in the infrastructure. FLEX can be used on a dedicated channel or can be mixed on a system with other protocols (POCSAG, GOLAY, ERMES, etc.). Paging operators are expected to populate the market with upwardly compatible FLEX pager units—a process Motorola likens to the radio broadcast industry's shift from monophonic FM to stereo.

At maximum speed, Motorola claims that FLEX can handle 600,000 alphanumeric pagers per channel—a 300 percent increase over the capacity of POCSAG (Post Office Code Standardization Advisory Group), the paging industry's current standard. Carriers can mix traffic using POCSAG or GOLAY (another Motorola-developed paging protocol) and FLEX in a single system. Another feature of FLEX is that it energizes the pager's electronics only when data are to be received, significantly improving its battery life.

Motorola began delivering pagers that accept the FLEX protocol late in 1993. Two major Motorola competitors, NEC Corp. and Glenayre Technologies, Inc., have licensed the FLEX technology. NEC plans to begin selling FLEX pagers in the U.S. when carriers install the infrastructure, some of which will be built by Glenayre. (NEC America and Glenayre Technologies have informally demonstrated ERMES hardware in the U.S., and NEC Japan and its U.K. subsidiary have built about 2,000 ERMES pagers to test in Europe.)

In addition to FLEX, Motorola has also developed ReFLEX and InFLEXion. ReFLEX is designed as a two-way paging and messaging protocol. By adding a response channel to a traditional paging system, ReFLEX allows carriers to provide two-way messaging within a competitive cost structure. Tango, the first two-way pager, which hit the market in the third quarter of 1995, is based on ReFLEX technology. The InFLEXion protocol is based on linear modulation and enables high-speed advanced voice messaging and data services. InFLEXion features system registration capability, which provides increased system capacity and regional and local frequency reuse. InFLEXion products are being developed for the Narrowband PCS channels and have been available through service providers since early 1996.

pACT, another relatively new two-way messaging and paging standard, is expected to compete with InFLEXion. Like cellular, pACT uses base-station transceivers that send and receive messages over the same network. This Bellcore-developed protocol uses frequency reuse and cell splitting to provide network efficiency and to raise the network's capacity as the number of subscribers increases.

pACT operates in FCC-licensed spectrum for Narrowband PCS that range from 901 to 901.5 MHz for the transmit channel and 940 to 940.5 MHz for the receive channel. pACT systems are also aware of the location of all devices on the network because each messaging device registers and authenticates itself from a single base station. Since each base station has a separate channel frequency, the Narrowband PCS spectrum frequency can be reused, allowing pACT networks to support millions of users.

Another paging standard is the Telocator Data Protocol (TDP). Adopted by the Personal Communications Industry Association (PCIA), TDP is the standard by which information will be entered and transmitted over the current paging infrastructure, as well as the new Broadband and Narrowband PCS systems.

Telocator is the former name of the PCIA, and TDP represents the next generation of that organization's efforts to expand the technology to meet new and emerging paging requirements. In fact, TDP succeeds the Telocator Alphanumeric Protocol (TAP), which is currently being used for alphanumeric paging. TAP was designed solely to deliver short text messages in a one-way paging system. With TDP, wireless carriers can implement a wide array of new features, allowing users to send data files and lengthy documents, and to transmit the same message to multiple recipients. It can also defer message delivery. TDP allows paging networks to store and reconcile messages, and acknowledges messages.

As an open standard, TDP is designed to ensure interoperability of wireless solutions from multiple carrier, infrastructure, and software vendors. Revisions and updates to the PCIA-sanctioned paging protocols will be administered by the PCIA's technical committees, which include representatives of wireless carriers, infrastructure manufacturers, and software developers.

All major wireless carriers are expected to adopt the TDP for advanced messaging. Typical of the potential applications for TDP, Lotus Development Corp. has announced a paging gateway to its Lotus Notes over the new paging protocol.

IT'S ERMES IN EUROPE

Europe, of course, has its own paging standard, the European Radio Messaging Service (ERMES), which allows users to roam from country to country. The International Telecommunications Union (ITU) recommended ERMES as the paging standard for international use in October 1994 and it was launched into service as an open standard in 1995.

With 39 signatories from 22 countries, the ERMES MoU Steering Group has become one of the biggest paging groups in the world. ERMES has already been adopted in the Middle East, Eastern Europe, and Southeast Asia. Saudi Arabia claims the world's largest ERMES paging network, with an initial capacity of 720,000 subscribers.

In the Arabian Gulf, the Gulf Cooperation Council States have agreed to adopt the ERMES standard, following their policy of adopting ITU recommended standards. The countries of the Arab League are expected to follow. Saudi Arabia launched its ERMES network in December 1995.

In the U.K., British Telecommunications Plc, Cable & Wireless Plc's Mercury Communications unit, Vodafone Group Plc, Hutchison Telecom, London Pager Co., Paging Network Ltd., and Message Telecommunications Ltd. have applied for licenses to operate ERMES paging systems.

France has three ERMES operators. Infomobile was the first company in the world to bring a high-speed paging system into commercial operation and now has more than 200 base stations in operation. TDR operates an ERMES alphanumeric Calling Party Pays system in the Paris area. FTMR is the third operator in France.

Europhivo has launched its ERMES service in Hungary and a second Hungarian operator, Easycall, is in commercial service in Budapest. Other ERMES operators include the Netherlands PTT, which launched its nationwide "TravelText" service in March 1996.

The ERMES Manufacturers' Marketing Association (EMMA) was formed in March 1995 and has held a series of seminars in China, Taiwan, South Korea, Hong Kong, Indonesia, Thailand, and the Philippines to help promote ERMES as a global paging standard. New members include Celcom of Malaysia, the first in the fast-growing Asia/Pacific region, and Europhivo of Hungary, the first ERMES signatory from Eastern Europe. Etisalat of the United Arab Emirates also is a full signatory of ERMES.

More than 20 companies in several countries were scheduled to launch ERMES commercial service in 1996.

MINIATURE CARDS FOR ORIGINAL EQUIPMENT MANUFACTURERS

Another organization that hopes to create an industry de facto standard is the Miniature Card Implementers Forum (MCIF). The miniature card is a low-cost, removable, reusable, digital storage medium for handheld consumer electronics devices such as cellular phones, personal digital assistants (PDAs), and digital cameras. Its technology is expected to allow consumers to use a single, transportable medium for their data needs.

The MCIF specification describes a card that can be used to store and exchange image, text, and data. By adopting the specification, the MCIF hopes to foster a compatible exchange of data between handheld consumer devices and personal computers.

The MCIF has at least 38 members, including Compaq Computer, Hewlett-Packard, Intel, Microsoft, Nokia, Eastman Kodak, Mitsubishi Electronics, Hyundai Electronics, and Philips Electronics. In supporting the memory cards, these companies are competing with about the same number of members of the Solid State Floppy Disk Card (SSFDC) alliance. The so-called Mini Card, 38 mm wide by 33 mm long by 3.5 mm high (1.5 × 1.3 × .14 inches), is smaller than a PCMCIA memory card and is targeted at consumer applications. Both the Mini and SSFDC cards can be inserted into adapters that fit into PCMCIA slots.

In future cellular phone applications, miniature cards will store audio data and function as message answering and retrieval centers. Even with the phone turned off, the card stores messages for the user to retrieve later. In pagers, the card enables personal voice messaging services.

Actual voice messages can be stored locally on the memory card rather than being stored in a remote database as with 800-number message-retrieval centers. Miniature cards can also be designed into telephone answering machines, allowing random access to recorded messages—a feature that is not available on current home phone answering systems. In handheld computers, miniature cards can store application programs and expand the computer's memory.

As usual, price and application will play at least as important a part in the Mini/SSFDC card battle as technological advancement.

LINKING OFFICE MACHINES

Another group, the Salutation Consortium, an international organization formed in 1994 to create a standard interface for intelligent peripherals, office machines, and services to exchange basic device characteristics and data formats, has updated its Salutation Architecture to provide access to Internet data from office machines. The new system (Version 2) also supports infrared (IR) links between handheld devices and desktop computers, and allows personal digital assistant (PDA) users to "beam" an address to a fax machine via IR directly from the address book on the PDA.

Meanwhile, the Infrared Data Association (IrDA) is now attempting to define interoperability for the cordless IR market for future consumer applications. According to IrDA, "The acceleration of digital convergence heightens the need to expand IrDA connectivity to these consumer devices."

The need to look more closely at standards in mobile communication satellites became clear when the Telecommunications Industry Association formed the Communications Interoperability Subcommittee (CIS) of the Satellite Division of the TIA in 1996. The group grew out of the Satellite Industry Task Force, which works with government agencies to promote satellites in the global and national information infrastructure.

The CIS's purpose is to address the protocols and standards needed to ensure effective global interoperability.

A PUSH FOR SPEAKeasy

The successful demonstration of the U.S. Department of Defense's multiband, multimode radio, "SPEAKeasy," led to the creation of yet another standards/interoperability group in 1996, tentatively called the Open Architecture Modular Multifunctional Information Transfer System (MMITS) Forum. The group's goal is to adopt an open architecture for a wireless system that includes network functionality, message processing, and multimedia interfaces that extend beyond the bounds of traditional radios.

Essentially, the forum hopes to explore market requirements, develop and influence emerging standards to meet these requirements, and promote the use of multimode/multiband wireless communications systems by both vendors and users.

Organizations represented at the initial MMITS meeting in April 1996 include BellSouth, Motorola, Raytheon (all U.S. companies), Tadiran (Israel), Thomson-CSF (France), the United Kingdom Ministry of Defence, and several U.S. government agencies, including the Federal Communications Commission (FCC), the National Telecommunications and Information Administration (NTIA), the Federal Aviation Administration (FAA), the U.S. Army's Training and Doctrine Command, and the U.S. Navy's Space and Naval Warfare Systems Command (SPAWAR).

With only a few meetings, MMITS was still feeling its way, but the DOD's Open System Joint Task Force, which plans to be a key participant in MMITS, has defined an open system "as having many suppliers, many users, an architecture with a long life, and the ability to easily incorporate technology upgrades."

The SPEAKeasy modular reprogammable radio program is sponsored by the DOD's Advanced Research Projects Agency (ARPA), the Army's Communications Electronics Command, and the Air Force's Rome Laboratory.

7   *Wireless Unplugged*

Is using a cellular or personal communications service (PCS) phone harmful to your health?

After years of debate and research, it does appear that, under certain circumstances, digital portable phones may interfere with cardiac pacemakers and hearing aids. It is still not clear whether or not cellular phone users are risking brain cancer from extended use of their phones.

Industry-sponsored research by the Wireless Technology LLC (WTL) has determined that some digital phones caused interference in more than half of the 975 pacemaker patients tested. Some digital phone technologies were better (or worse) than others. Time division multiple access (TDMA), for example, referred to in the study as North American Digital Cellular, caused 28 percent of the interference. Code division multiple access (CDMA) contributed to 15 percent of the interference, while less than 2 percent of the interference was caused by GSM (Global System for Mobile Communications) phones.

The WTL study recommends that pacemaker-dependent patients use analog rather than digital phones. It also suggests that none of the data developed thus far indicate that bystanders with pacemakers are at any risk of interference from wireless phone use.

At some point, the U.S. Supreme Court will consider whether manufacturers of medical devices that malfunction can be sued in state court. One of the issues to be resolved is whether or not pacemaker manufacturers can be sued if one of their products malfunctions due to electromagnetic interference (EMI) from a wireless phone.

Meanwhile, the early data from the University of Oklahoma's Center for the Study of Wireless Electromagnetic Compatability found that hearing aid users do experience interference when they're within 2 feet of a digital phone.

Another study by Boston-based Epidemiology Resources, Inc., indicates that portable telephones are probably not a health hazard in the short run, but it leaves open the question of whether long-term use of phones can cause cancer or other medical problems. The ER study compared mortality rates among a quarter of a million wireless phone users over a three-year period and found no increase in deaths among the portable phone users. Still, public health officials who examined the data believe the research is insufficient to conclude that there are no long-term effects, either from handheld wireless phones or from cellular towers.

Congress got into the act when it asked the U.S. General Accounting Office (GAO) to prepare a report on the status of research on the safety of cellular phones. The GAO's report, *Status of Research on the Safety of Cellular Telephones*, published in late 1994, found no link between cellular phones and brain cancer. The report states, "Existing research does not provide enough evidence to determine whether portable cellular telephones pose a risk to human health." The report suggests that controlled laboratory studies on animals and living cells are needed to determine if radiation from portable cellular telephones poses a human health risk.

THE FCC's ROLE

Even at the highest levels of government, it is not clear who has health-related watchdog responsibility for telecommunications products. The Federal Communications Commission (FCC) has always had a limited role in electromagnetic radiation matters. It is not a health organization and must rely on the advice of outside expert groups. But it does have an obligation under the 1969 National Environmental Policy Act to ensure that its licensing and regulatory actions do not create adverse health or environmental effects. And the Telecommunications Act of 1996 requires the FCC to develop health guidelines for the use of wireless devices.

The FCC will likely update its safety standards for electromagnetic radiation (EMR) exposure to bring them more closely into line with those adopted in 1992 by ANSI and IEEE. Applying this standard to the next generation of cellular systems, power levels would have to be between 0.31 and 0.39 watts to meet ANSI/IEEE specifications.

The FCC has also authorized total effective radiated power levels up to 500 watts for cellular towers, depending on geographic area and tower height. Most cellular towers in large cities operate at about 10 watts per channel. The Telecommunications Industry Association (TIA), which represents equipment manufacturers and works with ANSI to develop U.S. telecommunications standards, has encouraged the government to put into law what is already the de facto standard for land mobile manufacturers.

There are also important differences in international research. Whereas most European and Australian studies focus on GSM at 900 MHz with 2 watt handsets and 8 watt mobile phones, U.S. studies target all wireless technologies that will be used for cellular and PCS services in North America. Nevertheless, the Oklahoma University Center is using test results from European and Australian studies, among others, in its own studies.

To create actual signal structures of cellular calls, the OU Center used wireless phones instead of various simulated radio frequency (RF) signals. A Hearing Aid Test Design Group has been formed to develop the test protocol for the study and to act as peer reviewers to assess the test results and conclusions. The group is made up of representatives of 45 U.S. and international wireless equipment manufacturers and service providers, hearing aid companies, and government agencies.

But hearing aids are only a part of the problem. The cellular industry has been questioned for years about electromagnetic interference and radiation (EMI/EMR) caused by people holding their cellular phones next to their heads for extended periods of time.

One of the biggest hits came in January 1993 when a Florida widower claimed on the "Larry King Live" television talk show on CNN that his wife died from a brain tumor caused by her cellular phone. That show, his lawsuit against a cellular phone manufacturer and retailer, and the media blitz that followed threw a scare into some cellular phone users—and the cellular industry.

The industry took another shot in April 1996 when the London *Sunday Times* reported that RF radiation from handheld communications devices may pose more of a threat to users than the telecommunications equipment manufacturers have led the public to believe. The *Sunday Times* article, which ran under the headline, "Danger: Mobile Phones Can 'Cook' Your Brain," noted two unpublished studies in the U.S. and one in England that suggest a potentially dangerous link between mobile phone emissions and possible health problems. U.S. officials involved in wireless communications–related epidemiological studies described the research mentioned in the *Sunday Times* article as inaccurate and misleading. The publicity got worse when the London *Times* reported in a follow-up article that Hitachi and Mitsubishi had applied for patents for devices to shield consumers from RF radiation.

The proliferation of wireless products worldwide has prompted the United Nations to assign one of its agencies, the Geneva-based World Health Organization, the task of investigating whether wireless devices are a health risk. The WHO has begun a five-year study of the health effects of electromagnetic fields in frequencies up to 300 GHz, which would cover several electrical products—hair dryers, microwave ovens, and electrical power lines—as well as wireless communications.

Actually, there have already been hundreds of studies made over the past 40 years, none of which has turned up any real evidence of human brain damage. Mercifully, for the industry, a federal court agreed, dismissing the Florida suit against NEC Corp., which produced the phone, and GTE Mobilnet, which provided the service, because of a lack of scientific or medical research into the issue.

The wireless communications industry hailed the judge's decision as a rejection of "junk science." Others were less sanguine, saying, "The industry needs to do some biology."

Part of the problem is that most research in the field to date has focused on low-frequency emissions, such as those produced by overhead power lines. Until recently, very little work has been done at the high-frequency end of the spectrum where cellular phones operate. The most recent extensive review of electromagnetic field (EMF) research was completed in 1991 by a 120-person panel of experts assembled by the Institute of Electrical and Electronics Engineers (IEEE), the largest technical society in the world with more than 300,000 members. Of the 321 studies reviewed by that panel, 29 dealt with cancer, but none of these included human experiments. The IEEE Committee on Man and Radiation's official position is that, based on present knowledge, prolonged exposure to RF fields from portable and mobile telephone devices at or below the recommended levels is not hazardous to human health.

The number of medical equipment problems attributed to EMI has been relatively small—less than 50 reported cases in the U.S. in a recent one-year period, according to the IEEE Electromagnetic Compatability Society. But there are documented cases in which some apnea monitors failed to operate properly when exposed to weak RF signals, and some hospital patients have reportedly died as a result of these equipment failures. The U.S. has no current standards requiring manufacturers of medical equipment to shield their devices against EMI. IEEE and others who have studied EMI simply suggest that hospitals develop their own guidelines and try to use lower power devices until better information on EMI is available.

With 6 million people in the U.S. wearing hearing aids, it's not a small issue. Pacific Bell and Ericsson have been working with Self Help for Hard of Hearing People, Inc., to develop interim solutions that will enable hearing aid wearers to use wireless digital phones without experiencing interference.

Testing of human subjects is being conducted at the Hough Ear Institute in Oklahoma City, and by other research groups. The objective of this research is to determine the extent of the interference problems to hearing aid users.

Concerned industry leaders, the FCC, and the National Telecommunications and Information Administration (NTIA) have for years tried to convince the Environmental Protection Agency (EPA) to develop RF exposure guidelines. Unfortunately, the EPA terminated an in-house research program in 1986, citing budget problems and other priorities.

To its credit, the wireless personal communications industry has been pretty much up front about the issue. In early 1991, in its petition to the FCC requesting frequency assignments for its proposed Data PCS service, Apple Computer specifically asked the commission to consider the health consequences in making its decision. On a larger scale, the Cellular Telecommunications Industry Association (CTIA) has spent millions of dollars supporting various research programs and generated thousands of pages of documentation in an effort to get this issue behind its industry members. The Personal Communications Industry Association (PCIA) has formed an Electromagnetic Compatibility Task Force to survey the industry on interference between wireless technologies and medical devices. Eventually, the PCIA hopes to formulate protocols for PCS equipment testing.

MORE STUDIES ARE UNDERWAY

Two independent studies conducted by doctors at the Mount Sinai Medical Center in Miami and another at the Mayo Clinic have found evidence that digital phones interfered with cardiac pacemakers, particularly when the phones are held near pacemaker pulse generators, which are usually implanted near the collarbone.

The Mount Sinai studies tested two analog phones and three TDMA phones and found interference in the digital, but not the analog, models. The studies found no problems when the phones were placed near the ear. However, there was "significant" interference when the phones were held near the pulse generator. The Mayo Clinic studies involved 30 patients and used two analog phones, rated at 3 watts and 0.6 watts, and two similarly rated TDMA phones. The tests found "no significant interference" from the analog phones but several instances of interference from the 3-watt TDMA phone.

Formerly known as the Scientific Advisory Group on Cellular Telephone Research, or SAG (the name has been changed twice as the group's work was expanded to cover research for all wireless technologies), the WTL has dispersed millions of dollars in research funds.

Essentially, the WTL is trying to determine whether there is a public health risk from interference between cardiac pacemakers and wireless handheld telephones, and the extent of the risk, if one is identified. The studies include pacemaker-dependent and non-pacemaker-dependent patients of all ages who receive regular examinations and who agree to be tested at clinical sites. All types of pacemakers are included, and the test covers a variety of cellular phones, including analog and digital models.

The WTL study team includes scientists and physicians from the Food and Drug Administration, the University of Oklahoma, the Mayo Clinic, Mt. Sinai Hospital in Miami, the George Washington University Medical Center, the Health Instruments Manufacturers Association, and the CTIA. While the clinical study is in vivo, investigating how people use cellular telephones and whether this use produces interference, a coordinated laboratory in vitro study is being directed by the University of Oklahoma Center to determine the potential for electromagnetic interference.

The WTL has also awarded $1.6 million to three scientific teams to help develop standardized exposure systems for use in future WTL-sponsored experiments. Most of this funding was assigned to C. K. Chou of City of Hope National Medical Center in Duarte, California. Chou and his staff are expected to spend the first year of their three-year contract studying how humans react to long-term exposure to wireless communications devices.

The WTL has also funded specific, state-of-the-art research projects directly relevant to investigating the potential health effects from cellular telephones and other wireless communications technologies, and has produced a 100-page document reviewing virtually everything the WTL has learned in the very complex area of wireless technology research, and it shows the rationale for WTL's research goals and priorities.

Another approach by the WTL has been to review cellular customer usage information to help assess what exposure systems in human epidemiology studies best approximate real world use. Most of the information has been developed from business records. However, the data have been useful in providing the WTL and others researching EMR with a picture of cellular phone usage patterns. Among the more interesting details from the data gathered so far:

- Call duration is independent of system classification or geography. The percentage of calls two minutes or less was 79 percent in St. Louis, 80 percent in Dallas, and 81 percent for a rural system in Maine. The average call length for various urban and rural areas of Vermont and New Hampshire was one minute and 24 seconds.
- The distance between the cell site antennas and the phone determines the amount of power radiating from the phone; therefore, urban and suburban areas that have a greater density of antennas are areas where phones have a tendency to operate at less than full power.
- Heavy phone users tend to be in urban or suburban areas. Six hundred Southwestern Bell Mobile Systems (SBMS) customers were chosen at random from the Dallas area. Based on their billing records, their average usage was classified by the number of minutes per month. SBMS then looked at 14 days of billing tapes and recorded the originating cell location for all 600 customers' calls. The location was then specified as urban, suburban, or rural. When calls averaged more than 150 minutes per month, 48 percent were from urban areas, 28 percent from the suburbs, and 24 percent from rural areas. Usage patterns were far more similar in the 51–100 minute range, with urban users accounting for 39 percent of the calls, suburban 38 percent, and rural 23 percent. Under 50 minutes per month, 65 percent of the users were from urban areas, 29 percent from the suburbs, and 6 percent from rural areas.

THE GSM ISSUE

Virtually every major country in the world is studying the wireless health issue. According to many European cellular phone users, the GSM-based phones have been a problem almost from the day they were introduced. The GSM issue surfaced in the U.S. when a new organization called HEAR-IT NOW (Helping Equalize Access Rights in Telecommunications Now) asked the FCC to investigate the "safety and interference dangers" of GSM.

Along with its petition, HEAR-IT NOW, a coalition consisting of the Wireless Communications Council, Self Help for the Hard of Hearing, and the Alexander Graham Bell Association for the Deaf, turned over to the FCC copies of technical papers and news reports published in Europe and Australia describing how GSM may cause cardiac pacemakers to skip beats and automobile airbags to deploy without warning. According to HEAR-IT NOW, "GSM technology in wireless communications devices such as portable telephones would have disastrous consequences for those persons who are hard of hearing, and present serious safety concerns for millions of Americans."

With no GSM systems yet in operation in the U.S., HEAR-IT NOW's claims received very little attention in the general press, but did garner heavy coverage in the trade and business publications where they were viewed with some skepticism, partly because Jim Valentine, HEAR-IT NOW's spokesman and (at the time) president of Vienna, Virginia–based North American Wireless, had a contract with QUALCOMM, Inc., for the production of CDMA-based PCS handsets.

The CTIA called Valentine's charges irrelevant since GSM normally operates at power levels almost four times higher than the planned levels for GSM phones in the U.S. Nevertheless, the CTIA offered several suggestions for digital phone users with hearing aids. Among them:

- Use an analog wireless phone.
- Switch ears; don't use the digital phone in the ear with the hearing aid.
- Use a digital phone that comes with a standard plug-in extension device (a Walkman-like miniature speaker/microphone combination that fits in or around the ear, thus allowing the digital unit to be kept away from the ear and eliminating interaction).

HEAR-IT NOW criticized the CTIA for pretending the problems are minor and only a battle over market share. Valentine said, "The battle we are in today is one of public access, public safety, and preventing horrendous product liability lawsuits and strong government action that could cripple the entire wireless communications industry. All technologies should have to meet the same rules of public access and be free of the public safety dangers that we have witnessed and heard reported."

With GSM the technology of choice of several cellular carriers and PCS licensees in the U.S., HEAR-IT NOW says that millions of people who wear hearing aids will not be able to use GSM-type cellular or PCS phones and would suffer discomfort and pain if they were even near someone using a GSM device.

MORE BAD NEWS

HEAR-IT NOW has plenty of anecdotal evidence on GSM phones. The London-based Royal National Institute for Deaf People, for example, reported in its October 1994 *Factsheet*: "No hearing aid can be used with a hand-held digital (GSM) mobile telephone," adding that, "Analog mobile telephones do not cause this interference." In New Zealand, the National Audiology Center said that of the 29 hearing aid users it studied in 1993, 27 detected interference when using a 2-watt GSM telephone. The National Telecom Agency of Denmark issued a statement in June 1994 that said, "Researchers determined that 62 percent of hearing aid wearers will find that usage of a GSM telephone in either ear creates interference." That was tempered when Ole Lauridsen, the corporate director of R&D at Tele Denmark Research, wrote to FCC Chairman Reed Hundt in 1995. Lauridsen said, "In my little country of Denmark, 250,000 people (4.8 percent of the population) are currently using GSM telephones on two competitive, nationwide networks and not one single complaint has been received by the Danish Telecom inspector from hearing aid users, car owners, hospitals, airports, medical equipment suppliers, consumer protection agencies, etc."

In May 1995, the National Acoustic Laboratories in Australia published the findings of a GSM hearing aid interaction study initiated by Telecom Research Laboratories, AUSTEL (the country's telecommunications regulatory body), the Deafness Forum of Australia, the Spectrum Management

Agency, and hearing aid suppliers, including Australian Hearing Services. According to the Australian study, the level of interaction varies depending on the type of hearing aid. The CTIA's comment on the Australian study is that it "demonstrated that it is possible to design high-immunity hearing aids, as well as design and use digital mobile telephones in ways to minimize the problem of interaction with hearing aids."

MORE ON EMI/EMR

The International Coordinating Committee on Telecommunications Research (ICCTR) has been formed with representatives from the U.S., Europe, Japan, and Australia to help track EMI and EMR developments around the world.

In Europe, EMR research is carried out under the umbrella of the European Community's Framework Programme for Research and Technological Development. Results from the European Cooperation on the Field of Scientific and Technical Research (COST) in telecommunications, started in 1993, don't even hint at cancer-causing or cancer-promoting effects of high-frequency electromagnetic waves used in mobile communications.

Sweden, a country with more than 12 years of commercial cellular experience and one of the highest market penetration levels in the world, has found no evidence that cellular phones cause cancer. Sweden's Lund University says its research indicates that EMR does not promote cancer from nonthermal exposure levels at 915 MHz, the range in which some cellular phones operate in that country. Abstracts of technical papers presented before the European Bioelectromagnetics Association early in 1993 found "no convincing laboratory evidence" indicating that EMR causes tumor promotion at nonthermal exposure levels. Nevertheless, Sweden continues to study the issue and has considered placing certain restrictions on products that generate EMR.

AND IN JAPAN . . .

The Telecommunications Technology Council of Japan's Ministry of Posts and Telecommunications (MPT) issued a report in 1990 on electromagnetic wave exposure, complete with RF protection guidelines and recommendations for measuring electromagnetic radiation. In 1993, Japan's Research and Development Center for Radio Systems (RCR) adopted a large part of the report as a voluntary standard for RF exposure protection for manufacturers and operators associated with the RCR.

The RCR guidelines were revised; however, an MPT health report issued in April 1996 offered very little in the way of new information. The new Radio Frequency Exposure Protection Study Group, chaired by Yoshifuni Amemiya, a professor at the Kanazawa Institute of Technology, began its own study in September 1995, in an effort to establish some guidelines for safe radio equipment use. The study group completed its report in March 1996 and sent it to the MPT. But this report simply reviews existing regulations, compares them to those of other countries, and suggests that a framework is needed for future studies.

The MPT report concludes that current regulations, issued by the RCR in 1993, were similar to those of ANSI and IEEE standards, and therefore sufficient, at least for the moment.

8

What's Next?

Just the word *wireless* carries a certain cache. Everyone wants a piece of the action. Indeed, it would be hard to find a communications carrier or equipment manufacturer (any kind of equipment, from radios, antennas, and test and measurement instrumentation to integrated circuits and discrete components) that isn't trying to make a dent in the wireless market. Even magazine publishers who made a killing (and some who missed the boat) in the computer industry are trying to duplicate that success in wireless service. (The Cellular Telecommunications Industry Association's mailing list for news releases and related materials covers 175 magazines, newspapers, and newsletters.)

The industry is on a roll. A 1995 venture capital survey by Coopers & Lybrand L.L.P., the international accounting and consulting firm, reported: "The quantity of investments remains high, with Internet and wireless communications-related products and services leading the way."

New corporate formations and strategic alliances are being announced almost weekly and will speed the development of innovative wireless products and services. As technologies converge, demand for better, more individually tailored services will increase. To access these markets, businesses that create and use information will build new digital infrastructures of wireless networks.

Information providers are rapidly taking innovative programs into the marketplace. Network operators are expanding into customized services. Information consumers are discovering they can explore their mobility options and improve their productivity and make decisions faster than ever before.

The wired community is being taken over by the wireless consumer. Even the Internet can now be accessed wirelessly. But the biggest change may come from the industry itself. The telecommunications giants that have pretty much had the market to themselves will have to be more nimble; they will have to be able to move much more quickly to keep up with smaller, more innovative organizations.

Market research and consulting firm Kenneth W. Taylor & Associates says the global telecommunications industry is now entering the "massive cutthroat competition and new leadership" phase of the market during which some of the early entrants in the wireless industry are going to start taking a beating from new entrants who have superior investment partners, technology-based innovation, and fast-reacting top management that can effectively cope with and survive the inevitable price wars.

Taylor expects the new firms to feed well on the expanding digital wireless communication industry and projects the market for digital wireless base station electronics equipment at $12 billion in the year 2000. That's a huge jump from the $2.9 billion in revenues reported for 1995. Taylor also sees revenues from worldwide digital wireless communication carrier services climbing from $11.8 billion in 1995 to $92.4 billion in 2000—a 51 percent compound annual growth rate (CAGR).

But Taylor's key message to the industry is to watch out for the new players. Many of them have had time to figure out ways to beat some of the more traditional competitors, usually by targeting the right product or service, and by entering the right geographical market with the right partners at the right time.

At the same time, some of the older communications firms will likely make new and strong commitments to digital wireless markets. Taylor's list of old but strong organizations is a long one. It includes Lockheed Martin, Hughes Electronics, MCI, AT&T, Sprint/Sprint Spectrum, Motorola, Lucent Technologies, Ericsson, Nokia, Nortel, Alcatel, IBM, Texas Instruments, Hewlett-Packard, TRW, Fujitsu, Rockwell International, NEC, Toshiba, Oracle, Siemens-ROLM, BellSouth, SBC Communications, Pacific Telesis, Apple Computer, Hyundai, Lucky Goldstar Electronics, Hitachi, Philips, Sony, Nintendo, Sega, and Unisys.

Flexibility in the way wireless products are bought and sold will also become more important. Consumers and businesses increasingly will want more options. Some will buy packages of services that include combinations of local and long-distance telephone service (some wireless, some not), along with cable TV and Internet services. Others have found through market research that a lot of people want to unbundle their product purchases and services. AT&T Wireless Services has already committed to this approach and others will likely follow.

In fact, some companies will simply change the way they have historically done business. IBM, for example, is now allowing selected hardware and software companies to sell its products, usually packaged with their own. AT&T will bundle or unbundle its products and services, whatever it takes to get the business.

Paging will probably have to redefine itself as a low-cost two-way data service. Specialized Mobile Radios (SMRs) may have to go through some renewal process and become more than a dispatch service by focusing on business communications networking. Mobile communications satellite service providers may have to reconsider their original game plan and re-think their initial customer base. Cordless and personal communications services (PCS) phones may eventually disappear into cellular phones.

How far can the convergence of communications and computers (and entertainment) be pushed? For consumer electronics manufacturers used to creating, bundling, and introducing products with more bells and whistles than most people want or will ever use, there is no end to the possibilities. A single product with cellular voice and data (including fax and Internet access), satellite links, paging (or more advanced messaging), global positioning system (with a precision altimeter in case someone has to find you in a high-rise building during a fire), AM/FM radio, TV audio, and a scanner would bring new meaning to the term *multimode*. Matsushita has already developed a portable video phone with a 2.5-inch screen and minicamera; it will be available in Japan beginning in 1997.

We can make jokes about mental telepathy being the cheapest form of wireless communication but Sony is taking it seriously. The consumer electronic giant is studying extrasensory perception (ESP) and oriental medicine (which is based on body energy) under a program called ESPER. The Sony lab is trying to explain ESP scientifically and how information travels. For example, Sony has been investigating how the brain waves and skin temperature of people with ESP change when their powers are being used.

What's the connection with modern communications? Even Yoichiro Sako, who manages the Sony lab, isn't quite sure. But he believes strongly that artificial intelligence and phenomena that are very difficult for science to explain may find their way into the products of the next century.

NEW TELECOM LEGISLATION

Some of the biggest changes in the industry will result from the Telecommunications Act of 1996. Indeed, the new law will change the way Americans get their phone service. Long term, it provides an opportunity for wireless carriers to invade the turf of traditional wireline carriers.

Basically, the new law allows long-distance companies, regional Bell operating companies (RBOCs), and cable television companies to compete across the board. The law also opens the debate on the definition of "universal service," which will likely come to include access to information services. And it allows state regulators to open the local loop to competition. According to a study by A. T. Kearney, an international management consulting subsidiary of EDS, the Telecom Act will also accelerate the movement toward a global structure in which a limited number of consortia will dominate the world market for telecommunications. In fact, the study expects these consortia to incorporate information and entertainment companies to become $100 billion "convergence companies."

On balance, Kearney says the new Telecom Act provides a degree of competitive advantage to operators of wireless systems, but this advantage won't emerge until after 1998 when the new wireless systems are built out.

For voice services, the new legislation leaves intact the existing deregulation (and federal preemption of station regulation) regarding entry and price of persons holding federal commercial mobile radio service (CMRS) licenses. This includes existing cellular and new PCS providers. The Federal Communications Commission (FCC) believes preemption extends to CMRS spectrum used for fixed local loop telephone service. In other words, state regulators will not be able to prevent wireless local loop competition.

Under the new Telecom Act, cellular and PCS providers can provide bundled packages of local and long-distance services without offering the presubscription option that incumbent landline carriers will be required to offer. The Telecom Act exempts CMRS providers from equal access and presubscription obligations regarding long-distance carriers, and preempts antitrust decrees such as the one signed by AT&T as part of its purchase of McCaw Cellular Communications. However, according to Kearney, the FCC may require unblocked access to long-distance carriers through access codes or 800 numbers if the commission determines that customers are being denied access to the long-distance carriers of their choice and that this denial is contrary to the public interest.

In fact, two RBOCs—Southwestern Bell Mobile and Ameritech—entered the cellular long-distance market immediately after the Telecom Act was passed. Others are right behind.

What changed is the way the Bells are doing business. Before the landmark Telecom Act was signed into law, RBOC cellular subscribers were required by law to select a separate long-distance carrier when they signed on for local cellular service and were billed for the services separately. Now, RBOC customers receive one monthly bill for both local and long-distance cellular service from the one service provider.

One of the more controversial aspects of the act is that it limits what state and local zoning boards can do to prohibit the installation of transmission towers for wireless services. According to the Kearney study, the courts may ultimately have to decide how to balance the Telecom Act's recognition of legitimate zoning interests with the need to site wireless communications towers; however, zoning officials may not base zoning decisions on the alleged health effects of radio frequency emissions if the transmitters on those towers comply with FCC emission standards. This is an important provision in the law because it should make it easier for new PCS licensees to build their systems and begin to compete with local exchange carriers (LECs).

The Telecom Act requires the FCC to issue explicit rules that spell out how deregulation will happen and, in fact, the FCC has already adopted rules that offer guidelines for state and federal regulators for breaking open the $100 billion local market controlled by the seven Baby Bell phone companies.

In their final form, the new rules are expected to leave little room for interpretation of the new Telecom Act by local telephone, long-distance, and cable companies, as well as state regulators. However, there will be lengthy legal challenges from the Baby Bells and others that could delay this new era of telecom competition, starting with the new FCC rules calling for steep discounts from those who want to resell local service leased from incumbent carriers.

Under the new rules, the Bells will have to give up some revenues and offer to lease parts or all of their phone networks to new players at low prices. They will also have to give up some fees they now charge cellular and other wireless companies to complete calls. Analysts are predicting the FCC will order the Bells to cut the fees they charge wireless phone companies to terminate calls on the wired local networks—from around 0.5¢ to 0.7¢ a minute from the current level of about 3¢.

As part of its rule-making process, the FCC will be required to give television viewers the right to erect antennas to receive most video signals, including direct broadcast satellites (DBS), as well as multichannel, multipoint distribution system (MMDS) operators, which can provide more than 100 channels of compressed digital video to subscribers. According to the *House Report* on the Telecom Act, this section is intended to preempt not only state and local zoning laws but also homeowner association covenants and other similar arrangements. A. T. Kearney's bottom-line analysis of the Telecommunications Act of 1996: "Wireless wins big!"

LIBERALIZATION IN EUROPE

In June 1993, the European Council of Telecommunications Ministers adopted a resolution that sets a timetable for full liberalization of voice telephony by January 1, 1998. After liberalizing terminal equipment and all value-added services, the European Union (EU) has paved the way for a single market for telecommunications.

Most European telecom companies have been preparing to compete in this liberalized environment by integrating the EU's regulatory changes into their strategic plans. Most of these companies have begun to move toward a cost-based pricing structure. Several significant price reductions have gone into effect in the last few years, including lower rates for leased lines, toll-free (called freephone in Europe), Integrated Services Digital Network (ISDN), and virtual private network (VPN) services. At the same time, most of the major telecom companies have been upgrading their network infrastructures to develop high-speed trans-European services.

Britain, Sweden, Finland, and Denmark were among the first to open their telecom markets. Denmark, for instance, where Tele Danmark AS held a legal monopoly over most voice services and telephone lines, opened its phone market to new players in June 1996, 18 months ahead of the EU deadline for liberalization. The German Parliament voted at about the same time to open Germany's telecom industry to competition by the 1998 deadline.

Japan is also moving toward telecommunications reform, but not very quickly. Plans call for splitting up the Nippon Telegraph & Telephone Corp. (NTT) into one long-distance and two regional carriers by March 1999. The government would sell its share in the long-distance company and local companies would continue to be regulated until competition develops in local markets. However, some Japanese industry leaders believe that breaking up NTT would weaken Japan's position as a world leader in telecommunications research and development (R&D). Another concern in Japan is that passage of the Telecommunications Act in the U.S. has enhanced competition in the U.S., further improving the market position of American companies over Japan with new products and services.

Regulatory reform is also underway in Latin America but, like Japan, it has been a slow process. Brazil has probably made the most progress, but competitive telephone markets are not expected until at least the year 2000 in Argentina, Brazil, Chile, the Dominican Republic, Colombia, Mexico, Peru, and Venezuela.

FIRST-GENERATION PHONES

Cellular phones have come a long way in the past few years. As described in a briefing by Analog Devices, which makes the integrated circuits (ICs) for analog and digital signal processing for a wide range of communications applications, the global system for mobile communications (GSM) phones produced in 1993 typically include three digital signal processors (DSPs) and a system controller. A codec mixed-signal converter transformed incoming analog voice into a digital signal, which a separate speech transcoder compressed. A channel coder/decoder provided signal protection, and an equalizer enabled recovery of the received information. A baseband converter then transformed the digital signal back into analog form, and a multiple-component radio frequency (RF) transmitter circuit sent the analog signal to a wireless base station. A complete phone also required a keyboard, display, and 16-bit microcontroller, as well as a subscriber identification module (SIM) with subscriber information.

These handsets, now considered first-generation models, operated at 5 volts, and featured multiple ICs and multiple support components. They were also large with short talk and standby times. The 225 cc (13.73 cubic inches) Ericsson GH197, for example, weighed 296 grams (10.45 ounces), but could manage only one hour of talk time and seven hours of standby battery life. Although size, weight, and talk time lagged industry projections, GSM became widely accepted. The design goals were to reduce cost, size, and power consumption and increase talk and standby times.

In 1995, second-generation designs began to emerge that improved significantly on earlier models. Moving to 3 volt operation reduced power consumption, which inherently improved battery life. Increased silicon integration levels reduced the number of parts, not only shrinking the size of the handsets, but also enhancing reliability and manufacturability.

UMTS AND THE THIRD GENERATION

Most phone manufacturers are now developing third-generation phones that feature fully integrated chip sets while further reducing cost and power consumption.

Programs such as the Universal Mobile Telecommunications System (UMTS); the Future Public Land Mobile Telecommunications System (FPLMTS); and International Mobile Telecommunications 2000 (IMT 2000) are concepts under development to take personal communications into the new information society, with a full range of digital wireless services for the mass market.

FPLMTS is described by the International Telecommunications Union (ITU) as "third-generation systems which aim to unify the diverse systems we see today into a seamless radio infrastructure capable of offering a wide range of services around the year 2000 in many different radio environments." FPLMTS would integrate mobile and fixed services to allow access to virtually any public-switched telephone network (PSTN) in the world, from anywhere in the world.

UMTS was conceived by the European Telecommunications Standards Institute (ETSI) as a program that would offer digital multifunction, multimedia, multiapplication personal communications at rates up to 2 million bits per second, using terrestrial and satellite systems.

UMTS has several general objectives, which include the integration of residential, office, and cellular services into a single system based on one type of user equipment, a UMTS user number independent of the network or service provider, seamless global radio coverage, and speech quality comparable to existing fixed networks. Roaming with smart card-based multimode handsets will likely compete with some of the objectives of UMTS, but they may also help speed the implementation of UMTS services.

UMTS is expected to complement FPLMTS with detailed European standards and test specifications. UMTS may even support a wider range of services than FPLMTS and internetwork with existing European systems such as GSM, DECT, DCS-1800, TETRA, HIPERLAN, or satellites. However, a draft standard for UMTS isn't expected to be published until 1998. By then, UMTS and FPLMTS concepts may already be merged into IMT 2000, a system architecture based on hybrid networks that combine wireless, satellite, and cable technologies. IMT 2000 is expected to provide high-speed, multimedia, fully integrated mobile and fixed communications services. The U.K. government is seriously considering requiring companies to cooperate if they expect to win licenses to operate FPLMTS networks.

The European Union initiated the Research and Development in Advanced Communications Technologies in Europe (RACE) program in 1988 to support the introduction of integrated broadband communications. Since 1992, RACE has focused on system integration and prototyping of new services and applications.

Several system-level RACE programs are underway. One of the larger programs is TSUNAMI (Technology in Smart Antennas for Universal Advanced Mobile Infrastructure), which is a joint venture of ERA Technology, Motorola, France Telecom, Bosch Telecom, CSF, Dassault, Orange Personal Communications, and CSEM, and several European universities to develop an adaptive antenna compatible with the DCS-1800 standard, which they hope to demonstrate on the Orange DCS-1800 network by the end of 1998.

Other RACE programs include ATDMA and CODIT (TDMA and CDMA radio access research projects), MONET (a network architecture project covering service requirements, security issues, database architectures, and intelligent network designs), SAINT (satellite integration), MAVT (low-rate video coding), GIRAFFE (receiver front-end subsystems), and MBS (a long-range study of future mobile broadband systems).

The GSM MoU Association isn't impressed with either FPLMTS or UMTS, claiming that they offer very few benefits over and above those offered by GSM. According to the GSM association, the radio system preselection criteria specified by the ITU can be met by most existing second-generation systems.

The GSM group is now working on a "new" UMTS, guided not so much by technology but by the marketplace. The aim is to develop a strategy that will justify the huge research, standardization, and deployment cost to develop a third-generation system.

One area under investigation—and much in demand by many network operators—is the integration of mobile and fixed services. For example, GSM equipment developers and service providers believe that personal numbering and mobility management should be integrated functions for fixed and mobile systems, and that messaging services for mobile users should be available at fixed sites. Little work has been done in these areas by the ITU or ETSI.

At the same time, significant improvements will be needed in current radio technologies to squeeze more and more information into less and less spectrum. Research is progressing, but meaningful results aren't expected until late in 1997 or 1998.

Trying to maintain a leadership role, the GSM MoU Association has laid out four major tasks to develop a "true third-generation" system:

- Provide a market and business focus for third-generation work.
- Elaborate a clearer vision.
- Review the work of the ITU and ETSI.
- Provide and promote strategic guidance for standardization.

The UMTS Forum, an advisory group to the European Commission and the Third-Generation Interest Group (3GIG) within the GSM MoU Association, is already working on these items. FAMOUS, the term used to refer to a series of conferences held between the telecommunications leadership of the U.S., the European Union, and Japan, may also participate in the development of third-generation systems.

Another group, the Evolution of Land Mobile Radio (Including Personal) Communications, which operates within the European Cooperation in the Field of Scientific and Technical Research (COST) Action program, spends most of its time defining and, to some extent, developing new products.

SMART CARDS

Smart cards, which are already an important part of GSM outside the U.S., will make important inroads in the U.S. if, for no other reason, than they differentiate GSM-based DCS-1800 phones from cellular and other PCS services.

The smart cards, or SIMs, give mobile phone subscribers security, authentication, and ease of billing. Anyone with their own SIM card can use anyone else's GSM phone by simply inserting their card into the phone. The card owner, not the phone owner, will be billed for the call.

A network authentication test follows the insertion of the card into the phone and entering a four- to eight-digit assigned code. This is the phone's local personal identification number (PIN) check. If the PIN is invalid (for example, if it has been reported stolen), service can be automatically and immediately discontinued by the carrier.

Most SIM cards offer several features, or at least options, including abbreviated dialing numbers, last number dialed, and information on call charges with currency exchange. But the most important feature of the SIM cards is that they allow anyone to make a call anywhere in the world on a GSM phone. This is particularly convenient to global travelers because the U.S. GSM-based PCS-1900 operates in a different frequency band than most other countries. The phones may not work in every country, but the SIM card will.

In the future, multifunctional cards will be designed for use as telephone and bank cards.

Wired magazine says that home offices are out; being officeless is in. In other words, the technology is now available to conduct your business from anywhere, anytime.

It's not that people aren't trying. An item in a *New York Times* "sign-of-the-times" column tells about an executive from a Silicon Valley software company who was stopped for driving erratically by a California State Highway Patrol officer. When the officer asked about the open laptop computer and wireless modem on the passenger seat, the driver admitted that he was checking his e-mail and probably wasn't paying as close attention to the road as he should have been.

"Sir, what is the maximum safe speed for computing?" the officer asked.

"I guess that would be zero," the driver said.

Mobility is key. Anywhere, anytime communication continues to be the hallmark of any future wireless product design. But is all this mobility and freedom to communicate anywhere, anytime a good thing?

Not for everyone. Japanese commuters are already complaining about the "social manner problem" of cellular users. "They receive calls and speak loudly anywhere—in restaurants, tearooms, and on trains," notes one correspondent, leading the national railroad in Japan (now divided into several private companies called JR) to make regular announcements that riders should not use cellular phones while seated. So far, however, this is a suggestion, not a rule.

How about sitting in the middle seat on a five-hour flight and having to listen to your seat mates—one on each side—chatting away on their in-flight phones? The prospect becomes even more distressing as more commercial aircraft are equipped to receive—as well as send—phone calls.

Is this a good thing? Some people aren't so sure. Are people really going to pay several hundred dollars or more for a portable satellite phone, plus $3 a minute to use it? Absolutely. The market goes well beyond early adopters and developing and underdeveloped countries where telephone service is minimal at best. Even in the most developed countries, companies will gladly subscribe to a service that will allow them to keep in constant touch with key executives as they travel the world.

Will some wireless equipment manufacturers and service providers vanish before they can find a place in this huge market? Obviously, the answer is yes. Several manufacturers have already disappeared with little or no advance notice, others are struggling, and a few would-be PCS licensees have dropped from view when they couldn't come up with enough millions to keep up with the high cost of winning a license.

What does all this do to the Information Superhighway espoused by Vice President Al Gore and others? If anything, it puts it into overdrive. Half the households in the U.S. already have cordless phones. You don't have to look very far to see someone with a pager. Cellular phone users are everywhere. Laptop and notebook-size computers as well as new PDA-type devices are now loaded with new communications features.

Even within the telecommunications community there is disagreement over how fast this industry can grow and what business users and consumers expect in terms of new products and services. Most service subscribers will pay little attention to what technology is being used as long as they receive good coverage and good voice quality at a good price. However, surveys indicate that corporate users' expectations have increased. With wireless communications available to them, they expect increased productivity from fewer missed calls. Indeed, the dependence on fixed phones will drop rapidly—to the point where some users will take out their wired phones altogether.

Market research can only take the industry so far. You can't always ask people what they want. Products and services will have to be developed with a certain market foresight and then tested in the real world. As one industry executive put it, "Traditional market research techniques are not a reliable indicator. You have to put the services into people's hands and charge them money."

Any way you look at it, it's going to be a bumpy ride. Reed Hundt, the chairman of the FCC, could have been speaking for an entire industry when he said, "I don't know how far this industry is going. No one does."

Glossary of Wireless Terms and Acronyms

"A" Carrier: The nonwireline cellular company that operates in radio frequencies from 824 to 849 MHz.

Advanced intelligent networks (AIN): Systems that allow a wireless user to make and receive phone calls while roaming in areas outside the user's "home" network. These networks, which rely on computers and sophisticated switching techniques, also provide many personal communications service (PCS) features such as "one person/one phone."

Advanced Mobile Phone System (AMPS): The U.S. standard for analog cellular telephones.

Airtime: Time spent talking on a cellular phone, which is usually billed to the subscriber on a per-minute basis.

Allocation: The designation of a band of frequencies for a specific radio service or services. The Federal Communications Commission (FCC) and the National Telecommunications and Information Administration (NTIA) are responsible for frequency allocations in the U.S.

Alphanumeric: A message (for example, on an alphanumeric pager) displaying both letters ("alphas") and numbers ("numerics").

Analog: The traditional method of transmitting voice signals where the radio wave is based on electrical impulses that occur when speaking into the phone. Most cellular companies today transmit in analog.

"B" Carrier: The wireline cellular carrier, usually the local telephone company, which operates on the frequencies 869–894 MHz.

Sources: Cellular Telecommunications Industry Association (CTIA), National Semiconductor, and InterDigital Communications Corp.

Bandwidth: The total range of frequencies required to transmit a radio signal without undue distortion. The required bandwidth of a radio signal is determined by the amount of information in the signal being sent.

Base station: The fixed transmitter/receiver device with which a mobile radio transceiver establishes a communication link in order to gain access to the public switched telephone network.

Broadband: A communications channel with a bandwidth greater than 64 kilobits per second that can provide high-speed data communications via standard telephone circuits.

Business PCS: A communication system that adds wireless capability to an in-building or campus communications network. Also known as Wireless PBX or Enterprise PCS.

Cell: The geographic area served by a single low-power transmitter/ receiver. A cellular system's service area is divided into multiple "cells."

Cellular digital packet data (CDPD): Introduced in 1992 by McCaw Cellular, IBM, and a group of eight other major cellular companies, CDPD uses the idle time in the analog cellular telephone system to transmit packet-size data at rates up to 19.2 kilobits per second.

Centrex: The switching system of a local telephone operator.

Channel: A single path of the spectrum band taken up by a radio signal, usually measured in kilohertz (kHz). Most analog cellular phones use 30 khz channels. Motorola's Narrow AMPS uses a 10 kHz channel.

Circuit-switched data: Circuit-switched data involve keeping a circuit open between users for the duration of a connection.

Code division multiple access: CDMA is a digital technology that uses a low-power signal "spread" across a wide bandwidth. With CDMA, a phone call is assigned a code instead of a certain frequency. Using the identifying code and a low-power signal, a large number of callers can use the same group of channels. Some estimates indicate CDMA's capacity increase over analog may be as much as 20 to 1. The Telecommunications Industry Association (TIA) has awarded CDMA interim standard approval (IS-95).

Commercial Mobile Radio Service: The regulatory classification that the Federal Communications Commission (FCC) uses to govern all commercial wireless service providers, including personal communications service (PCS), cellular, and Enhanced Specialized Mobile Radio (ESMR).

CT-1: Cordless telephone–first generation, or any variety of North American, European, and Japanese analog cordless telephones.

CT-2: Cordless telephone–second generation, a digital cordless telephone standard; generally used in a residential cordless phone, a Telepoint application, or in a small office WPBX system. CT-2 handsets can initiate but cannot receive calls.

DCS-1800: Digital communications service 1,800 MHz is a variant of Global System for Mobile Communications (GSM).

Digital: A method of transmitting a human voice using the computer's binary code, 0's and 1's. Digital transmission offers a cleaner signal than analog technology. Cellular systems providing digital transmission are currently in operation in several locations for both trial and commercial service.

Digital Advanced Mobile Phone System (DAMPS): Digitally enhanced AMPS based on the ID-54 standard. Also referred to as a TDMA.

Digital Cordless Telephone U.S. (DCTU): A version of DECT proposed for the U.S. PCS market.

Digital European Cordless Telecommunications (DECT): A digital cordless telecommunications system intended initially for WPBX applications, but may be used in the consumer market. DECT supports both voice and data communications.

Downlink: The transmission of radio frequency (RF) signals from a satellite to an earth station.

Dual-mode phone: A phone that operates on both analog and digital networks.

Earth station: The electronic ground equipment used with a parabolic-shaped antenna or "dish" to process RF signals to and from a satellite.

Enhanced Specialized Mobile Radio (ESMR): The next generation of SMR, ESMR takes advantage of digital technology combined with cellular system architecture to provide greater capacity than existing SMR systems. (*See also* **Specialized Mobile Radio**).

European Telecommunications Standards Institute: (ETSI): One of the European organizations responsible for establishing common, industry-wide standards for telecommunications.

Federal Communications Commission (FCC): The government agency responsible for the allocation of radio spectrum for communication services in the U.S.

Footprint: The area of the earth's surface covered by a satellite signal.

Frequency reuse: Because of their low power, radio frequencies assigned to one channel in a cellular system are limited to a single cell. However, carriers are free to reuse the frequencies in other cells in the system without causing interference.

Geostationary satellite: A satellite whose speed is synchronized with the speed of the earth's rotation so it is always in the same spot over the earth (geosynchronous orbit). Most geosynchronous satellites operate 22,300 miles above the equator.

Gigahertz (GHz): A frequency equal to one billion Hertz, or cycles per second.

Global positioning system (GPS): A network of satellites developed by the U.S. Department of Defense that provides precise location determination to special receivers.

Global System for Mobile Communications (GSM): Originally called the Groupe Speciale Mobile, it is the European digital cellular transmission standard, which has been adopted by several other countries around the world for cellular and personal communications services (PCS).

Handoff: Cellular systems are designed so that a phone call can be initiated while driving in one cell and continued no matter how many cells are driven through. The transfer to a new cell, known as a handoff, is designed to be transparent to the cellular phone user. During a cellular conversation, when the user reaches the edge of the service area of a cell, computers in the network assign another tower in the next cell to provide the phone with continuing service.

Hertz: The unit of measuring frequency signals (one cycle per second).

HIPERLAN: The Higher-Performance Radio Local Area Network is a European standard for short-range (about 50 meters) high-performance radio local area network. HIPERLAN operates in the 5.1–5.3 GHz band. Another band from 17.1 to 17.3 has been designated for HIPERLAN use, but detailed specifications are not expected to be finalized until at least the end of 1996.

Industrial-scientific-medical (ISM): The unlicensed radio band in North America and some European countries. It is also referred to as Part 15.247, the FCC regulation that defines the parameters for use of the ISM bands in the U.S., including power output, spread spectrum, and noninterference. Commonly used ISM bands include 902–928 MHz, 2,400–2,483 MHz, and 5,725–5,850 MHz.

Infrastructure equipment: The fixed transmitting and receiving equipment in a communications system, usually consisting of a base station, base station controllers, antennas, switches, management information systems, and other equipment that make up the backbone of the system that sends and receives signals from mobile or handheld subscriber equipment and/or the public switched telephone network (PSTN).

Integrated Services Digital Network (ISDN): A switched network providing end-to-end digital connectivity for simultaneous transmission of voice and data over multiplexed communications channels.

Interoperability: The ability to migrate communications transmissions among a variety of local, regional, and national networks. Switching between the different networks would be transparent to the user.

IS: Interim standard.

IS-41: The protocol for roaming within the U.S., as designated by the Telecommunications Industry Association (TIA).

IS-54: The dual-mode (analog and digital) cellular standard in North America. In the analog mode, IS-54 conforms to the AMPS standard.

IS-95: The code division multiple access (CDMA) standard for U.S. digital cellular service, as designated by the TIA.

IS-136: The time division multiple access (TDMA) standard, as designated by the TIA.

Japan Digital Cellular (JDC): A digital cellular standard developed by NTT of Japan, operating in Japan at 800 and 1,500 MHz.

JTAC: Japanese variant of the Total Access Communications System (TACS) analog standard; developed by Motorola for Japan.

Local multipoint distribution service (LMDS): A new "wireless cable" service operating in the 28 GHz band. LMDS uses low-power transmitters, configured in a cellular-like arrangement, to transmit video to receivers in homes and businesses.

Major trading area (MTA): A personal communications service (PCS) area designated by Rand McNally and adopted by the FCC to determine the 51 MTAs in the U.S.

Megahertz (MHz): A measurement of frequency equaling one million cycles per second. One cycle per second is one Hertz.

Metropolitan statistical area (MSA): An MSA denotes one of the 306 largest urban population markets as defined by Rand McNally and designated by the FCC as a guide to determine coverage areas for cellular networks. Two cellular operators are licensed in each MSA.

Microwave: Radio frequency (RF) signals between 890 MHz and 20 GHz. Point-to-point microwave transmission is commonly used as a substitute for copper or fiber cable.

Mobile telephone switching office (MTSO): The central computer that connects a cellular phone call to the public telephone network. The

MTSO controls the entire system's operations, including monitoring calls, billing, and handoffs.

National Telecommunications and Information Administration (NTIA): An agency of the U.S. Commerce Department, the NTIA is the president's adviser on communications policy and is responsible for administering all federal government use of the radio spectrum, including military communications.

Nonwireline cellular company: The FCC licensed two cellular systems in each market—one for the local telephone company, and the second, the "A" carrier, for other applicants. The distinction between A and B (the wireline cellular carrier) was meaningful only during the FCC's licensing process. Once a system is constructed, it can be sold to anyone. Thus, in some markets today, both the A and B systems are owned by telephone companies—one happens to be the local phone company for the area and the other is a phone company that decided to buy a cellular system outside its home territory.

Nordic Mobile Telecommunications System (NMT): A European analog cellular standard operating at 450 MHz and 900 MHz.

Part 15: An FCC ruling that defines the parameters for use of the industrial-scientific-medical (ISM) bands in the U.S. (such as low power output, spread spectrum, noninterference, etc.).

Personal digital assistant (PDA): A portable computing device capable of transmitting data. PDAs can be used for paging, data messaging, electronic mail, receiving stock quotations, personal computing, and facsimile, and as a personal electronic organizer.

Personal HandyPhone System (PHS): Japan's designation for its digital cordless telephony standard.

POPS: A cellular industry term for population. If the coverage area of a cellular carrier includes a population base of one million people, it is said to have one million POPS. The financial community uses the number of potential users as a measuring stick to value cellular carriers.

Public branch exchange (PBX): A telephone switching system designed to both control and route calls in large multiphone environments, such as offices. Most PBXs can handle custom features for users' specific telecom requirements.

Public switched telephone network (PSTN): The land-based telecommunications system through which cellular calls are routed.

PTT: The European government organizations responsible for postal and telecommunications services within their respective countries are usually called PTTs, which stands for Post, Telephone, and Telecommunications.

Radio frequency (RF): In terms of cellular applications, RF is the part of the electromagnetic spectrum between the audio and high-range frequencies (between 500 kHz and 300 GHz). Cellular transmission frequencies are found in two locations in the microwave segment of the spectrum—between 824–849 MHz (megahertz) and 869–894 MHz.

RCR: Research and Development Center for Radio Systems, the Japanese organization that provides industry-wide telecommunications standards.

Reseller: A middleman who buys blocks of time from a cellular carrier at discounted wholesale rates and then resells them at retail prices.

Roaming: Using a cellular phone outside your usual service area; in a city other than the one in which you live, for example.

Rural service area (RSA): The FCC divided the less populated areas of the country into 428 RSAs and licensed two service providers per RSA.

Specialized Mobile Radio (SMR): A private, mobile dispatch radio service usually used by businesses, such as taxi services.

Spectrum: The complete range of electromagnetic waves, which can be transmitted by natural sources such as the sun and man-made devices such as cellular phones. Electromagnetic waves vary in length and therefore have different characteristics. Longer waves in the low-frequency range can be used for communications, while shorter waves of high frequency show up as light. Spectrum with even shorter wavelengths and higher frequencies are used in X-rays.

Spread spectrum: Originally developed by the military because it offered secure communications, spread spectrum radio transmissions essentially "spread" a radio signal over a very wide frequency band in order to make it difficult to intercept and difficult to jam.

Standby time: The amount of time a fully charged wireless phone can be left on before its battery runs down.

Telecommunications Industry Association (TIA): The U.S.-based organization established to provide industry-wide standards for telecommunications equipment used in North America.

Telepoint: A cordless telephone system in which a subscriber can make but not receive phone calls in public areas that have been equipped with Telepoint base stations. The system is not mobile; the user must remain essentially in a fixed location throughout the duration of the call. Both service and equipment are less expensive than cellular.

Terrestrial communications: A system where all transmitters and receivers are on the ground.

TETRA: Trans-European Trunked Radio Access, the European digital cellular land mobile radio system.

Time division multiple access (TDMA): The cellular industry established a TDMA digital standard in 1989. TDMA increases the channel capacity by chopping the signal into pieces and assigning each one to a different time slot. Current technology divides the channel into three time slots, each lasting a fraction of a second. Thus, a single channel can be used to handle three simultaneous calls.

Total access communications system (TACS): An analog cellular system used mainly in Europe; it has also been implemented in some areas in Japan, Britain, China, and other regions of the world.

UMTS: Universal Mobile Telephone Service.

Universal Digital PCS (UD-PCS): An alternative system proposed by Bell Communications Research (Bellcore) for in-building cordless phone service.

Uplink: The transmission of a radio frequency (RF) signal from an earth station to a satellite.

Wireless local-area network (WLAN): A network that allows the transfer of data and the ability to share resources, such as printers, without the need to physically connect each node, or computer, with wires.

Wireless local loop: Wireless systems that can be used to replace cooper to connect telephones and other communications devices with the public switched telephone network (PSTN).

International Wireless Communications Spectrum Guide

Advanced Mobile Phone System (AMPS)—Rx: 869–894 MHz, Tx: 824–849 MHz

Personal Communications Service (PCS)—Rx: 1,930–1,990 MHz, Tx: 1,850–1,910 MHz

Nordic Mobile Telephone (NMT-450)—Rx: 463–468 MHz, Tx: 453–458 MHz and (NMT-900)—Rx: 935–960 MHz, Tx: 890–915 MHz

Enhanced Total Access Communications System (ETACS)—Rx: 917–950 MHz, Tx: 872–905 MHz and (NTACS)—Rx: 915–925 MHz, Tx: 860–870 MHz

Personal HandyPhone System (PHS)—1,895–1,907 MHz

Digital European Cordless Telecommunications (DECT)—1,880–1,990 MHz

Cordless Telephone/Second-Generation (CT-2)—864–868 MHz and (CT-2+)—930–931 MHz and 940–941 MHz

Digital Cordless Telephone (DCT)—1,900 MHz

Note: Rx refers to the receiving frequency; Tx is the transmission frequency.

Global System for Mobile Communications (GSM)—Rx: 935–960 MHz, Tx: 890–915 MHz

IS-54 Time Division Multiple Access (TDMA)—Rx: 869–894 MHz, Tx: 824–849 Mhz

IS-136 Enhanced TDMA—Rx: 869–894 MHz, Tx: 824–849 MHz

IS-95 Code Division Multiple Access (CDMA)—Rx: 869–894 MHz, Tx: 824–849 MHz

Japanese Cordless Telephone (JCT)—254–380 MHz

Personal Access Communications System (PACS)—1,900 MHz

Digital Cellular System (DCS-1800 or DCS-1900)—Rx: 1,805–1,880 MHz, Tx: 1,710–1,785 MHz

Personal Digital Cellular (RCR-27)—Rx: 810–826 MHz, Tx: 940–956 MHz and Rx: 1,477–1,501 MHz, Tx: 1,429–1,453 MHz

Cellular Digital Packet Data (CDPD)—Rx: 869–894 MHz, Tx: 824–849 MHz

ARDIS—Rx: 851–869 MHz, Tx: 806–824 MHz

RAM Mobile Data (Mobitex)—(North America) Rx: 935–941 MHz, Tx: 896–902 MHz; (Europe/Asia)—403–470 MHz

IEEE 802.11 Wireless Local Area Networks—(North America/Japan) 2,400–2,483 MHz; (Japan) 2,470–2,499 MHz

Directory of Wireless Communications Organizations

ADC Kentrox
14375 Northwest Science Park Dr.
Portland, OR 97229
(503) 643-1681
Fax: (503) 641-3341

Advanced Micro Devices
One AMD Place
P.O. Box 3453
Sunnyvale, CA 94088
(408) 749-5439

Advanced Wireless
  Communications, Inc.
7435 Indio Way
Sunnyvale, CA 94086
(408) 735-8833

Advantis
3401 W. Dr. Martin Luther
  King Jr. Blvd.
Tampa, FL 33607
(813) 878-4207

AIM USA
634 Alpha Dr.
Pittsburgh, PA 15238
(412) 963-9047

Air Communications, Inc.
274 San Geronimo Way
Sunnyvale, CA 94086
(408) 749-9883
Fax: (408) 749-8089

Aironet Wireless Communications, Inc.
P.O. Box 5292
Akron, OH 44334
(216) 665-7900

Alcatel Network Systems
1225 N. Alma Rd.
Richardson, TX 75081
(214) 996-5000
Fax: (214) 996-5409

Alexander Resources Co.
4854 E. Onyx Ave.
Scottsdale, AZ 85253
(602) 948-8225
Fax: (602) 948-1081

Allen Telecom Group
30500 Bruce Industrial Parkway
Cleveland, OH 44139
(216) 349-8695
Fax: (216) 349-8692

Alpha Industries
20 Sylvan Rd.
Woburn, MA 01801
(617) 935-5150
Fax: (617) 935-2359

ALPS Electric (USA) Inc.
3553 North First St.
San Jose, CA 95134
(408) 432-6544

American Electronics Association
AEA Japan Office
Yonbancho 11-4
Suite 101
Chiyoda-ku
Tokyo 102, Japan
(03) 3237-7195
Fax: (03) 3237-1237

American Mobile Satellite Corp.
10802 Parkridge Blvd.
Reston, VA 22091
(703) 758-6000
Fax: (703) 758-6111

American Mobile Telecommunications
   Association
1150 18th St., NW
Suite 250
Washington, DC 20036
(202) 331-7773

American National Standards Institute
11 W. 42nd St.
New York, NY 10036
(212) 642-4900

American Personal Communications
1025 Connecticut Ave., NW
Suite 904
Washington, DC 20036
(202) 296-0005

Ameritech Mobile Communications
2000 West Ameritech Center Dr.
Hoffman Estates, IL 60195
(708) 234-9700

Amtech Systems Corp.
17304 Preston Rd.
E-100
Dallas, TX 75252
(214) 733-6060
Fax: (214) 733-6699

ANADIGICS, Inc.
35 Technology Dr.
Warren, NJ 07059
(908) 668-5000
Fax: (908) 668-5068

Andrew Corp.
10500 West 153rd St.
Orland Park, IL 60462
(708) 349-3300
Fax: (708) 349-5444

APCO International
2040 South Ridgewood Ave.
South Daytona, FL 32119
(800) 949-2726

Apple Computer
2025 Mariani Ave.
Cupertino, CA 95014
(408) 974-6790

Applied Engineering
P.O. Box 5100
Carrollton, TX 75011
(214) 241-0055

AP Research
19672 Stevens Creek Blvd.
Suite 175
Cupertino, CA 95014
(408) 253-6567

ARDIS
300 Knightsbridge Parkway
Lincolnshire, IL 60069
(708) 913-1215
Fax: (708) 913-4768

Arianespace Inc.
700 13th St., NW
Suite 230
Washington, DC 20005
(202) 628-3936

ARIA Wireless Systems, Inc.
140 Mid County Dr.
Orchard Park, NY 14127
(716) 662-0874
Fax: (716) 662-0823

Arthur D. Little, Inc.
Acorn Park
Cambridge, MA 02140
(617) 864-5770

Aspect Telecommunications
1730 Fox Dr.
San Jose, CA 95131
(408) 441-2200

Astronet Corp.
37 Skyline Dr.
Suite 4100
Lake Mary, FL 32746
(407) 333-4900

AT/Comm, Inc.
America's Cup Bldg.
Marblehead, MA 01945
(617) 631-1721
Fax: (617) 631-9721

AT&T Consumer Electronics
5 Woodhollow Rd.
Parsippany, NJ 07054
(201) 581-3000

AT&T Easylink Services
400 Interspace Parkway
Parsippany, NJ 07054
(201) 331-4000

AT&T Microelectronics
555 Union Blvd.
Allentown, PA 18103
(800) 372-2447

AT&T Network Wireless Systems
67 Whippany Rd.
Whippany, NJ 07981
(201) 606-4206

AT&T Paradyne
8545 126th Ave. North
P.O. Box 2826
Largo, FL 34649
(813) 532-2200
Fax: (813) 532-5436

AT&T Wireless
(McCaw Cellular Communications)
P.O. Box 97060
Kirkland, WA 98083
(206) 827-4500

Audiovox Corp.
150 Marcus Blvd.
Hauppauge, NY 11788
(516) 231-7750

Aydin Corp. (West)
30 Great Oaks Blvd.
San Jose, CA 95119
(408) 629-0100

Belgacom USA
301 Riverside Ave.
Westport, CT 06880
(203) 221-5150
Fax: (203) 222-8401

Bell Atlantic NYNEX Mobile Systems
180 Washington Valley Rd.
Bedminster, NJ 07921
(908) 306-7508
Fax: (908) 306-6927

Bell Communications Research (Bellcore)
290 West Mt. Pleasant Ave.
Livingston, NJ 07039
(800) 523-2673

Bell Mobility
20 Carlson Court
Etobicoke, Ontario M9W 6V4, Canada
(416) 674-2220

BellSouth Cellular
1100 Peachtree St., N.E.
Suite 1000
Atlanta, GA 30309
(404) 249-0800
Fax: (404) 249-0782

BIS Strategic Decisions
One Longwater Circle
Norwell, MA 02061
(617) 982-9500
Fax: (617) 878-6650

Blaupunkt
Robert Bosch Corp.
2800 S. 25th Ave.
Broadview, IL 60153
(708) 865-5200

Boonton Electronics Corp.
791 Route 10
Randolph, NJ 07869
(201) 584-1077

BT (C.B.P.) Ltd.
Annandale House
1 Hanworth Rd.
Sunbury-on-Thames
Surrey, UK
44-932-765766

BT North America
2560 North First St.
P.O. Box 49019, MS-F25
San Jose, CA 95161
(800) 872-7654

Cable Television Laboratories
400 Centennial Parkway
Louisville, CO 80027
(303) 661-9100
Fax: (303) 661-9199

California Microwave, Inc.
Wireless Network Division
985 Almanor Ave.
Sunnyvale, CA 94086
(408) 732-4000

Canon, Inc.
3-30-2 Shimomaruko, Ohta-ku
Tokyo 146, Japan
81-3-5482-8067
Fax: 81-3-5482-5129

Cantel Mobile Systems
40 Eglington Ave. East
Toronto, Ontario M4P 3A2, Canada
(416) 440-1300

Casio, Inc.
570 Mt. Pleasant Ave.
Dover, NJ 07801
(201) 361-5400

CDMA Development Group
650 Town Center Dr.
Suite 820
Costa Mesa, CA 92626
(714) 545-9400
Fax: (714) 545-8600

Cellsat, Inc.
532 S. Gertruda, Ave.
Redondo Beach, CA 90277
(310) 316-6301

Cellular One
5001 LBJ Freeway
Suite 700
Dallas, TX 75244
(214) 443-9901

Cellular Telecommunications Industry
    Association
1250 Connecticut Ave., NW
Suite 200
Washington, DC 20036
(202) 785-0081

CenCall Communications
3231 S. Zuni
Englewood, CO 80110
(303) 761-4707

Center for the Study of Wireless
    Electromagnetic Compatibility
University of Oklahoma
202 W. Boyd
Suite 23
Norman, OK 73019
(405) 325-2429

Centre Suisse d'Electronique et de
    Microtechnique SA
Jaquet-Droz 1
P.O. Box 41
CH-2007
Neuchatel, Switzerland
41 38 205 111
Fax: 41 38 205 720

Chevalier (OA) Limited
2303-2305 Great Eagle Centre
23 Harbour Rd.
Wanchai
Hong Kong
(852) 827-2827

Chugoku Cellular Telephone Co.
Asahi Seimei Hiroshima Bldg., 7th Floor
4-12 Embisu, Naka-ku
Hiroshima City 730, Japan
(082) 242-0120

Cincinnati Microwave, Inc.
One Microwave Plaza
Cincinnati, OH 45249
(513) 489-5400
Fax: (513) 247-4109

C. Itoh & Co.
Communications Business Dept.
2-5-1 Kita-Aoyama, Minato-ku
Tokyo 107, Japan
(03) 3497-3186
Fax: (03) 3497-3177

Clarion Corp. of America
661 W. Redondo Beach Blvd.
Gardena, CA 90247
(310) 327-9100

Claris Corp.
5201 Patrick Henry Dr.
Box 58168
Santa Clara, CA 95052
(408) 987-7000
Fax (408) 987-7558

Columbia Communications Corp.
4733 Bethesda Ave.
Suite 610
Bethesda, MD 20814
(301) 907-8800
Fax: (301) 907-2420

Columbia Institute for Tele-Information
Columbia Business School
809 Uris Hall
New York, NY 10027
(212) 854-4222
Fax: (212) 932-7816

Columbia PCS Inc.
201 N. Union St.
Suite 410
Alexandria, VA 22314
(703) 518-5073

Comaro Wireless Technologies, Inc.
5 Jenner
Suite 100
Irvine, CA 92718
(714) 450-4000
Fax: (714) 450-8000

Communications Industry Association
   of Japan
8th Floor, Sankei Bldg., Annex
7-2, Ohtemachi, 1-chome
Chiyoda-ku, Tokyo 100, Japan
(03) 3231-3156
Fax: (03) 3246-0495

Compaq Computers Co.
20555 FM 149
P.O. Box 692000
Houston, TX 77269
(713) 370-0670

Comptek Telecommunications
110 Broadway
Buffalo, NY 14203
(716) 842-2700

COMSAT Mobile Communications
22300 COMSAT Dr.
Clarksburg, MD 20871
(301) 428-2253
Fax: (301) 601-5894

Comsearch
11720 Sunrise Valley Dr.
Reston, VA 22094
(703) 476-2672
Fax: (703) 476-2787

Constellation Communications, Inc.
10530 Rosehaven St.
Suite 200
Fairfax, VA 22030
(703) 352-1733

Coral Systems
1500 Kansas Ave.
Suite 2E
Longmont, CO 80501
(303) 772-5800
Fax: (303) 772-8230

Cox Enterprises, Inc.
1400 Lake Hearn Dr.
Atlanta, GA 30319
(404) 843-5000

Creative Strategies Research
   International
46 Old Ironsides Dr.
Suite 490
Santa Clara, CA 95054
(408) 748-3400

CruisePhone, Inc.
1100 Park Central Blvd.
Suite 1800
Pomano Beach, FL 33064
(305) 974-9601

CTIA Foundation for Wireless
   Telecommunications
1250 Connecticut Ave., NW
Suite 200
Washington, DC 20036
(202) 785-0081

Cylink
910 Hermosa Court
Sunnyvale, CA 94086
(408) 735-5800
Fax: (408) 735-6643

Cyplex
18 Clinton Dr.
Hollis, NH 03049
(603) 882-8104

Data Broadcasting Corp.
1900 S. Norfolk St.
Suite 150
San Mateo, CA 94402
(415) 571-1800

Datacomm Research Co.
920 Harvard St.
Wilmette, IL 60091
(708) 256-1763

DataPro Information Services Group
600 Delran Pkwy.
Delran, NJ 08075
(609) 764-0100
Fax: (609) 764-2814

Dauphin Technology, Inc.
377 East Butterfield Rd.
Suite 900
Lombard, IL 60148
(708) 971-3400

DDI Corp.
Ichibancho FS Bldg.
8 Ichibancho, Chiyoda-ku
Tokyo 102, Japan
81-3-3221-9526
Fax: 81-3-3221-9527

Decision Resources, Inc.
Bay Colony Corporate Center
1100 Winter St.
Waltham, MA 02154
(617) 487-3737

Delco Electronics Corp.
One Corporate Center
Kokomo, IN 46904
(317) 451-0657
Fax: (317) 451-0659

DeTeMobil
Oberkassler Str. 2-53227
Bonn, Germany
492289364360
Fax: 492289364365

Digital Microwave Corp.
170 Rose Orchard Way
San Jose, CA 95134
(408) 943-0777
Fax: (408) 944-1648

Digital Ocean, Inc.
11206 Thompson Ave.
Lenexa, KS 66219
(913) 888-3380

Digital Telecom
1355 Peachtree St.
Suite 650
Atlanta, GA 30309
(404) 607-0053
Fax: (404) 892-2585

Digital Wireless Corp.
One Meca Way
Norcross, GA 30093
(404) 564-5540
Fax: (404) 564-5541

Dream IT, Inc.
1255 Cedar Ridge Lane
Colorado Springs, CO 80919
(719) 598-9000

DSC Communications Corp.
1000 Coit Rd.
Plano, TX 75075
(214) 519-3000
Fax: (214) 519-2322

DSP Telecom, Inc.
666 Plainsboro Rd.
Suite 525
Plainsboro, NJ 08536
(609) 799-5397

Edge Media
P.O. Box 762
Acton, MA 01720
(508) 263-1866

EDS Personal Communications Corp.
1601 Trapelo Rd.
Waltham, MA 02154
(617) 890-1000
Fax: (617) 890-0367

Electromagnetic Energy Association
1255 23rd St., NW
Suite 850
Washington, DC 20037
(202) 452-1070
Fax: (202) 833-3636

Electronic Industries Association
2500 Wilson Blvd.
Arlington, VA 22201
(703) 907-7500
Fax: (703) 907-7501

Electronic Industries Association
of Japan
Tokyo Chamber of Commerce and
Industry Bldg.
2-2, Marunouchi 3-chome
Chiyoda-ku, Tokyo 100, Japan
(03) 3213-5861

Ellipsat International, Inc.
1120 19th St., NW
Suite 480
Washington, DC 20036
(202) 466-4488

EMI Communications Corp.
P.O. Box 4872
Syracuse, NY 13221
(315) 433-0022

EO, Inc.
800A East Middlefield Rd.
Mountain View, CA 94043
(415) 903-8100

E-Plus Mobilfunk
Thyssen Trade Center
Hans-Guenther-Sohl-Strasse 1
4000 Duseldorf 1, Germany
(01149) 211-967-7590

Ericsson Business Communications Inc.
5757 Plaza Dr.
Cypress, CA 90630
(714) 236-6500

Ericsson GE Mobile
Communications, Inc.
15 E. Midland Ave.
Paramus, NJ 07652
(201) 265-6600
(404) 325-7555

Ericsson Radio Systems, Inc.
740 East Campbell Rd.
Richardson, TX 75081
(214) 238-3222
Fax: (214) 952-8783

ETE, Inc.
12526 High Bluff Dr.
Suite 300
San Diego, CA 92130
(619) 793-5400

European Community Delegation
2300 M St., NW
Washington, DC 20037
(202) 862-9500
Fax: (202) 429-1766

European Public Paging Association
c/o Ericsson Radio Messaging AB
Box 830, Gardsfogdevagen 18A
161 24 BROMMA
Sweden
46 8 7575965
Fax: 46 8 4044994

European Road Transport Telematics
Implementation Coordination
Organization (ERTICO)
Avenue Henri Jaspar 113
Brussels, B-1060 Belgium
32-2-538-0262
Fax: 32-2-538-0273

European Space Agency
8-10 rue Mario Nikis
75738 Paris CEDEX 15
France
33 1 53697416
Fax: 33 1 53697690

Eurosat Distribution Ltd.
Head Office, London
1 Oxgate Centre, Oxgate Lane
Edgeware Rd., London NW2 7JG, UK
(081) 452-6699

Eutelsat
33, Avenue du Maine, 75755
Tout Maine Montpasrnasse
Paris, France
33 1 45 38 47 57
Fax: 33 1 45 38 46 64

Ex Machina, Inc.
45 East 89th St.
#39-A
New York, NY 10128
(718) 965-0309

Federal Communications Commission
1919 M. St., NW
Washington, DC 20055
(202) 632-7557
Fax: (202) 632-1587
Office of Engineering and Technology
(202) 653-8117
Spectrum Allocations
(202) 652-8108
Common Carrier Bureau
(202) 634-7058

Final Analysis, Inc.
7500 Greenway Center Dr.
Suite 1240
Greenbelt, MD 20770
(301) 474-0111

Fingertip Technologies, Inc.
620 Newport Center Dr.
Suite 650
Newport Beach, CA 92660
(714) 759-9399

France Telecom
6 Place d'Alleray
75505 Paris CEDEX 15
France
33 1 44449393
Fax: 33 1 44448034

Frost & Sullivan/Market Intelligence
106 Fulton St.
New York, NY 10038
(212) 233-1080

Frost & Sullivan International
Sullivan House
4, Grosvenor Gardens
London SW1W ODH, UK
(071) 730-3438

Fujitsu Network Transmission Systems
2801 Telecom Parkway
Richardson, TX 75082
(214) 690-6000
Fax: (214) 497-6981

Fujitsu Personal Systems, Inc.
5200 Patrick Henry Dr.
Santa Clara, CA 95054
(408) 982-9500

Gandalf Mobile Systems Inc.
2 Gurdwara Rd.
Nepean, Ontario K2E 1A2, Canada
(613) 723-6500

GEC Plessey
1500 Green Hills Rd.
P.O. Box 660017
Scotts Valley, CA 95067
(408) 439-6049
Fax: (408) 438-5576

GeoWorks
2150 Shattuck Ave.
Berkeley, CA 94704
(510) 644-0883
Fax: (510) 644-0928

GEO Systems
227 Granite Run Dr.
Lancaster, PA 17601
(717) 293-7500

Geotek Communications, Inc.
20 Craig Rd.
Montvale, NJ 07645
(201) 930-9305
Fax: (201) 930-9614

Glenayre Technologies, Inc.
4800 River Green Parkway
Duluth, GA 30136
(404) 623-4900
Fax: (404) 623-0210

Global One
12490 Sunrise Valley Rd.
Reston, VA 22096
(703) 689-6040
Fax: (703) 689-7592

Globalstar L.P.
3200 Zanker Rd.
P.O. Box 640670
San Jose, CA 95134
(408) 473-4436
Fax: (408) 473-5750

GO Corp.
919 East Hillsdale Blvd.
Suite 400
Foster City, CA 94404
(415) 345-7400

Goldstar
1850 W. Drake Dr.
Tempe, AZ 85283
(602) 752-2200

GPS International Association
206 East College St.
Grapevine, TX 76051
(800) 269-1073

Granite Communications, Inc.
9 Townsend West
Suite 1
Nashua, NH 03063
(603) 881-8666

GRE America
425 Harbor Blvd.
Belmont, CA 94002
(415) 591-1400
Fax: (415) 591-2001

Great Plains Software
1701 Southwest 38th St.
Fargo, ND 58103
(701) 281-0550

GRiD Systems Corp.
7 Village Circle
Westlake, TX 76262
(817) 491-5200

GSM MoU Association
Avoca Court
Temple Rd.
Blackrock Co
Dublin, Ireland
353 1 2695922
Fax: 353 1 2695958

GTE PCS Group
600 N. Westshore Blvd.
Suite 600
Tampa, FL 33609
(813) 282-6154

GTE Mobile Communications
245 Perimeter Center Parkway
P.O. Box 105194
Atlanta, GA 30348
(404) 391-8011
Fax: (404) 391-6788

GTE Spacenet
1700 Old Meadow Rd.
McLean, VA 22102
(703) 848-1391

Hand Held Products
7510 E. Independence Blvd.
Suite 100
Charlotte, NC 28227
(704) 537-1444
Fax: (704) 532-4191

Hayes Microcomputer Products, Inc.
5835 Peachtree Corners East
Norcross, GA 30092
(404) 840-9200

Hazeltine Corp.
450 East Pulaski Rd.
Greenlawn, NY 11740
(516) 262-8499
Fax: (516) 262-8002

HEAR-IT NOW
1050 Connecticut Ave., NW
Suite 1100
Washington, DC 20036
(202) 861-1725

Herschel Shosteck Associates, Ltd.
Wheaton Plaza, South Office Bldg.
11160 Viers Mill Rd.
Suite 709
Wheaton, MD 20902
(301) 589-2259
Fax: (301) 588-3311

Hewlett-Packard Co.
5301 Stevens Creek Blvd.
Santa Clara, CA 95052
(408) 727-0700

HighwayMaster
16479 Dallas Parkway
Suite 300
Dallas, TX 75248
(214) 732-2500

Hitachi, Ktd.
4-6 Kanda-Surugadai, Chiyoda-ku
Tokyo 101-10, Japan
81-3-5295-5511
Fax: 81-3-3256-5498

Hong Kong Call Point
19th Floor
Century Plaza Three
Tai Koo Shing
Hong Kong
(852) 803-3663

Hong Kong Post Office
Telecommunications Branch
5th Floor
Sincere Bldg.
Central Hong Kong
(852) 852-9688

Hong Kong Telecom CSL Limited
19th Floor
City Plaza Phase Three
Hong Kong
(852) 803-8231

Hughes Network Systems
11717 Exploration Lane
Germantown, MD 20876
(301) 428-5500

Hughes Spaceway
P.O. Box 92424
Los Angeles, CA 90009
(310) 364-4840

Hutchison Paging Limited
9th Floor
Manlong House
611-615 Nathan Rd.
Kowloon
Hong Kong
(852) 710-6828

Hutchison Telecom
27th Floor
Great Eagle Centre
23 Harbour Rd.
Hong Kong
(852) 828-3230

Hutchison Telephone Co. Ltd.
22/F, Citicorp Centre
18 Whitfield Rd.
Causeway Bay
Hong Kong
(852) 807-9765

IBM Personal Computer Co.
1000 N.W. 51 St.
Boca Raton, FL 33432
(407) 443-2000

ICD/Link Resources
5 Speen St.
Framingham, MA 01701
(508) 872-8200

ICO Global Communications
1 Queen Caroline St.
Hammersmith, London W6 9BN, UK
(44) 181 600 1000
Fax: (44) 181 600 1199

IDB Mobile Communications, Inc.
6903 Rockledge Rd.
Suite 500
West Bethesda, MD 20817
(301) 214-8700

Illinois Superconductor Corp.
1840 Oak Ave.
3rd Floor
Evanston, IL 60201
(708) 866-0435

Independent Telecommunications
   Network, Inc.
8500 West 110th St.
Suite 600
Overland Park, KS 66210
(913) 491-1600

Industrial Computer Systems, Inc.
27972 Meadow Dr.
Evergreen, CO 80439
(303) 674-0700

Industry Canada
Spectrum Information Technology &
   Telecommunications Division
235 Queen St.
Ottawa, Ontario K1A 0H5, Canada
(613) 998-0368
Fax: (613) 952-1203

In-Flight Phone Corp.
122 West 22nd St.
Suite 100
Oak Brook, IL 60521
(708) 573-2660

Information Technology Association
   of America
1616 North Fort Myers Dr.
Suite 1300
Arlington, VA 22209
(703) 522-5055

InfraLAN Technologies, Inc.
12 Craig Rd.
Acton, MA 01720
(508) 266-1500

Inmarsat-P
1 Queen Caroline St.
London W6 9BN, UK
(44) 181 600 1000

In-Stat Inc.
7418 East Helm Dr.
Scottsdale, AZ 85260
(602) 483-4440

Institute of Electrical and Electronics
  Engineers
445 Hoes Lane
Piscataway, NJ 08855
(908) 981-0060

Institute of Electronics and
  Communication Engineers of Japan
Hamazaki Bldg., 4F, 40-14
Hongo 2-chome, Bunkyo-ku
Tokyo 113, Japan
03 817 5831

Institute of Navigation
1800 Diagonal Rd.
Suite 480
Alexandria, VA 22314
(703) 683-7101

Instrument Society of America
67 Alexander Dr.
P.O. Box 12277
Research Triangle Park, NC 27709
(919) 549-8411

Integration Systems, Inc.
625-B Purissima St.
Half Moon Bay, CA 94014
(415) 726-2620

Intel Corp.
2625 Walsh Ave.
Santa Clara, CA 95052
(408) 765-4483

Intelsat
3400 International Dr.
Washington, DC 20008
(202) 944-6963

Interactive Television Association
1019 19th St., NW
Suite 1000
Washington, DC 20036
(202) 408-0008
Fax: (202) 408-0111

InterDigital Communications Corp.
2200 Renaissance Blvd.
Suite 105
King of Prussia, PA 19406
(215) 278-7800

International Digital
  Communications, Inc.
5-20-8 Asakusabashi, Taito-ku
Tokyo 111-61, Japan
81-3-5820-0061
Fax: 81-3-5820-5370

International Mobile Satellite
  Organization
99 City Rd.
London EC1Y 1AX, UK
(44) 71 728 1000

International Semiconductor
  Cooperation Center
Urban Toranomon Bldg.
16-4, Toranomon 1-chome
Minato-ku, Tokyo 105, Japan
(03) 3597-8273

International Telecommunications
Users Group
18 Westminster Palace Gardens
London W1, UK
(44) 799-2446

International Telecommunication Union
Place des Nations
CH-1211 Geneva 20, Switzerland
(41) 22 730 61 61
Fax: (41) 22 730 64 44

Intuit, Inc.
155 Linfield Ave.
P.O. Box 3014
Menlo Park, CA 94026
(415) 322-0573

Iridium, Inc.
1401 H St., NW
Suite 800
Washington, DC 20005
(202) 326-5600
Fax: (202) 842-0006

ITS America
400 Virginia Ave., SW
Suite 800
Washington, DC 20024
(202) 484-IVHS
Fax: (202) 484-3483

Japan Business Group
1137 N. Harvey
Oak Park, IL 60302
(708) 383-9525
Fax: (708) 383-9529

Japan Electronic Industry
Development Association
Kikai Shinko Kaikan
5-8 Shibakoen 3-chome
Minato-ku, Tokyo 105, Japan
(03) 3433-1922
Fax: (03) 3433-2003

Japan Ministry of Posts and
Telecommunications
1-3-2 Kasumigaseki
Chiyoda-ku, Tokyo 100-90, Japan
(03) 3504-4086

Japan Radio Co.
Akasaka Twin Tower (Main)
2-17-22 Akasaka, Minato-ku
Tokyo 107, Japan
81-3-3584-8836
Fax: 81-3-3584-8878

Japan R&D Center for Radio Systems
Bansui Bldg.
1-5-16 Toranoman
Minato-ku
Tokyo 105, Japan
(03) 3592-1101
Fax: (03) 3592-1103

K and M Electronics, Inc.
11 Interstate Dr.
West Springfield, MA 01089
(413) 781-1350

Kansai Cellular Telephone Co.
Umeda Center Bldg., 13th Floor
2-4-12 Nakazaki-nishi, Kita-ku
Osaka City 530, Japan
(06) 375-8666
Fax: (06) 375-8200

Kansai Digital Phone Co.
Crystal Tower, 23rd Floor
1-2-27 Shiromi, Chou-ku
Osaka 540, Japan
(06) 949-5082
Fax: (06) 945-4718

Kenwood USA Corp.
2201 E. Dominquez St.
Long Beach, CA 90810
(310) 639-9000

Kokusai Denshin Denwa Co.
1-8-1 Otemachi, Chiyoda-ku
Tokyo 100, Japan
81-3-3275-4365
Fax: 81-3-3275-4229

Korea Mobile Telecom
Central Research Center
58-4, Hwaam-Dong, Yoosung-Gu
Taejon, Korea 305-348
82-42-865-0564
Fax: 82-42-865-0633

Korea Telecom
100 Sejong-ro, Chongro-Gu
110777 Seoul, Korea
82 2 7503827
Fax: 82 2 7503830

Kyushu Cellular Telephone Co.
Fuji Bldg.
1-12-9 Watanabe-dori, Chou-ku
Fukuoka City 810, Japan
(092) 713-6420

LanAir
Atidim Technology Park
Bldg. 3
Tel Aviv 61131, Israel
972 3 6459162
Fax: 972 3 6487146

Latvian Mobile Telephone Co.
39 Unijas St.
1039 RIGA
Latvia
371 2 569183
Fax: 371 7 828253

Leo One USA
150 N. Meramec
St. Louis, MO 63105
(314) 746-0567

Lexicus
345 Forest Ave.
Suite 45
Palo Alto, CA 94301
(415) 323-4771
Fax: (415) 323-4772

Link Resources Corp.
79 Fifth Ave.
New York, NY 10003
(212) 627-1500

LOCATE
17 Battery Place
Suite 1200
New York, NY 10004
(212) 509-5595

Loral Aerospace Corp.
7375 Executive Place
Suite 101
Seabrook, MD 20706
(301) 805-0591

Loral Microwave-NARDA
435 Moreland Rd.
Hauppauge, NY 11788
(516) 231-1700

Loral Qualcomm Satellite Services, Inc.
3825 Fabian Way
Palo Alto, CA 94303
(415) 852-5601

Lotus Development Corp.
Mobile Computing Division
One Rogers St.
Cambridge, MA 02142
(800) 448-2500
Fax: (617) 693-5561

LXE
303 Research Dr.
Norcross, GA 30092
(404) 447-4224

M/A-COM
100 Chelmsford St.
Lowell, MA 08351
(508) 453-3100
Fax: (508) 656-2900

Magnavox Nav-Com, Inc.
9 Brandywine Dr.
Deer Park, NY 11729
(516) 667-2235

Marconi Communications Inc.
11800 Sunrise Valley Dr.
Reston, VA 22091
(703) 620-0333
Fax: (703) 620-0415

Maritime Cellular Networks, Inc.
560 Village Blvd.
Suite 150
West Palm Beach, FL 33409
(407) 689-3050

Marubini Corp.
Telecommunications Dept.
1-4-2 Ohtemachi, Chiyoda-ku
Tokyo 100, Japan
(03) 3282-2363
Fax: (03) 3282-4835

Maryland Semiconductor, Inc.
22250 Comsat Dr.
P.O. Box 179
Clarksburg, MD 20871
(301) 353-8400
Fax: (301) 365-6845

Matra Marconi Space
Gunnels Wood Road, Stevenage
Herts SG1 2AS, UK
44-(0)1438 736698
Fax: 44-(0)1438 736069

MCI Communications Corp.
1801 Pennsylvania Ave., NW
Washington, DC 20006
(202) 872-1600

Mentor Engineering, Inc.
609 14th St., NW #503
Calgary, Alberta T2N 2A1, Canada
(403) 777-3760
Fax: (403) 777-3769

Mercury Communications Ltd.
90 Long Acre
London WC2E 9NP, UK
44-71-836-2449

Mercury One-2-One
Imperial Place
Maxwell Rd.
Borehamwood, Hertfordshire, UK
WD6 1E6 1EA
44-181-214-2320
Fax: 44-181-214-2322

Metricom, Inc.
980 University Ave.
Los Gatos, CA 95030
(408) 399-8200
Fax: (408) 399-5147

Metriplex, Inc.
25 First St.
Cambridge, MA 02141
(617) 494-9393

Micron Communications
2805 E. Columbia Rd.
Boise, ID 83706
(208) 368-3971

Micronet Communications, Inc.
720 Avenue F
Suite 100
Plano, TX 75074
(214) 422-7200

Microsoft Corp.
One Microsoft Way
Redmond, WA 98052
(206) 936-4375
Fax: (206) 936-7329

Mikros System Corp.
3490 U.S. Highway 1
Bldg. 5
Princeton, NJ 08540
(609) 987-1513

Minnesota Department of Transportation
Research & Strategic Initiatives
117 University Ave.
Room 253
St. Paul, MN 55155
(612) 296-4935

Mitsubishi Electronics America, Inc.
800 Biermann Court
Mt. Prospect, IL 60056
(708) 699-4317
Fax: (708) 824-7221

Mitsubishi International Corp.
1500 Michael Dr.
Suite B
Wood Dale, IL 60191
(708) 860-4200
Fax: (708) 860-8398

Mitsui & Co.
Telecom Business & Project Dept.
1-2-1 Ohtemachi, Chiyoda-ku
Tokyo 100, Japan
(03) 3285-7286
Fax: (03) 3285-9441

MobileComm
1800 E. County Line Rd.
Suite 300
Ridgeland, MS 39157
(601) 977-0888

Mobile Communications Holdings, Inc.
1120 19th St., NW
Suite 460
Washington, DC 20036
(202) 466-4488

Mobile Datacom Corp.
22001 Comsat Dr.
Clarksburg, MD 20871
(301) 428-2103

Monicor Electronic Corp.
2964 N.W. 60th St.
Ft. Lauderdale, FL 33309
(305) 979-1907

Motorola Cellular Infrastructure Group
1501 West Shure Dr.
Arlington Heights, IL 60004
(708) 632-5000
Fax: (708) 632-4123

Motorola, Inc.
Cellular Subscriber Group
600 North US Highway 45
Libertyville, IL 60048
(708) 523-5000

Motorola/EMBARC Communication
    Services
1500 N.W. 22nd Ave.
Boynton Beach, FL 33426
(407) 364-2000
Fax: (407) 364-3683

MTA/EMCI
1130 Connecticut Ave., NW
Suite 325
Washington, DC 20036
(202) 835-7800

Multimedia Telecommunications
    Association (formerly North
    American Telecommunications
    Association)
2000 M St., NW
Suite 550
Washington, DC 20036
(202) 296-9800
Fax: (202) 296-4993

Murata Electronics North America, Inc.
2200 Lake Park Dr.
Smyrna, GA 30080
(404) 436-1300
Fax: (404) 436-3030

Murata/Muratec
5560 Tennyson Pkwy.
Plano, TX 75024
(214) 403-3300
Fax: (214) 403-3400

NASA
400 Maryland Ave., SW
Washington, DC 20546
(202) 358-1983

National Association of State
    Telecommunications Directors
Iron Works Pike
P.O. Box 11910
Lexington, KY 40578
(606) 231-1939

National Cable Television Association
1724 Massachusetts Ave., NW
Washington, DC 20036
(202) 775-3550

National Paging & Personal
    Communications Association
2117 L St., NW
Suite 175
Washington, DC 20037
(202) 554-6722

National Semiconductor
2900 Semiconductor Dr.
Santa Clara, CA 95052
(408) 721-3892

National Telecommunications &
    Information Administration
U.S. Department of Commerce
Washington, DC 20230
(202) 377-1866
Fax: (202) 501-6198

NCR (AT&T)
WaveLAN Products
1700 South Patterson Blvd.
Dayton, OH 45479
(800) 225-5627

NEC America
Mobile Radio Division
383 Omni Dr.
Richardson, TX 75080
(800) 421-2141
Fax: (214) 907-4563

Netherlands Institute for Conformance
    Testing of Telecommunications
    Equipment (NKT)
P.O. Box 30605
2500 GP The Hague, Netherlands
070-3410582

Nettech Systems, Inc.
324 Wall St.
Princeton, NJ 08540
(609) 680-0100

NEXTEL Communications
201 Route 17 North
Rutherford, NJ 07070
(201) 438-1400

Nippon Idou Tsushin Corp.
Marukin Banco Bldg.
6 Rokubancho, Chiyoda-ku
Tokyo 102, Japan
(03) 3263-5314
Fax: (03) 3263-2133

Nokia Mobile Phones
2300 Tall Pines Drive
Suite 120
Largo, FL 34641
(813) 536-4443
Fax: (813) 530-3599

Norcom Networks Corp.
3650 131st Ave.
Suite 510
Bellevue, WA 98006
(206) 649-8868

North American Wireless, Inc.
1919 Gallows Rd.
Suite 950
Vienna, VA 22182
(703) 760-3100
Fax: (703) 760-3230

Northern Telecom
2221 Lakeside Blvd.
Richardson, TX 75208
(214) 684-8821

NovAtel Communications Ltd.
6732 8th St. Northeast
Calgary AB T2E 8M4, Canada
(403) 295-4999
Fax: (403) 295-0230

NTT Mobile Communications
    Network, Inc.
Shin-nikko Bldg.
4th Floor, East Tower
2-10-1 Toranomon, Minato-ku
Tokyo 105, Japan
(03) 5563-7230
Fax: (03) 5563-7238

Odyssey Telecommunications
    International, Inc.
One Space Park
Redondo Beach, CA 90278
(310) 812-5227
Fax: (310) 812-8232

Oki Telecom
437 Old Peachtree Rd.
Suwanee, GA 30174
(404) 995-9800

Omnipoint Corp.
7150 Campus Dr.
Colorado Springs, CO 80920
(719) 591-0823

ON Technology Corp.
One Cambridge Center
Cambridge, MA 02142
(617) 374-1400

Oracle Corp.
500 Oracle Parkway
Redwood Shores, CA 94065
(415) 506-7000

Orange Personal Communications
    Services
St. James Court
Great Park Rd.
Almondsbury
Bristol BS12 4QJ, UK
01454 624600
Fax: 01454 618501

Orbital Communications Corp.
21700 Atlantic Blvd.
Dulles, VA 20166
(800) ORBCOMM

Orbitel Mobile Communications
The Keytech Centre
Ashwood Way
Basingstoke, Hamshire RG23 8BG, UK
(44) 256-843468

Pacific Communication Sciences, Inc.
10075 Barnes Canyon Rd.
San Diego, CA 92121
(619) 535-9500
Fax: (619) 535-9235

Pacific Link Communications Limited
30/F, 26 Harbour Rd.
China Resources Bldg.
Hong Kong
(852) 879-8688
Fax: (852) 879-7770

PA Consulting Group
279 Princeton Rd.
Hightstown, NJ 08520
(609) 426-4700

PacTel Corp.
PacTel Corporate Plaza
2999 Oak Rd.
Walnut Creek, CA 94596
(510) 210-3645

Paging Network Inc.
4965 Preston Park Blvd.
Suite 600
Plano, TX 75093
(214) 985-4100

Paging Services Council
2233 Wisconsin Ave., NW
Suite 500
Washington, DC 20007
(202) 333-0700

Palm Computing
4410 El Camino Real
Suite 108
Los Altos, CA 94022
(415) 949-9560

PanAmSat
1 Pickwick Plaza
Suite 270
Greenwich, CT 06830
(203) 622-6664
Fax: (203) 622-9163

Panasonic Communications &
    Systems Co.
Two Panasonic Way
Secaucus, NJ 07094
(201) 348-7000
Fax: (201) 392-6305

Pannon GSM Telecommunications Co.
Vaci u. 37
1134 Budapest, Hungary
(361) 270 41 30

Paragraph International
1035 Pearl St.
Suite 104A
Boulder, CO 80302
(303) 443-8777

Pastel Development Corp.
113 Spring St.
New York, NY 10012
(212) 941-7500

PCMCIA
1030 East Duane Ave.
Suite G
Sunnyvale, CA 94086
(408) 720-0107

PCS Development Corp.
P.O. Box 272
Greenville, SC 29602
(803) 467-1627

Peninsula Wireless Communications
1150 Morse Ave.
Sunnyvale, CA 94089
(408) 747-1900

PenMagic Software Inc.
310-260 West Esplanade
North Vancouver, BC V7M 3G7,
    Canada
(604) 988-9982

Performance Systems International, Inc.
510 Huntmar Park Dr.
Herndon, VA 22070
(703) 904-7187

Persoft, Inc.
465 Science Dr.
Madison, WI 53711
(608) 273-6000
Fax: (608) 273-8227

Personal Communications Limited
9933364 Harcourt House
39 Glouchester Rd.
Wanchai
Hong Kong
(852) 860-8282

Personal Communications Industry
  Association
1501 Duke St.
Alexandria, VA 22314
(703) 836-3528
Fax: (703) 836-3528

Personal Communications Information
  Services, Inc.
1110 N. Glebe Rd.
Suite 500
Arlington, VA 22201
(703) 528-7300

Personal Technology Research
296 Newton Ave.
Waltham, MA 02154
(617) 893-2600

Philips Semiconductors
811 East Arques Ave.
Sunnyvale, CA 94088
(800) 447-1500

Photonics Corp.
2940 N. First St.
San Jose, CA 95134
(408) 955-7930

Pinpoint Communications, Inc.
14651 Dallas Parkway
Suite 600
Dallas, TX 75240
(214) 789-8900
Fax: (214) 789-8989

Pioneer Electronics
2265 E. 220th St.
Long Beach, CA 90810
(310) 835-6177

Plexsys International Corp.
610 Herndon Parkway
Suite 700
Herndon, VA 22070
(703) 904-4000
Fax: (703) 904-0980

Pocket Communications, Inc.
2550 M St., NW
Suite 200
Washington, DC 20037
(202) 490-4300

Portugal Telcom
Av. Foneca Pereira de Melo 40-7
F-108 Lisbon, Portugal
351 1 350 4753
Fax: 351 1 352 3444

PowerTek Industries, Inc.
14550 East Fremont Ave.
Englewood, CO 80112
(303) 680-9400

Premier Telecom Products Inc.
600 Industrial Parkway
Industrial Airport, KS 66031
(913) 791-7000

Primary Access Corp.
10080 Carroll Canyon Rd.
San Diego, CA 92131
(619) 536-3000

Probe Research
3 Wing Dr.
Suite 240
Cedar Knolls, NJ 07927
(201) 285-1500

Proxim, Inc.
295 North Bernardo Ave.
Mountain View, CA 94043
(415) 960-1630
Fax: (415) 964-5181

Psion
118 Echo Lake Rd.
Watertown, CT 06795
(203) 274-7521

P&T Luxembourg
2 rue Emile Bian
L-2999 Luxembourg
Belgium
352 49915503
Fax: 352 491221

PTT Telecom
Maanweg 174
2516 AB The Hague, Netherlands
31 70 34 33 356
Fax: 31 70 38 16 581

Pulse Engineering
12220 World Trade Dr.
San Diego, CA 92128
(619) 674-8222
Fax: (619) 674-8373

Pyramid Research, Inc.
14 Arrow St.
Cambridge, MA 02138
(617) 868-4725
Fax: (617) 868-5574

Q-bit Corp.
2575 Pacific Ave. NE
Palm Bay, FL 32905
(407) 727-1838

QUALCOMM, Inc.
6455 Lusk Blvd.
San Diego, CA 92121
(619) 587-1121
Fax: (619) 487-8276

Racotek, Inc.
7401 Metro Blvd.
Suite 500
Minneapolis, MN 55439
(612) 832-9800

Radiance Communications, Inc.
2338 A Walsh Ave.
Santa Clara, CA 95051
(408) 980-5360

RadioLAN
99 W. Tasman Dr.
Suite 102
San Jose, CA 95134
(408) 526-9170
Fax: (408) 526-9174

RadioMail Corp.
2600 Campus Dr.
San Mateo, CA 94403
(415) 286-7800

RAM Mobile Data
10 Woodbridge Ctr. Dr.
Woodbridge, NJ 07095
(908) 602-5500
Fax: (908) 602-1262

Raytheon Co.
528 Boston Post Rd.
Sudbury, MA 01776
(508) 440-2678

RDC Communications Ltd.
1735 Beach Park Blvd.
Foster City, CA 94404
(415) 345-4018
Fax: (415) 345-4694

Reliability Center for Electronic
  Components of Japan
6th Floor, Shin Daiichi Bldg.
4-13, Nihonbashi 3-chome
Chuo-ku, Tokyo 103, Japan
(03) 3272-2736

Republic of China
General Chamber of Commerce
Fu Hsing South Rd.
6th Floor, Sect. 1
Taipei, Taiwan
(2) 701-2671

Republic of the Philippines
Department of Transportation and
 Communications
National Telecommunications
 Commission
865 Vibal Bldg.
EDSA Corner Times St., Q.C.

Retix
2401 Colorado Ave.
Santa Monica, CA 90404
(310) 828-3400

Rockwell International
4311 Jamboree Rd.
Newport Beach, CA 92660
(714) 833-4600

Rogers Cantel Mobile
 Communications Inc.
Suite 2600
Commercial Union Tower
P.O. Box 249
Toronto Dominion Centre
Toronto, Ontario M5K 1J5, Canada
(416) 777-0880

Rural Cellular Association
2711 LBJ Freeway
Suite 560
Dallas, TX 75234
(800) 722-1872
Fax: (214) 243-6139

Russian Ministry of
 Telecommunications
7 Tverskaya St.
103375 Moscow
Russia
7 95 2927218
Fax: 7 95 2302097

Samsung
3655 North First St.
San Jose, CA 95134
(800) 446-0262

Sanyo
21350 Lassen St.
Chatsworth, CA 91311
(818) 998-7322

Satellite Broadcasting and
 Communications Association
225 Reinekers Lane
Suite 600
Alexandria, VA 22314
(703) 549-6990

Saudi Ministry of Post, Telegraph and
 Telephone
11112 Riyadh
Saudi Arabia
966 1 463 7072
Fax: 966 1 4032048

Scientific Advisory Group on Cellular
 Telephone Research
1711 N St., NW
Suite 200
Washington, DC 20036
(202) 833-2800

Semiconductor Industry Research
 Institute of Japan
23rd Floor, Fukoku Seimei Bldg.
2-2, Uchisaiwaicho 2-chome
Chiyoda-ku, Tokyo 100, Japan
(03) 3593-7243
Fax: (03) 3593-7250

Sharp Electronics Corp.
Sharp Plaza
Mahwah, NJ 07430
(201) 529-8200
Fax: (201) 529-9697

Shinsegi Telecom, Inc.
Kumsegi Bldg. 16, Ulchiro-1Ga
Chung-Gu, Seoul, 100-191, Korea
82-2-3708-1935

Shintom West
20435 South Western Ave.
Torrance, CA 90501
(310) 328-7200

Siemens Stromberg-Carlson
900 Broken Sound Parkway
Boca Raton, FL 33487
(407) 955-6054

Sierra Wireless, Inc.
8999 Nelson Way
Burnaby, BC V5A 485, Canada
(604) 668-7328

Sign-On Systems
9440 Santa Monica Blvd.
Suite 705
Beverley Hills, CA 90210
(310) 274-7477

Simware, Inc.
20 Colonnade Rd.
Ottawa, Ontario K2E 7M6, Canada
(613) 727-1779

Singapore Telecommunications
28-00 Comcentre
31 Exeter Rd.
0923 Singapore
65 838 3682
Fax: 65 733 1350

Skyphone
BT Aeronautical
43 Bartholomew Close
London EC1A 7HP, UK
44-171-492-4213
Fax: 44-171-492-4633

SkyTel Corp.
1350 I St., NW
Suite 1100
Washington, DC 20005
(202) 408-7444

Skywire L.P.
2620 Thousand Oaks Blvd.
Memphis, TN 38118
(901) 363-9535

Slate Corp.
15035 North 73rd St.
Scottdale, AZ 85260
(602) 443-7322

Smart Card Forum
3030 N. Rocky Point Dr. West
Suite 670
Tampa, FL 33607
(813) 286-2339
Fax: (813) 281-8752

Smart Card Industry Association
1420 K St., NW
Suite 400
Washington, DC 20005
(202) 371-1600
Fax: (202) 789-0335

Snider Telecom
P.O. Box 4189
Little Rock, AR 72214
(501) 661-7600

Software Corp. of America
100 Prospect St.
Stamford, CT 06901
(203) 359-2773
Fax: (203) 359-3198

Software Publishers Association
Pen Special Interest Group
1730 M St., NW
Suite 700
Washington, DC 20036
(202) 452-1600

Solectek Corp.
6370 Nancy Ridge Dr.
San Diego, CA 92121
(619) 450-1220
Fax: (619) 457-2681

Sony Corp. of America
One Sony Dr.
Park Ridge, NJ 07656
(201) 930-7066
Fax: (201) 930-6623

Southwestern Bell Mobile Systems
18111 Preston Rd.
Suite 900
Dallas, TX 75252
(214) 733-2001
Fax: (214) 733-2012

Spacecom Systems
One Technology Plaza
7140 S. Lewis Ave.
Tulsa, OK 74136
(918) 488-4848

Spar Aerospace Limited
Satellite & Communications
   Systems Div.
21025 Trans-Canada Highway
Ste-Anne-de-Bellevue, Quebec H9X
   3R2, Canada
(514) 457-2150

SpectraLink Corp.
1650 38th St.
Suite 202E
Boulder, CO 80301
(303) 440-5330

Spectrix Corp.
108 Wilmot Rd.
Suite 450
Deerfield, IL 60015
(708) 317-1770
Fax: (708) 317-1517

Spectrum Ericsson
45 Crossways Park Dr.
Woodbury, NY 11797
(516) 822-9810

Spectrum Information
   Technologies, Inc.
1615 Northern Blvd.
Manhasset, NY 11030
(516) 627-8992

Sprint Cellular
O'Hare Plaza
8725 W. Higgins Rd
Suite 650
Chicago, IL 60631
(312) 399-2225
Fax: (312) 693-7432

Stanford Telecommunications, Inc.
2421 Mission College Blvd.
Santa Clara, CA 95056
(408) 980-5684
Fax: (408) 727-1482

Starsys Global Positioning
4400 Forbes Blvd.
Lanham, MD 20706
(301) 459-8832

Stat S.A.
36-38 Agias Paraskevis Str.
12132 Athens, Greece
30 1 570700
Fax: 30 1 5707070

Strategies Unlimited
201 San Antonio Circle
Suite 205
Mountain View, CA 94040
(415) 941-3438

Sumitomo Corp.
Satellite Communications
   Business Dept.
1-2-2 Hitotsu-bashi, Chiyoda-ku
Tokyo 100, Japan
(03) 3217-5087
Fax: (03) 3217-7018

Sumitomo Electric U.S.A., Inc.
Park Avenue Tower
65 East 55th St., 16th Floor
New York, NY 10022
(212) 308-6444
Fax: (212) 308-6575

Sun Microsystems
2550 Garcia
Mountain View, CA 94043
(415) 960-1300

Supreme Telecommunications Ltd.
Prabhu Darshan
230 Lady Jamshedji Rd.
Dadar (W), Mumbal 400 028
India
91 22 4457677
Fax: 91 22 4449048

Swedish Mobile Telecommunications
    Association
Box 1416, S-111 84
Stockholm, Sweden
46 8 2476 48
Fax: 46 8 21 84 96

Swiss Telecom PTT
Viktoriastrasse 21
CH 3030 Bern, Switzerland
41 31 338 2421
Fax: 41 31 338 7383

Symbionics Networks Ltd.
St. John's Innovation Park
Crowley Rd.
Cambridge CB 4WS, UK
44 1223-421025
Fax: 44 1223-421031

Symbol Technologies, Inc.
2145 Hamilton Ave.
San Jose, CA 95128
(408) 369-2634
Fax: (408) 446-4630

Tadiran Telecommunications
c/o Oreet Marketing Communications
15 Kineret St.
51201 Rene Beyak
Israel
972 3 5706527
Fax: 972 3 5706526

Tandy Corp.
700 One Tandy Center
Forth Worth, TX 76102
(817) 390-3300

TDMA Forum
P.O. Box 71
Massillon, OH 44648
(216) 833-8690
Fax: (216) 833-4002

Technologic Partners
419 Park Ave. South
Suite 500
New York, NY 10016
(212) 696-9330

TekNow, Inc.
4909 E. McDowell
Phoenix, AZ 85008
(602) 266-7800

Tektronix, Inc.
P.O. Box 500
Beaverton, OR 97077
(503) 629-4057

Telecom Denmark A/S
Telegrade 2
DK-2630 Taastrup
(45) 42-52-91-11
Fax: (45) 42-52-80-76

Telecom Finland
P.O. Box 54
00511 Helsinki, Finland
358 0 20401
Fax: 358 0 20402032

Telecom Italia
Cia Paolo di Dono 44
00143 Roma, Italy
39 6 3687 4930
Fax: 39 6 3687 4909

Telecommunications Division
Dept. of Transport, Energy and
  Communications
Ely Court
7 Ely Place
Dublin 2, Ireland
(353) 1 671-5233

Telecommunications Industry
  Association
2500 Wilson Blvd.
Suite 300
Arlington, VA 22201
(703) 907-7700
Fax: (703) 907-7727

Telecommunications Research
  Associates
505 Bertrand Ave.
St. Marys, KS 66536
(800) 872-4736

Tele Denmark
Norregade 21
0900 Copenhagen K, Denmark
45 33993183
Fax: 45 33145625

Teledesic Corp.
2300 Carillon Point
Kirkland, WA 98033
(206) 803-1400

Teledyne Microwave
1290 Terra Bella Ave.
Mountain View, CA 94043
(415) 960-8601

Telefonia
Beatrix de Bobadilla 3a
28040 Madrid, Spain
34 1 5849999
Fax: 34 1 5849643

Teleglobe
1000 de la Gauchetiere St.
Montreal H3B 4X5, Canada
(514) 868-1367
Fax: (514) 868-7764

Telecom SA Ltd.
Corporate Business (SK) P/Bag X74
Pretoria, South Africa
27 12 3111531
Fax: 27 2 3238255

Telepartner International
135 South Rd.
Farmington, CT 06032
(800) 935-3270
Fax: (203) 674-8243

Telesciences
351 New Albany Rd.
Moorestown, NJ 08057
(609) 866-1000
Fax: (609) 866-2439

Telesystems SLW, Inc.
85 Scarsdale Rd.
Suite 201
Don Mills, Ontario M38 2R2, Canada
(416) 441-9966

Telia Sweden
Marbackagatan 11
S-123-86 Farsta, Sweden
46 8 713 1964
Fax: 46 8 713 7362

Tellabs, Inc.
30 North Ave.
Burlington, MA 01803
(617) 273-1400
Fax: (617) 273-4160

Telular Corp.
920 Deerfield Pkwy.
Buffalo Grove, IL 60089
(708) 465-4500
Fax: (708) 465-4501

The Freedonia Group
3570 Warrenville Center Rd.
Suite 201
Cleveland, OH 44122
(216) 921-6800

The Gartner Group
5201 Great America Parkway
Suite 219
Santa Clara, CA 95054
(408) 748-1111

The PCS Group
15851 North Dallas Parkway
Suite 600
Dallas, TX 75248
(214) 450-5930
Fax: (214) 450-5931

3Com Corp.
5400 Bayfront Plaza
P.O. Box 58145
Santa Clara, CA 95052
(408) 764-5000
Fax: (408) 764-5004

Timex Corp.
Timex Data Link Business Group
Park Rd. Ext.
Middlebury, CT 06762
(203) 573-5000

TMI Communications
1601 Telesat Ct.
P.O. Box 9826
Ottawa, Ontario K1G 5M2, Canada
(613) 742-4000
Fax: (613) 742-4100

Tokai Digital Phone Co.
1-13-26 Nishiki, Naka-ku
Nagoya-shi 460, Japan
(052) 222-5151
Fax: (052) 222-5150

Tokyo Digital Phone Co.
Sumitomo Sarugaku-cho Bldg.
2-8-8 Sarugaku-cho, Chiyoda-ku
Tokyo 101, Japan
(03) 5280-2222
Fax: (03) 5259-7129

Toshiba America, Inc.
9740 Irvine Blvd.
Irvine, CA 92713
(714) 583-3000

Totally Wireless
1062 S. Saratoga-Sunnyvale Rd.
San Jose, CA 95129
(408) 366-5960
Fax: (408) 253-2093

Traveling Software
18702 North Creek Parkway
Bothell, WA 98011
(206) 483-8088

TRW Wireless Communications
Building J7
495 Java Dr.
P.O. Box 3510
Sunnyvale, CA 94088
(408) 752-2867

TU-KA Cellular Kansai
KF Center Bldg., 8th Floor
3-6-2 Bingo-machi, Chou-ku
Osaka City 541, Japan
(06) 266-5822
Fax: (06) 266-9775

TU-KA Cellular Tokia, Inc.
Hazama Bldg., 3rd Floor
2-4-16 Nishiki-ku
Nagoya City 460, Japan
(052) 222-3111
Fax: (052) 222-3110

TU-KA Cellular Tokyo Co.
Shiba Dai-Mon Center Bldg.
5th Floor
1-10-11 Shiba Dai-Mon, Minato-ku
Tokyo 105, Japan
(03) 5400-6100
Fax: (03) 5400-6150

U.S. Department of Commerce
Trade Statistics Division
Room 2217
14th St. and Constitution Ave., NW
Washington, DC 20230
(202) 377-4211

U.S. GPS Industry Council
1100 Connecticut Ave., NW
Suite 535
Washington, DC 20036
(202) 296-1653

U.S. Paging Corp.
1680 Route 23 North
Wayne, NJ 07470
(201) 305-6000

U.S. Robotics Communications
  Corp.
4505 South Wasatch Blvd.
Salt Lake City, UT 84124
(801) 273-6635

U.S. Telephone Association
1401 H St., NW
Suite 600
Washington, DC 20005
(202) 326-7300

US WEST NewVector Group
3350 161st Ave. S.E.
Bellevue, WA 98008
(206) 747-4900
Fax: (206) 450-8695

United Telecommunications, Inc.
2330 Shawnee Mission Parkway
Westwood, KS 66213
(913) 624-2641

Universal Paging Corp.
20 Broad Hollow Rd.
Melville, NY 11747
(516) 385-4100

UTAM, Inc.
1155 Connecticut Ave., NW
Washington, DC 20036
(202) 429-6565

UTC—The Telecommunications
  Association (formerly Utilities
  Telecommunications Council)
1140 Connecticut Ave., NW
Suite 1140
Washington, DC 20036
(202) 872-0030

Venture Development Corp.
One Apple Hill
Natick, MA 01760
(508) 653-9000

Virginia Polytechnic Institute and
  State University
Mobile & Portable Radio Research
  Group
Blacksburg, VA 24061
(703) 231-6834

Vodafone Group Plc
The Courtyard
2-4 London Rd.
London RG13 1JL, UK
44 1635 33251
Fax: 44 1635 37242

Vodafone Pty Ltd.
799 Pacific Highway
Chatswood NSW 2067
Australia
P.O. Box 1066
Chatswood NSW 2057
61 2 415 7000
Fax: 61 2 415 7071

VITA
1600 Wilson Blvd.
Suite 500
Arlington, VA 22209
(703) 276-1800

VTech Communications (UK) Limited
Unit B5, Armstrong Mall
Southwood Summit Centre
Famborough, Hants GU14 0NR, UK
44 1252 370595
Fax: 44 1252 373440

Watkins-Johnson
3333 Hillview Ave.
Palo Alto, CA 94304
(415) 813-2958

Westinghouse Electric Corp.
P.O. Box 17319
Baltimore, MD 21203
(410) 765-4146
Fax: (410) 765-7941

Wi-LAN, Inc.
30012 Manning Close, N.E.
T2E 7N6 Calgary
Alberta, Canada
(403) 273-9133

Windata, Inc.
10 Bearfoot Rd.
Northboro, MA 01532
(508) 393-3330

WinStar Telecommunications Group
7799 Leesburg Pike
Suite 401 South
Tysons Corner, VA 22043
(703) 917-9117
Fax: (703) 917-6557

Wireless Cable Association International
1140 Connecticut Ave., NW
Suite 810
Washington, DC 20036
(202) 452-7823

Wireless Communications Alliance
1639 Lewiston Dr.
Sunnyvale, CA 94087
(408) 737-7720

Wireless Communications Division
Office of Telecommunications
International Trade Administration
U.S. Department of Commerce
Room 1009
Washington, DC 20230
(202) 482-4466

Wireless Dealers Association
11833 N. 55th Place
Phoenix, AZ 85254
(800) 624-6918
Fax: (800) 820-2284

Wireless LAN Alliance
409 Sherman Ave.
Palo Alto, CA 94306
(415) 328-5555
Fax: (415) 328-5016

Wireless LAN Interoperability Forum
1111 W. El Camino Real
Suite 109-171
Sunnyvale, CA 94087
(415) 960-1630

Wireless Logic, Inc.
4010 Moorpark Ave.
Suite 105
San Jose, CA 95117
(408) 246-1538

Wireless Spectrum Technology, Inc.
P.O. Box 1417
Northbrook, IL 60065
(708) 480-0066

Wireless Telecom, Inc.
3025 South Parker Rd.
Suite 1000
Aurora, CO 80014
(303) 338-4200

Wireless Telecom Group, Inc.
(NOISE/COM)
E. 49 Midland Ave.
Paramus, NJ 07652
(201) 261-8797

WiSE Technologies
130 Knowles Dr.
Los Gatos, CA 95030
(408) 376-0250
Fax: (408) 376-0506

WordPerfect Corp.
1555 North Technology Way
Orem, UT 84057
(801) 222-4050
Fax: (801) 222-3977

Wynd Communications Corp.
4251 S. Higuera
Bldg. 800
San Luis Obispo, CA 93401
(800) 549-6000

Xircom, Inc.
2041 Landings Dr.
Mountain View, CA 94043
(415) 691-2500
Fax: (415) 691-1064

Yankee Group
200 Portland St.
Cambridge, MA 02114
(617) 367-1000
Fax: (617) 367-5760

Zenith Data Systems
2150 East Lake Cook Rd.
Buffalo Grove, IL 60089
(800) 553-0331

Zsigo Wireless Data Consultants, Inc.
2875 Northwind Dr.
Suite 232
East Lansing, MI 48823
(717) 337-3995

WORLD CELLULAR MARKET, BY COUNTRY

The following is a country-by-country table listing information on international cellular systems, including standard used, startup date, number of subscribers, system operator, system supplier, and other details as of August 1996. The table is updated three times a year and is available by subscription from the National Technical Information Service, or via the World Wide Web at www.ita.doc.gov/industry/tai/telecom/telecom.html.

The information was compiled by the U.S. Department of Commerce, International Trade Administration, Office of Telecommunications. The principal information source is the overseas staff of the U.S. and Foreign Commercial Service reporting from U.S. embassies around the world, supplemented by trade publications and industry research. The latter includes *International Cellular Newsletter*, published by Kagan World Media, Ltd.; the *Financial Times*; *Mobile Communications International*; *Cellular Marketing*; *Cellular Business*; *PICA Journal*; the Cellular Telecommunications Industry Association; MTA/EMCI; *Cellular/Mobile Communications Directory*; Asociacion Mexicana de Concesionaires de Radiotelefonia Celular, A.C. (AMCEL); Northern Business Information; and International Technology Consultants.

Note that the number of subscribers is based on the most recent figures available for each network.

Country	Standard	Start-Up	Subscribers	As of	System Operator	U.S. Partners	Vendor
Albania	GSM	5/96			GSM of Albania		Alcatel
Algeria	NMT-900	12/89	1,500	12/95	Algerian PTT		Nokia
Andorra	NMT-450	7/90	235	5/96	Servei de Telecomunicacions d'Andorra		Ericsson
	GSM	3/95	3,587	5/96	Servei de Telecomunicacions d'Andorra		Motorola, Siemens
Angola	AMPS	2/94	1,994	12/95	Angola Telecom		Motorola
Anguilla	AMPS	1995			Cable & Wireless		Plexsys
Antigua and Barbuda	AMPS	12/89			Boatphone Antigua		Plexsys
Argentina	NAMPS	11/89	200,000	3/95	Compania de Radiocomunicaciones Moviles S.A.	BellSouth 42.46%, Motorola 57%	Motorola
	AMPS-TDMA	3/93	140,000	1/96	MoviStar		Ericsson
	AMPS	10/94	120,000	6/96 CTI ttl	Compania de Telefonos del Interior	GTE 23%, AT&T 10%	Lucent, Astronet, Plexsys

Country	Standard	Launch	Subscribers	Date	Operator	Ownership	Equipment
Argentina	AMPS	10/94			Compania de Telefonos del Interior	GTE 23%, AT&T 10%	Lucent, Astronet, Plexsys
	AMPS-TDMA	3/96			Compania de Comunicaciones Personales del Interior		Ericsson
	AMPS-TDMA	3/96			Telefonica Comunicaciones Personal		Ericsson
Aruba	AMPS-TDMA	1993?			SETAR		Nortel
	NAMPS	End 95			Caribbean Wireless Communications		Motorola
Australia	AMPS	12/86	1,801,000	12/95	Telstra Mobilenet		Ericsson
	GSM	4/93	250,000	12/95	Telstra Mobilenet		Ericsson
	GSM	5/93	350,000	3/96	Optus	BellSouth 24.5%	Nokia, Nortel Matra
	GSM	4/93	220,000	3/96	Vodafone		Ericsson
Austria	C-450	11/84	31,562	5/96	Mobilkom		Motorola
	TACS	7/90	245,998	5/96	Mobilkom		Motorola
	GSM	12/93	166.308	5/96	PTV		Alcatel, Motorola, Nortel, Siemens
	GSM	7/1/96			Oe-Call		Siemens

Country	Standard	Start-Up	Subscribers	As of	System Operator	U.S. Partners	Vendor
Azerbaijan	GSM	10/96			Azercell		Ericsson
	TACS	2/94	2,100	2/96	Baksell		Motorola
Bahamas	AMPS-TDMA	1988	3,000	12/94	BATELCO		Nortel
Bahrain	TACS	9/87	24,500	10/94	Bahrain Telecommunications Company		NEC
	GSM	5/95			Bahrain Telecommunications Company		Ericsson
Bangladesh	AMPS	8/93	3,800	12/95	Bangladesh Telecom		Motorola
Barbados	AMPS	1990	2,500	12/94	BARTEL		Nortel
Bashkortostan					Unknown		
Belarus	NMT-450	4/93	5,400	2/96	BelCel		Ericsson
Belgium	NMT-450	7/87	43,500	5/96	Belgacom Mobile		Alcatel, Bell Telephone, Nokia, Ericsson
	GSM	1/94	231,000	5/96	Belgacom Mobile	AirTouch 25%	Alcatel, Motorola, Siemens
	GSM	1996			Mobistar		Alcatel, Motorola, Nortel Matra

Country	System	Start	Subscribers	Date	Operator	Ownership	Equipment
Belize	AMPS-TDMA	11/1/93	1,852	1/95	Belize Telecom, Ltd.	MCI 23.5%	Nortel
Benin	AMPS	1995	1,050	12/95	OPT		Unknown
Bermuda	AMPS-TDMA	2/87	5,000	12/94	Bermuda Cellular		Nortel
Bolivia	AMPS	11/91	8,100	12/95	Telefonica Celular de Bolivia	Millicom 89%	Motorola
	AMPS	2Q96			ENTEL		Unknown
Botswana	Unknown	1996?			Botswana Telecom and foreign partner		Undecided
Brazil	AMPS	1990	85,000	12/95	Telebrasilia		Nortel; tender for DAMPS pending
	AMPS	8/90	120,000	12/95	Telerj		NEC
	AMPS	8/93	600,000	12/94	Telesp		NEC
	AMPS	1993			Telesp		NEC
	AMPS	1993			Telesp		NEC
	AMPS	5/92	34,000	12/94	Telepar		Lucent, Plexsys
	AMPS	1992	12,000	12/94	Telest		Nortel
	AMPS	1993	80,000	12/95	CRT		Ericsson
	AMPS-TDMA	1993	7,800	12/94	Sercomtel		Ericsson
	AMPS	1994			Telem		
	AMPS	1993	10,000	12/94	CTBC		Ericsson

Country	Standard	Start-Up	Subscribers	As of	System Operator	U.S. Partners	Vendor
Brazil	AMPS	1993	95,000	12/95	Telemig		Nortel
	AMPS	1993			Telepar		NEC
	AMPS	1993	29,000	12/94	Telebahia		NEC
	AMPS	1993	19,700	12/94	Telegoias		Nortel
	AMPS	1993	20,000	12/94	Telesc		Ericsson
	AMPS	1994	16,600	12/94	Teleceara		Ericsson
	AMPS	1993			Telesp		Motorola
	AMPS	1993			Telesp		Lucent
	AMPS	1993			CETERP		Nortel
	AMPS	1993			Telesp		Motorola
	AMPS	1994	6,000	12/94	Telpe		Ericsson
	DAMPS	1996?			Two licenses in each region to be tendered 10/96		Undecided
Brunei	AMPS	2/89	7,700	10/94 AMPS	Jabatan Telekom Brunei		Motorola
	AMPS	1992?	11,000	1994	Jabatan Telekom Brunei		NEC
	GSM	End 94			JTB/Sultan		Lucent

Country		Date	Subscribers	Launch	Operator	Partner	Vendor
Bulgaria	NMT-450	12/93	19,700	2/96	Radio Telecommunications Company		Ericsson
	GSM	9/95	5,350	2/96	MobilTEL AG	US WEST	Siemens
	GSM	Late 96			Mobikom		Undecided
Burkina Faso	Undecided	1997?			Onatel		Undecided
Burma (Myanmar)	AMPS	11/92	2,500	6/95	Myanmar Posts & Telecoms		Ericsson
Burundi	AMPS	9/93	500	Mid-96	Telecel-Burundi	Telecel International Ltd.	Plexsys
Cambodia	AMPS	10/92	5,000	12/93	Camtel		Motorola
	ETACS	Unknown	2,000	Mid-94	DPT/TRI		Ericsson
	NMT-450	Unknown	300	4/94	CamShin		Nokia
	NMT-900	Unknown	2,000	4/94	DPT/Casacom		Nokia
	AMPS	Unknown	2,000	Mid-94	Cambodia Cellular		Unknown
	Unknown	Unknown			SingTel Cambodia		Unknown
	GSM	Late 96			Cambodia GSM Ltd.	Millicom 61.5%	Unknown
Cameroon	GSM	1994	2,485	12/95	Dirtel		Philips, Siemens
Canada	AMPS-DAMPS	7/85	946,300	6/94 B-band	Bell Cellular		Nortel

Country	Standard	Start-Up	Subscribers	As of	System Operator	U.S. Partners	Vendor
Canada	AMPS-TDMA	1/86	580,000	6/94	Cantel Cellular		Ericsson
	AMPS-CDMA	Unknown			BC Cellular		Nortel, NovAtel
	AMPS	Unknown			AGT Cellular		Unknown
	AMPS	Unknown			Islandtel Cellular		Unknown
	AMPS	Unknown			MT&T Cellular		Unknown
	AMPS	5/89			NB Tel Cellular		Nortel
	AMPS	Unknown			Newtel Cellular		Unknown
	AMPS	Unknown			MTS Cellular		Unknown
	AMPS	Unknown			SaskTel Cellular		Unknown
	AMPS	Unknown			EDTel Cellular		Unknown
	AMPS	Unknown			Kenora Cellular		Unknown
	AMPS	Unknown			Thunder Bay Cellular		Unknown
	AMPS	Unknown			Quebec Telephone Cellulaire		Unknown
Cayman Islands	AMPS	1987	1,595	12/94	Cable & Wireless (W.I.) Ltd.		Plexsys
Central African Republic	AMPS	1995	500	Mid-96	TELECEL-CAR	Telecel International Ltd.	Unknown

Country	Technology	Start	Subscribers	Date	Operator	Ownership	Vendor
Chile	AMPS-TDMA	3/89	80,000	3/96	CTC Cellular		Ericsson, NEC
	AMPS-TDMA	5/89	80,000	3/96	BellSouth Celular	BellSouth 100%	Nortel
	AMPS	6/91	35,000	11/94	Telecom Celular	Motorola 33%	Motorola
	AMPS	1991	35,000	11/94	VTR Celular Telecommunications	SBC 40%	Motorola, Plexsys
China	TACS	1989	3,435,000	12/95 total	Beijing Telecom. Administration		Motorola
	TACS	1989			Beijing Telecom. Administration		Ericsson
	TACS	1989			Shanghai Post & Telecom. Administration		Ericsson, Motorola
	TACS	1989			Guangdong Mobile Communications Corp.		Ericsson
	TACS	1992			Hebei PTA		Ericsson
	TACS	1992			Hainan PTA		Ericsson
	TACS	1992			Hohhot PTA		Ericsson
	TACS	1992			Chengdu/Chongqing		Ericsson, Nortel
	TACS	1992			Fuzhou PTA		Motorola
	TACS	1992			Harbin PTA		Motorola
	TACS	1992			Jinan, Qingdao		Motorola

Country	Standard	Start-Up	Subscribers	As of	System Operator	U.S. Partners	Vendor
China	TACS	1992	10,464	3/95	Taiyuan PTA		Motorola
	TACS	1992			Wuhan PTA		Motorola
	TACS	1992			Xiamen PTA		Motorola
	TACS	1992			Zhengzhou/Luoyang		Motorola
	TACS	1992			Xian PTA		Motorola
	AMPS	1993	7,000	7/94	Urumqi PTA		Lucent, Motorola
	AMPS	1993			Yinchuan PTA		Lucent, Motorola
	AMPS	1993			Xi'an PTA		Lucent, Motorola
	AMPS	1993			Baoji PTA		Lucent, Motorola
	TACS	1993?			Tianjin PTA		Ericsson, Motorola
	TACS	1993?	45,000	2/95	Guangxi PTA		Ericsson
	TACS	1993	10,000	4/95	Xuzhou PTA		Motorola
	TACS	1993?			Zhejiang PTA		Motorola
	Unknown	Unknown			Chongqing PTA		NovAtel
	GSM	12/94			Guangdong Mobile Communications Corp.		Ericsson, Nortel Matra, Siemens
	GSM	8/95			Zhejiang PTA		Alcatel, Motorola, Nokia, Siemens

Technology	Date	Number	Date	Operator	Vendor
GSM	10/94	10,000	7/95	Beijing Telecom. Administration	Motorola, Nokia
ETDMA	1994			Chengdu PTA	Alcatel, Hughes Network Systems
GSM	2/95			Shanghai Post & Telecom. Administration	Alcatel, Siemens
TACS	Unknown			Hunan PTA	Ericsson
AMPS-TDMA	1995			UniCom	Ericsson
AMPS	1995			UniCom	Ericsson
GSM	1995			UniCom	Italtel
GSM	7/95			UniCom	Motorola
GSM	1995			UniCom	Motorola
GSM	1995			UniCom	Unknown
GSM	1995			UniCom	Unknown
TACS	1995			Liaoning PTA	Motorola
TACS	1995			Heilongjiang PTA	Motorola
Unknown	Unknown	695	12/94	Tibet Autonomous Region PTA	Unknown
GSM	7/95			Hebei PTA	Nortel
AMPS	Unknown			Shaanxi PTA	Lucent

Country	Standard	Start-Up	Subscribers	As of	System Operator	U.S. Partners	Vendor
China	AMPS	Unknown			Xinjiang PTA		Lucent
	AMPS	Unknown			Yunnan PTA		Lucent
	AMPS	Unknown			Ningxia PTA		Lucent
	GSM	10/95			Shandong PTA		Ericsson
	GSM	End 95			Fujian PTA		Motorola, Nokia
	GSM	12/95			Hubei PTA		Italtel
	GSM	1995?			Guangxi PTA		Ericsson
	GSM	10/95			Liaoning PTA		Ericsson
	GSM	7/95			UniCom		Motorola, Siemens
	GSM	7/95	10,000	12/95	UniCom Shanghai Co.	Bell Atlantic, McCaw 23%	Motorola, Nokia, Siemens
	GSM	7/95			UniCom		Motorola, Siemens
	GSM	Mid-96			UniCom		Unknown
	GSM	Mid-96			UniCom		Unknown
	GSM	3/96			Henen PTA		Nokia

Technology	Date	Subscribers	Launch	Operator	Partner	Vendor
GSM	1/94			Guangzhou PTA		Nortel
GSM	5/94			Shenzhen PTA		Siemens
GSM	10/94			Zhuhai PTA		Italtel
GSM	End 95			Jiangsu PTA		Ericsson, Alcatel
Unknown	1995			Shaanxi PTA		Nortel
GSM	5/95			Hainan PTA		Italtel
GSM	3/95			Tianjin PTA		Motorola, Nortel
GSM	1996			UniCom		Unknown
GSM	1996			UniCom		Unknown
GSM	1996			UniCom	Ameritech	Unknown
GSM	1996			UniCom		Unknown
GSM	1996			UniCom		Unknown
GSM	1996			UniCom		Unknown
GSM	1996			UniCom		Unknown
GSM	1996			UniCom		Unknown
GSM	1996			UniCom		Unknown
GSM	1996			UniCom		Unknown
TACS	1995	27,000	5/95	Zhongshan PTA		Unknown
GSM	End 95			UniCom		Unknown
GSM	11/95			Heilongjiang PTA		Ericsson

Country	Standard	Start-Up	Subscribers	As of	System Operator	U.S. Partners	Vendor
China	GSM	1996			Ningxia PTA		Alcatel
	GSM	1996			Guizhou PTA		Alcatel
	GSM	1996			Shaanxi PTA		Alcatel
	GSM	1996			Hunan PTA		Motorola
	GSM	3Q96			Sichuan PTA		Motorola
	GSM	1996			Fujian UniCom		Motorola
	GSM	1996			Yantai UniCom		Motorola
	GSM	1996			Nantong UniCom		Motorola
	GSM	1996			Shenzhen UniCom		Motorola
	GSM	1997			Suzhou UniCom		Unknown
	TACS	1992			Jiangsu PTA		Motorola
	GSM	1996			Jiaxing PTB		Alcatel
	GSM	1996			Jilin PTA		Alcatel
	GSM	1996			UniCom		Unknown
	GSM	3Q96			Chongqing UniCom		Nortel
	GSM	1997			Heilongjiang UniCom		Unknown
Colombia	AMPS-TDMA	6/17/94	44,000	8/95	Celumovil	McCaw 35%, LCC	Nortel

Country	System	Date	Subscribers	Date	Operator	Ownership	Vendor
	AMPS-TDMA	7/94	47,916	12/95	Cocelco		Ericsson
	AMPS-TDMA	6/9/94	15,000	8/95	Celumovil de la Costa	McCaw 10%, LCC	Nortel
	AMPS-TDMA	4/94	81,339	12/95	Comcel S.A.		Nortel
	AMPS-TDMA	Mid-94	40,886	12/95	Occel S.A.		Nortel
	AMPS-TDMA	8/31/94	14,000	12/95	Celcaribe	Millicom 32.5%	Ericsson
Cook Islands (New Z.)	AMPS	10/96			PTT		Plexsys
Costa Rica	AMPS	5/89	0	6/96	Millicom Costa Rica	Millicom 70%	NovAtel
	AMPS	4/94	18,000	12/95	Instituto Costarricense de Electrodad		Plexsys
Cote d'Ivoire	GSM	2Q96			Comstar	International Wireless	Motorola, Siemens
	GSM	End 96			Loteny Telecom	Telecel International Ltd.	Unknown
	GSM	End 96			Societe Ivoirenne de Mobiles		Unknown
Croatia	NMT-450	10/90	38,500	2/96	Croatian Post & Telecommunications		Ericsson

Country	Standard	Start-Up	Subscribers	As of	System Operator	U.S. Partners	Vendor
Croatia	GSM	7/95			Croatian Post & Telecommunications		Siemens
	NMT-450	Unknown			Tomsktelekom & Nicola Tesla		Unknown
Cuba	AMPS	2/93	2,000	6/95	Cubacel		Ericsson
Curacao	AMPS	2/89	8,500	6/95	Setel Cellular		Ericsson
Cyprus	NMT-900	12/88	22,679	5/96	Cyprus Telecom Authority		Ericsson
	GSM	1995	34,725	5/96	Cyprus Telecom Authority		Ericsson
Czech Republic	NMT-450	10/91	50,000	3/96	Eurotel Prague	Bell Atlantic 24.5%, US WEST 24.5%	Nokia
	GSM	7/1/96			Eurotel Prague	Bell Atlantic 24.5%, US WEST 24.5%	Nokia
	GSM	9/96			RadioMobil		Motorola, Siemens

Country	Standard	Date	Subscribers	Date	Operator	Ownership	Equipment
Denmark	NMT-450	9/81	28,814	5/96	TeleDanmark Mobil		Ericsson
	NMT-900	12/86	275,015	5/96	TeleDanmark Mobil		Ericsson
	GSM	3/92	360,000	5/96	TeleDanmark Mobil		Ericsson
	GSM	3/92	360,000	5/96	Dansk Mobiltelefon	BellSouth 29%	Nokia
Djibouti	GSM	1996?			Undecided		Undecided
Dominica	AMPS	Unknown			BOATPHONE		Plexsys
Dominican Republic	AMPS	3/87	38,000	12/95	CODETEL	GTE 100%	Lucent, Motorola
	NAMPS	1992	30,000	12/95	TRICOM	Motorola 40%	Motorola, Nortel
	AMPS	1989	3,000	12/95	All Americas Cable & Radio		Nortel
Ecuador	AMPS-TDMA	1994	30,000	5/96	Conecel		Nortel
	AMPS	7/94	20,000	5/96	Otecel	McCaw 2%; LCC	Ericsson
					Gov't may license two more operators		Undecided
Egypt	MATS	5/87	12,000	12/95	Arab Republic of Egypt Nat'l. Telecom. Org.		Matsushita
	GSM	11/96			Arab Republic of Egypt Nat'l Telecom. Org.		Alcatel

Country	Standard	Start-Up	Subscribers	As of	System Operator	U.S. Partners	Vendor
El Salvador	AMPS-NAMPS	1/93	15,000	12/95	Telemovil El Salvador	Millicom 70%	Ericsson
	AMPS	Unknown			To be tendered 1995		Undecided
Estonia	NMT-450	3/91	20,036	2/96	Eesti Mobiil Telefon		Ericsson, Nokia
	NMT-900	7/92	2,000	12/94	Eesti Mobiil Telefon		Ericsson, Nokia
	GSM	1/95	1,950	2/96	Radiolinja Estonia AS		Nokia
	GSM	9/93	14,200	2/96	Eesti Mobiil Telefon	Millicom	Ericsson
Faroe Islands	NMT-450	1/89	2,048	5/96	Telefonwerk Foroya Lodtings		Ericsson
	NMT-900	1/92	967	5/96	Telefonwerk Foroya Lodtings		Ericsson
Fed. Rep. of Yugoslavia	NMT-900	Unknown			PTT		Alcatel, Ericsson
	GSM	12/96			Mobile Telecommunications Srbija		Ericsson
Fiji	GSM	8/94	2,400	12/95	Fiji Cellular Ltd.		Alcatel, Ericsson
Finland	NMT-450	12/81	194,711	5/96	Telecom Finland		Ericsson, Mitsubishi, Mobira

	NMT-900	12/86	441,522	Telecom Finland		Ericsson, Mobira, Nokia
	GSM	6/92	377,206	Telecom Finland		Ericsson, Nokia
	GSM	1992	190,000	Radiolinja Oy		Nokia, Siemens
France	RC2000	11/85	127,000	France Telecom Mobiles		Matra
	NMT-450	4/89	134,394	Societe Francaise du Radiotelephone	SBC 10%, BellSouth 4%	Alcatel, Mobira, Nokia
	GSM	7/92	884,000	France Telecom Mobiles		Alcatel, Ericsson, Matra, Motorola, Nortel
	GSM	3/93	455,029	Societe Francaise du Radiotelephone	SBC 10%, BellSouth 4%	Lucent, Alcatel, Motorola, Nokia, Philips, Sagem, Siemens
French Polynesia	GSM	Unknown		Tikiphone		Alcatel
Gabon	AMPS	1986	3,500	Unknown		Motorola
Gambia, The	TACS	1994	1,100	Unknown		Motorola

Country	Standard	Start-Up	Subscribers	As of	System Operator	U.S. Partners	Vendor
Georgia	AMPS	7/94			MegaCom Ltd.	Schomann International Corp.	Plexsys
	GSM	8/95			Diur La-Ole		Motorola, Siemens
Germany	C-Netz	5/86	650,000	5/96	DeTeMobil		Alcatel, Siemens
	GSM	7/92	1,680,000	5/96	DeTeMobil		Alcatel, Motorola, Philips, Siemens
	GSM	7/92	1,738,000	5/96	Mannesman Mobilfunk GmbH	AirTouch 34.5%	Ericsson, Siemens
Ghana	TACS	5/92	3,637	12/95	Millicom Ghana	Millicom 80%	Motorola
	AMPS	1/94	2,563	12/95	CelTel Ghana	Schelle Cellular	Lucent
	GSM	1996			Francis Walker Ghana Ltd.		Alcatel, Motorola
	GSM	6/96			ScanCom Ltd.		Unknown
Gibraltar	GSM	1/95	861	5/96	Gibtel		Ericsson
Greece	GSM	7/93	196,420	5/96	Panafon		Ericsson
	GSM	7/93	182,937	5/96	STET-Hellas	Nynex 20%	Ericsson, Italtel

Country	Standard	Date	Subscribers	Date	Operator	Ownership	Equipment
Greenland	NMT-900	11/92	1,500	6/95	TeleGreenland A/S		Unknown
Grenada	AMPS	1/90			Grentel Boatphone		Plexsys
Guadeloupe	AMPS	7/91			France Antilles Boatphone		Plexsys
Guam	AMPS	1992	11,000	6/95	Unknown		Unknown
Guatemala	NAMPS	10/90	29,999	12/95	Comunicaciones Celulares	Millicom 45%	Motorola
Guatemala	AMPS	1995			Tender canceled 9/95; to be retendered		Undecided
Guinea	AMPS	1993	700	Mid-96	Telecel-Guinea	Telecel International Ltd. 60%	Plexsys
Guinea	AMPS	1995	500	12/95	Spacetel-Guinea		Unknown
Guyana	AMPS	1992			GTT	??	Nortel
Haiti	AMPS	1997			Cellular Communications Int'l	Cellular Communications Int'l	Undecided
Honduras	AMPS	Mid-96			Inversiones Rocafuerte S.A.	Motorola 25% Millicom	Motorola
Hong Kong	TACS	1/84	40,000	8/95	Communications Services Ltd.		NEC
Hong Kong	AMPS	6/85	63,000	3/95	Hutchison Telecom	Motorola 30%	Motorola

Country	Standard	Start-Up	Subscribers	As of	System Operator	U.S. Partners	Vendor
Hong Kong	TACS	1/89	64,000	3/95	Hutchison Telecom		Ericsson, Motorola
	ETACS	8/89	61,800	8/95	Pacific Link		Ericsson
	TDMA	10/92	125,530	8/95	Pacific Link		Ericsson
	CDMA	9/28/95	20,000	5/96	Hutchison Telecom	Motorola	Motorola
	GSM	7/93	250,000	4/96	Communications Services Ltd.		Nokia
	GSM	1/93	180,000	3/96	SmarTone	McCaw 30%	Ericsson, Nokia
	GSM	5/29/95	10,000	8/95	Hutchison Telecom	Motorola	Motorola, Siemens
Hungary	NMT-450	10/90	73,548	2/96	Westel Radiotelefon	US WEST 49%	Ericsson, Nokia
	GSM	3/30/94	170,000	5/96	Westel 900	US WEST 49%	Ericsson
	GSM	3/94	100,000	5/96	Pannon GSM		Nokia
Iceland	NMT-450	7/86	21,898	5/96	Post og Simamalastofnunin		Ericsson
	GSM	8/15/94	14,327	5/96	Post og Simamalastofnunin		Ericsson
India	GSM	1995	21,891	5/96	Sterling Cellular	CCI 49%	Motorola, Siemens

Standard	Date	Subscribers	Operator	Investor	Equipment
GSM	8/95	21,891	Bharti Cellular	Millicom 12%	Ericsson, Motorola
GSM	End 95	11,375	Hutchison Max		Ericsson, Motorola
GSM	Mid-95	10,200	BPL Mobile	LCC 13%	Motorola, Siemens
GSM	7/31/95	3,680	Modi Telstra		Nokia
GSM	7/95	3,154	Usha Martin Telecom		Motorola, Siemens
GSM	1995	3,175	Mobile Telecom Services	AirTouch	Ericsson
GSM	6/95	2,800	Skycell Communications	Millicom 24.5%, BellSouth 24.5%	Motorola, Nokia
GSM	1997		AT&T-Birla	AT&T	Ericsson
GSM	End 96		AT&T-Birla	AT&T	Ericsson
GSM	1997		Escotel-First Pacific		Lucent
GSM	1997		Escotel-First Pacific		Lucent
GSM	1997		Escotel-First Pacific		Lucent
GSM	1997		Cellular Communications India	AirTouch	Ericsson

Country	Standard	Start-Up	Subscribers	As of	System Operator	U.S. Partners	Vendor
India	GSM	1997			Reliance-NYNEX	NYNEX	Unknown
	GSM	1997			Reliance-NYNEX	NYNEX	Unknown
	GSM	1997			Reliance-NYNEX	NYNEX	Unknown
	GSM	1997			Reliance-NYNEX	NYNEX	Unknown
	GSM	1997			Reliance-NYNEX	NYNEX	Unknown
	GSM	1997			Reliance-NYNEX	NYNEX	Unknown
	GSM	1997			Reliance-NYNEX	NYNEX	Unknown
	GSM	1997			Koshika Telecom		Alcatel
	GSM	1997			Koshika Telecom		Alcatel
	GSM	1997			Koshika Telecom		Alcatel
	GSM	1997			Koshika Telecom		Alcatel
	GSM	1997			Bharti Telecom-STET		Unknown
	GSM	1997			Tata Communications-Bell Canada		Nokia
	GSM	1997			US WEST-BPL	US WEST	Nokia
	GSM	1997			US WEST-BPL	US WEST	Motorola
	GSM	1997			US WEST-BPL	US WEST	Motorola
	GSM	1997			Fascel Ltd.		Nokia

Country	Technology	Date	Subscribers	Operator	License	Partner	Vendor
	GSM	1997		Modicom		Vanguard	Unknown
	GSM	1997		JT Mobile			Ericsson
	GSM	1997		Hinduja-HCL			Unknown
	GSM	1997		Aircell Digital			Siemens
	GSM	1997		Modicom		Vanguard	Unknown
	GSM	1997		JT Mobile			Ericsson
	GSM	1997		Aircell Digilink			Siemens
	GSM	1997		Hexacom India			Ericsson
	GSM	1997		Aircell Digilink			Siemens
	GSM	1997		Bharti Telnet			Unknown
	GSM	1997		Hexacom India			Ericsson
Indonesia	NMT	4/86	30,000	PT Mobile Selular Indonesia	1/96	Bell Atlantic	Ericsson, Nokia
	AMPS		8,100	PT Telekomindo Prima Bhakti	1/96		Unknown
	AMPS	8/91	75,000	PT Komselindo	6/96		Motorola
	NAMPS			PT Metro Seluler Nusantara			Unknown
	GSM	11/94	108,430	PT Satelindo	3/96		Alcatel

Country	Standard	Start-Up	Subscribers	As of	System Operator	U.S. Partners	Vendor
Indonesia	GSM	7/94	35,000	3/96	PT Telkomsel		Ericsson, Motorola, Siemens
	GSM	Mid-96			PT Excelcomindo	Nynex 23%	Ericsson
Iran	GSM	6/94	0	Mid-93	Telecom Company of Iran		Nokia
	GSM	Unknown			Celcom		Tendered
	GSM	3/95			KIFZO		Unknown
Ireland	TACS	12/85	138,367	5/96	Eircell		Ericsson
	GSM	6/93	38,000	5/96	Eircell		Ericsson
	GSM	4Q96			Esat Digifone		Nortel
Israel	NAMPS	3/86	250,000	12/95	Motorola Telephone Cellular Communications Ltd.	Motorola	Motorola
	AMPS-TDMA	12/27/94	300,000	5/96	CellCom Israel Ltd.	BellSouth 33%	Nortel
	Unknown	1997			Third license to be awarded 1996 or 1997		Undecided
Italy	RTMS-450	9/85	8,300	5/96	Telecom Italia Mobile		Italtel
	ETACS	4/90	3,659,800	5/96	Telecom Italia Mobile		Ericsson, Italtel

Country	Standard	Date	Subscribers	Date	Operator	Ownership	Equipment suppliers
	GSM	10/92	702,000	5/96	Telecom Italia Mobile		Ericsson, Italtel, Marconi, Siemens
	GSM	12/95	200,000	5/96	Omnitel-Pronto Italia	Bell Atlantic 11.6%, AirTouch 11.7%, CCI 10.3%	Nokia
Jamaica	AMPS	8/91	22,500	12/94	Telecom Jamaica		NEC
Japan	NTT	12/79	2,260,000	3/95	NTT DoCoMo		NEC
	NTT	12/88	278,000	2/95	Nippon Idou Tsushin (IDO)		NEC
	JTAC	4/89	750,000	2/95	Daini Denden		Motorola
	JTAC	10/91	600,000	12/95	Nippon Idou Tsushin (IDO)		Motorola
	PDC-800MHz	1993	3,082,000	6/95	NTT DoCoMo		Ericsson, Motorola, NEC
	PDC	6/1/94	138,400	6/95	TU-KA Cellular Tokai	Motorola 8%, GTE 3%, US WEST 2%, Nynex 1%	Motorola, NEC
	PDC	7/7/94	111,700	6/95	TU-KA Cellular Tokai	GTE 2%	Motorola, NEC
	PDC	4/1/94	235,600	9/95	TU-KA Cellular Kansai	GTE 4%	Motorola, NEC

Country	Standard	Start-Up	Subscribers	As of	System Operator	U.S. Partners	Vendor
Japan	PDC-1.5GHz	4/1/94	164,000	6/95	Tokyo Digital Phone	AirTouch 15%	Ericsson, Toshiba
	PDC	7/26/94			Central Japan Digital Phone	AirTouch 23%	Ericsson
	PDC	5/94	120,000	5/95	Kansai Digital Phone	AirTouch 13%	Ericsson
	PDC	1/4/96	90,000	2/96	Digital TU-KA Kyushu	AirTouch 4.5%, GTE 4.25%	Ericsson
	PDC	2/96	25,000	1/96	Kyushu Cellular Telephone		Motorola
	PDC-1.5GHz	6/96			Digital TU-KA Chugoku	AirTouch 4.5%, GTE 4.5%	Unknown
	PDC	Unknown			Shikoku Cellular Telephone		Unknown
	PDC	Unknown			Okinawa Cellular Telephone		Unknown
	PDC	2/96			Chugoku Cellular Telephone		Motorola
	PDC	2/96			Hokuriko Cellular Telephone		Motorola

Country	Standard	Start	Subscribers	Date	Operator	Owner/Partner	Equipment
	PDC-1.5GHz	1997			Digital TU-KA Tohoku	AirTouch 4.5%, GTE 4.5%	Unknown
	PDC-1.5GHz	1994	832,800	6/95	Nippon Idou Tsushin (IDO)		Ericsson, NEC
	PDC-1.5GHz	1994	1,308,000	6/95	Daini Denden		Motorola, NEC
	PDC-1.5GHz	5/97			Digital TU-KA Hokkaido Co.	AirTouch 4.5%, GTE 4.5%	Unknown
	PDC-1.5GHz	6/97			Digital TU-KA Hokuriku	GTE 4.5%	Ericsson
Jordan	GSM	6/95	12,000	12/95	Jordan Mobile Telephone Services Co.	Motorola	Motorola, Siemens
Kazakhstan	AMPS	10/5/93	1,200	4/95	Kazakhstan Wireless Communications/Tolkyn	Belle Mead	NovAtel, Telular
	NAMPS	1994?	1,500	5/95	BECET		Motorola
Kenya	ETACS	1993	2,600	12/95	Kenya Posts & Telecoms Co.		NEC
	GSM	2Q96		12/95	Kenya Posts & Telecoms Co.		Siemens
Korea	AMPS	4/84	2,260,000	6/96	Korea Mobile Telecom. Corp.		Lucent, Motorola

Country	Standard	Start-Up	Subscribers	As of	System Operator	U.S. Partners	Vendor
Korea	CDMA	4/96			Shinsegi Mobile Telecom Co.	AirTouch 11.3%, SBC 8.3%, Qualcomm 2.6%	Tendered
	CDMA	3/15/96	60,000	6/96	Korea Mobile Telecom Co.		LG Information & Communications Ltd.
Kuwait	ETACS	12/91	120,000	1/95	Kuwait Mobile Telephone Systems Co.		Ericsson, NEC
	GSM	11/94	15,000	1/95	Kuwait Mobile Telephone Systems Co.		Motorola, Siemens
	GSM?	Unknown			Tentative approval for second operator		Undecided
Kyrghystan	Unknown	Unknown	400	6/95	Unknown		Unknown
Laos	AMPS	1/93	500	6/95 Total	Lao Posts & Telecom. Enterprise		Plexsys
	GSM	12/94			Lao Shinawatra Telecom Co. Ltd.		Ericsson

Country	System	Date	Subscribers	Date	Operator	Partner	Supplier
Latvia	NMT-450	10/92	10,100	2/96	Latvia Mobile Telephone		Mitsubishi
	GSM	1/95	7,600	2/96	Latvia Mobile Telephone Co.		Nokia
	GSM	Unknown			Latvia GSM	Millicom	Unknown
	GSM	1996–7			BalTel		Unknown
Lebanon	AMPS	7/91	10,000	1/95	Spacetel		Nortel, NovAtel
	GSM	5/95	30,000	5/95	FTML		Ericsson
	GSM	5/95	20,000	5/95	Libancel SAL		Motorola, Siemens
Lesotho	GSM	12/95			Vodacom Lesotho Pty.		Motorola, Siemens
Libya	GSM				ORBIT		Ericsson
Lithuania	NMT-450	2/92	8,550	2/96	UAB Comliet	Millicom 24.5%	Nokia
	NMT-900	1994			UAB Comliet		
	GSM	3/95	7,700	2/96	OmniTEL	Motorola 38%	Motorola, Siemens
	GSM	9/95	3,500	6/96	Mobilion Telekomunikacijos	Millicom 18.5%	Ericsson
Luxembourg	NMT-450	6/85	140	5/96	Enterprise des P&T Luxembourg		Ericsson

Country	Standard	Start-Up	Subscribers	As of	System Operator	U.S. Partners	Vendor
Luxemburg	GSM	6/93	35,000	5/96	Enterprise des P&T Luxembourg		Philips, Siemens
	GSM	1997		Unde-cided	To be tendered 2Q96		
Macau	TACS	11/88	35,300	12/95 Total	Compania de Telecom Macau		Ericsson
	GSM	7/95			Compania de Telecom Macau		Ericsson
Macedonia	GSM	8/96			PTT Macedonia		Ericsson
Madagascar	AMPS	7/25/94	1,900	Mid-96	TELECEL-Madagascar	Telecel International Ltd.	Motorola
Malawi	GSM	7/96			Telecom Networks Malawi		Alcatel
Malaysia	NMT	1/85	96,000	3/96	Jabatan Telekom Malaysia		Ericsson
	ETACS	8/89	598,000	6/95	Cellular Communications Network		Ericsson
	AMPS-TDMA	5/94	201,000	3/96	Mobikom		Ericsson
	GSM	3Q95	21,000	9/95	Cellular Communications Network		Lucent, Ericsson

Country	Standard				Operator	Ownership	Equipment
	GSM	8/95	65,000	10/95	Binariang	US WEST 20%	Ericsson
Malta	ETACS	7/90	12,291	5/96	Telecell Ltd.		Plexsys
Martinique	AMPS	7/91	4,000	6/95	France Antilles Boatphone		
Mauritius	ETACS	6/89	11,735	12/95	Emtel/Currimjee Jeewanjee Millicom	Millicom 46%	Motorola, NovAtel
	GSM	1/96			Mauritius Telecom		Alcatel
Mexico	AMPS	10/89	460,000	Mid-96	Radio Mobil Dipsa	SBC 10%	Ericsson, Motorola, Nokia, Toshiba
	AMPS-TDMA	11/89	220,000	Mid-96	IUSACELL co.ttl	Bell Atlantic 44%	Nortel
	AMPS	11/90	19,000	Mid-96	Baja Celular Mexicana	Motorola 42%	Motorola
	AMPS	11/90	18,000	Mid-96	Movitel del Noroeste	Motorola (Baja Celular), General Cellular, GTE 13%, McCaw 19%	Motorola, Nortel
	AMPS	11/90	25,000	Mid-96	Telefonia Celular del Norte	Motorola 80%, Motorola Sprint 20%	Motorola

Country	Standard	Start-Up	Subscribers	As of	System Operator	U.S. Partners	Vendor
Mexico	AMPS	11/90	25,000	Mid-96	Celular de Telefonia	Motorola 40%	Nortel, Motorola
	AMPS	8/90		See MX-92	Comunicaciones Celular de Occidente (IUSACELL)	Bell Atlantic 44%	Motorola
	AMPS	1990		See MX-02	Sist. Telefonica Portatiles Celular (IUSACELL)	Bell Atlantic 44%	Nortel
	AMPS	1990		See MX-02	Telecom. de Golfo (IUSACELL)	Bell Atlantic 44%	Nortel
	AMPS	1990	14,000	Mid-96	Portatel del Sureste	Associated Communi-cations 26%, LCC 23%	Motorola
Monaco	GSM	Unknown			Office des Telephone de Monaco		Alcatel
Mongolia	GSM	3/18/96			MobiCom		Alcatel
Montenegro	GSM	4Q96			Promonte GSM		Unknown
Montserrat	AMPS	12/92			Cable & Wireless (W.I.) Ltd.		Plexsys

Country	System	Date	Subscribers	Date	Operator	Equipment
Morocco	NMT–450	1989	9,000	2/95	Office National des Postes et Telecom.	Ericsson
	GSM	4/94	4,000	2/95	Office National des Postes et Telecom.	Motorola, Siemens
Mozambique	GSM	Unknown			Telecomunicacoes de Mocambique?	Undecided
Namibia	GSM	4/27/95	3,500	12/95	Mobile Telecommunications Co. Ltd.	Ericsson, Motorola, Siemens
Nauru	AMPS	8/94			Unknown	Lucent, Plexsys
Nepal	GSM	1997			Tendered 1/96	Undecided
Neth. Antilles	AMPS	1988			East Caribbean Cellular	Plexsys
Netherlands	NMT–450	1/85	16,536	5/96	Royal Dutch Post & Telecom	Ericsson, Philips
	NMT–900	1/89	281,464	5/96	Royal Dutch Post & Telecom	Ericsson, Philips
	GSM	7/1/94	329,000	5/96	Royal Dutch Post & Telecom	Alcatel, Ericsson, Nokia
	GSM	9/95	89,000	5/96	Libertel	Ericsson

Country	Standard	Start-Up	Subscribers	As of	System Operator	U.S. Partners	Vendor
New Caledonia	GSM				OPT		Alcatel
New Zealand	AMPS-TDMA	8/87	325,600	12/95	Telecommobile Communications	Ameritech, Bell Atlantic 49.9% total	Ericsson
	GSM				Telstra		Unknown
	GSM	7/93	30,000	12/95	Bell South New Zealand	BellSouth 80%	Nokia
Nicaragua	AMPS	4/93	7,000	5/96	NicaCell	Motorola	Motorola
	AMPS	Unknown			Teleglobo		Undecided
Nigeria	ETACS-800	11/92	30,000	12/95	Mobile Telecom Services	Digital Telecom	Ericsson, Motorola
	GSM	1995			EMIS Nigeria		Motorola, Siemens
	GSM	1995			International Wireless Inc./Comstar	Int'l Wireless Inc.	Motorola, Siemens
Norway	NMT-450	7/81	188,499	5/96	Telenor Mobil A/S		Ericsson, Mitsubishi, Nokia

Country	System	Date	Subscribers	Date	Operator	Ownership	Equipment
	NMT-900	12/86	292,454	5/96	Telenor Mobil A/S		Ericsson, Mitsubishi, Nokia
	GSM	5/93	318,000	5/96	Telenor Mobil A/S		Ericsson, Nokia
	GSM	9/93	275,000	2/96	Netcom GSM A/S	Ameritech 24.95%, Millicom	Motorola, Nokia, Siemens
Oman	NMT-450	5/85	7,500	12/95	General Telecom. Organization		Ericsson
	GSM	1996			General Telecom. Organization		Tendered
Pakistan	AMPS	11/90	32,500	12/95	Paktel		Ericsson
	AMPS	12/90	13,000	12/95	Pakcom	Millicom 59.3%	Ericsson
	GSM	8/94	7,000	12/95	Pakistan Mobile Communications Ltd.	Motorola 66%	Motorola, Siemens
Panama	AMPS-TDMA	6/28/96			BellSouth Corp. de Panama		Ericsson
	AMPS	Unknown			Intel S.A.		Undecided
Papua New Guinea	AMPS	12/1/95			Post and Telikom Corp.		Unknown
Paraguay	AMPS	8/92	15,807	12/95	Telefonica Celular del Paraguay	Millicom 72%	Motorola

Country	Standard	Start-Up	Subscribers	As of	System Operator	U.S. Partners	Vendor
Paraguay	AMPS	1996			Antelco		Unknown
Peru	AMPS-TDMA	4/90	40,000	5/96	Tele2000	Cellular International	Lucent, Nortel, NovAtel
	AMPS-TDMA	7/91	41,872	12/95	CPT-Celular		Nortel
	AMPS-TDMA	1993			ENTEL		Nortel
	Unknown	1996?			May be awarded 1995–6		Undecided
Philippines	NAMPS	3/91	240,000	3/96	Pilipino Telephone Corp.		Lucent, Motorola, NEC
	NAMPS	2/92	95,000	12/95	Express Telecommunications Co.	Millicom 40%	Motorola
	AMPS	2/91	27,000	12/95	Isla Communications		Unknown
	ETACS	2/94	137,000	3/96	Smart Information Technologies		Ericsson
	GSM	7/94	28,000	12/95	Isla Communications		Ericsson, Motorola, Siemens
	GSM	1995	51,000	3/96	Globe Telecom		
	GSM	1995			Smart Information Technologies		Unknown

Country	System	Date	Subscribers	Date	Operator	Partner	Equipment
Poland	NMT-450	6/19/92	95,000	5/96	Centertel	Ameritech 24.5%	Ericsson, Nokia
	GSM	4Q96			Polska Telefonia Cyfrowa	US WEST 22.5%	Ericsson, Siemens
	GSM	11/96			Polkomtel	AirTouch 19.25%	Nokia
Portugal	C-Netz	1/89	17,300	5/96	Telecom. Moveis Nacionais		Siemens
	GSM	12/92	190,000	5/96	Telecom. Moveis Nacionais		Motorola, Philips, Siemens
	GSM	10/92	213,000	5/96	Telecel	AirTouch 23%; LCC	Ericsson
Qatar	MATS	1982	3,500	1/95	Q-Tel		Ericsson
	GSM	2/94	5,000	1/95	Q-Tel		Motorola, Siemens
Reunion	GSM	Unknown			Societe Francaise du Radiotelephon		Alcatel
Romania	NMT-450	4/93	11,700	12/95	Telefonica Romania		Ericsson
	GSM	1Q96			Telefonica Romania		Unknown
	GSM	1997			Bids for two GSM licenses due 9/96		Undecided

Country	Standard	Start-Up	Subscribers	As of	System Operator	U.S. Partners	Vendor
Russia	NMT-450	6/91	10,723	3/96	Delta Telecom	US WEST 42.5%	Ericsson, Nokia
	NMT-450	12/91	17,980	3/96	Moscow Cellular Communications	US WEST 22%, Millicom 20%	Benefon, Ericsson, Motorola, Nokia
	AMPS-TDMA	1992	32,000	5/96	Vimpel Communications	FGI Wireless	Ericsson, Motorola, Plexsys, Telular
	AMPS	2/94	354	12/94	Personal Systems Network	Millicom 65%	Motorola
	GSM	8/94	2,000	12/94	Mobile TeleSystems		Motorola, Siemens
	GSM	1993?			Krakor		Unknown
	GSM	1993?	2,000	1Q94	Delta Telecom	US WEST	Ericsson, Nokia
	GSM	1/95	2,000	5/95	North-West GSM		Motorola, Nokia
	GSM	1993?			SMARTS		Nokia
	GSM	8/94	400	1/95	United Telecom Bashkortostan (RTDC)	US WEST	Ericsson
	GSM	5/22/95	1,023	3/96	NCC	US WEST thru RTDC	Alcatel

Technology	Date	Subscribers	As of	Operator	Owner	Equipment
GSM	1993?			US WEST	US WEST	Unknown
GSM	3/21/96			Ural Westcom	US WEST thru RTDC 49%	
GSM	1993?			US WEST	US WEST	Unknown
GSM	1993?			US WEST	US WEST	Unknown
GSM	1993?			US WEST	US WEST	Unknown
DAMPS	1993?	2,551	3/96	AKOS	US WEST	Hughes Network Systems
GSM	1993?			US WEST	US WEST	Unknown
GSM	1993?			US WEST	US WEST	Unknown
AMPS	1995?			Regional Cellular Network Co.	Millicom 45%	Motorola
AMPS	1995?			Belgorod Cellular Network	Millicom 60%	Unknown
AMPS	5/94	545	12/94	Chelyabisk Cellular	Millicom 51%	Ericsson
AMPS	1995?			Kursk Cellular Network	Millicom 60%	Unknown
NMT-450	1994			Bashinformsvyaz		Ericsson
NMT-450	End 94			RISS Telekom	SS Telecom, Comspan Int'l	Unknown
GSM	Unknown			Russian Telecom Development Corp.	US WEST	Alcatel, Motorola

Country	Standard	Start-Up	Subscribers	As of	System Operator	U.S. Partners	Vendor
Russia	GSM	Unknown			Russian Telecom Development Corp.	US WEST	Alcatel, Motorola
	GSM	11/94	896	3/96	Dontelecom (RTDC)	US WEST	Iskratel, Motorola, Siemens
	GSM	Unknown			Russian Telecom Development Corp.	US WEST	Unknown
	NAMPS	1995	3,000	5/95	St. Petersburg Telecom		Motorola
	GSM	1994?			Russkaya Telefinnaya Kompaniya		Unknown
	GSM	1994?			Russkaya Telefinnaya Kompaniya		Unknown
	GSM	1994?			Russkaya Telefinnaya Kompaniya		Unknown
	GSM	1994?			Russkaya Telefinnaya Kompaniya		Unknown
	NMT-450	4/94	2,500	4/96	Tver Cellular Communications	US WEST	Ericsson
	AMPS	Unknown			Unknown		Unknown
	AMPS				United Telecom Far East		Unknown

Technology	Year	Operator	Manufacturer
AMPS		United Telecom Far East	Unknown
AMPS		United Telecom Far East	Unknown
GSM	1995	Northwest GSM	Nokia
GSM	1995	Northwest GSM	Nokia
GSM	1995	Northwest GSM	Nokia
GSM	1994?	Chelyabinsk Svyazinform	Alcatel
GSM	1995	Italtel	Italtel
GSM	1995	Italtel	Italtel
GSM	1994?	Kilion-S	Unknown
GSM	1994?	Sakhalin Ltd.	Unknown
GSM	1994?	Gorivont-RT	Italtel
GSM	1994?	Vartelecom	Unknown
NMT-450	End 95	Moscow Cellular Communications	Ericsson
NMT-450	1994?	Telecom Finland	Unknown
NMT-450	1994?	Telecom Finland	Unknown
NMT-450	1994?	Telecom Finland	Unknown
NMT-450	1994?	Telecom Finland	Unknown
NMT-450	1995	Telecom Finland	Unknown
NMT-450	1994?	Perm Rossvyazinform	Unknown

Country	Standard	Start-Up	Subscribers	As of	System Operator	U.S. Partners	Vendor
Russia	NMT-450	1994?			Sverdlovsk Ross./Metropolitan Comms. Co.		Unknown
	NMT-450	1994?			Vartelecom	US WEST	Unknown
	NMT-450	1994?			TsUMS-22		Unknown
	NMT-450	7/17/95			Global Telesystems Group		Unknown
	NMT-450	1994?			Martelcom		Unknown
	NMT-450	1994?			Komi Rossvyazinform		Unknown
	NMT-450	1995			Tula Electrosvyaz		Unknown
	NMT-450	1995			Bryansk Rossvyazinform		Unknown
	NMT-450	1995			Kaluga Rossvyazinform		Unknown
	NMT-450	1995			Ryazan Electrosvyaz		Unknown
	NMT-450	1995			ITS		Unknown
	NMT-450	1995	533	3/96	Baykalwestcom	US WEST	Unknown
	NMT-450	1995			Krasnoyarsk Rossvyazinform		Unknown
	NMT-450	1995			Gorizont-RT		Unknown
	NMT-450	1995			Metropolitan Comms. Co.		Unknown
	NMT-450	1995			Nizhnevartovsk Neftegas		Unknown

Technology	Date	Company	Partner	Vendor
Unknown	9/95	Chuvashia Mobile		Lucent
CDMA	1997	Chelyabinsk Svyazinform		Qualcomm
Unknown	9/95	Global TeleSystems		Lucent
Unknown	9/95	Global TeleSystems		Lucent
Unknown	9/95	Global TeleSystems		Lucent
TDMA	Unknown	Orensot		Ericsson
Analog	Unknown	Dagestan Cellular Network Co.		Samsung
CDMA	1997?	IV Telecom		Samsung
AMPS	Unknown	Vostok	GTS Cellular 51%	Unknown
AMPS	Unknown	Vostok	GTS Cellular 51%	Unknown
AMPS	Unknown	Vostok	GTS Cellular 51%	Unknown
AMPS	Unknown	Vostok	GTS Cellular 51%	Unknown
AMPS	Unknown	Vostok	GTS Cellular 51%	Unknown
AMPS	Unknown	Vostok	GTS Cellular 51%	Unknown

Country	Standard	Start-Up	Subscribers	As of	System Operator	U.S. Partners	Vendor
Russia	AMPS	Unknown			Vostok	GTS Cellular 51%	Unknown
	AMPS	Unknown			Vostok	GTS Cellular 51%	Unknown
	AMPS	Unknown			Vostok	GTS Cellular 51%	Unknown
	AMPS	Unknown			Vostok	GTS Cellular 51%	Unknown
	DAMPS	1996			Orensot		Ericsson
	Unknown	Unknown			Dal Telecom		Unknown
	Unknown	Unknown			Sotovaya Co. Ltd.		Unknown
	NMT-450	5/95			CommStruct Int'l	GTS Cellular	Unknown
	NMT-450	4/96			RTDC	US WEST	Unknown
	AMPS	Unknown			VimpelCom		Unknown
Saipan	Unknown	11/91			Unknown		Unknown
Samoa (Am.)	AMPS	1987	2,900	6/95	Samoan Office of Communications		Motorola
	AMPS	Unknown			Undecided		Motorola

Country	Standard	Date	Subscribers	Launch	Operator	Supplier	Ownership
Samoa (W.)	Unknown	Unknown			Unknown		Undecided
Saudi Arabia	NMT-450	8/81	20,000	1/95	Saudi Telecom	Ericsson, Philips	
	GSM	1/96	250,000	1/96	Saudi Telecom	Lucent, Philips	
	GSM	Unknown			Royal Palace Network	Motorola, Siemens	
Senegal	RC2000	4/92	500	Mid-95	SONATEL	Unknown	
	GSM	1998			SONATEL	Undecided	
	GSM	1998				Undecided	
Seychelles	GSM	11/95			Cable & Wireless (Seychelles) Ltd.	Alcatel	
Singapore	AMPS	8/88	170,000	5/95	Singapore Telecom	NEC	
	ETACS	8/91	125,000	12/94	Singapore Telecom	Ericsson	
	GSM	3/8/94	102,000	6/95	Singapore Telecom	Ericsson	
	GSM	4/1/97			MobileOne Pte. Ltd.	Nokia	
Slovakia	NMT-450	10/91	14,270	2/96	Eurotel Bratislava	Nokia	Bell Atlantic 24.5%, US WEST 24.5%

Country	Standard	Start-Up	Subscribers	As of	System Operator	U.S. Partners	Vendor
Slovakia	GSM	1996–7			Eurotel	Bell Atlantic 24.5%, US WEST 24.5%	Undecided
	GSM	1996–7			To be awarded 8/96		Undecided
Slovenia	NMT-450	10/90	31,000	5/96	Mobitel		Ericsson
	GSM	4Q96			Mobitel		Ericsson
	GSM	1996			To be awarded 1996		Undecided
Soloman Islands	Unknown	Unknown			Cable & Wireless		Unknown
South Africa	C-Netz	5/86	10,000	6/95	Vodacom		Siemens
	GSM	6/1/94	330,000	12/95	Vodacom GSM		Alcatel, Motorola, Siemens
	GSM	6/1/94	190,000	12/95	Mobile Telephone Networks	SBC 15.5%	Ericsson
	GSM	Unknown			Third license may be issued		

Country	Standard	Date	Subscribers	Date	Operator	Ownership	Equipment
Spain	NMT-450	6/82	15,500	5/96	Telefonica Servicios Moviles		Ericsson, Philips
	TACS	4/90	1,148,500	5/96	Telefonica Moviles		Lucent, Motorola
	GSM	7/95	248,000	5/96	Telefonica Moviles		Lucent, Ericsson, Motorola
	GSM	10/95	130,000	5/96	Airtel S.A.	AirTouch 16.35%	Ericsson, Siemens
Sri Lanka	ETACS	7/89	23,000	5/96	Celltel Lanka	Millicom 76%	Motorola
	ETACS	1993?	15,000	5/96	Lanka Cellular Service		NEC
	AMPS	4/93	18,000	5/96	Mobiltel		Ericsson
	GSM	2/95	1,600	5/96	MTN Network Ltd.		Alcatel
St. Kitts and Nevis	AMPS	1989			St. Kitts & Nevis Boatphone		Plexsys
St. Lucia	AMPS	12/90			St. Lucia Boatphone		Plexsys
St. Martin/Bartholemy	AMPS	9/91			St. Martin Mobiles		Plexsys
St. Vincent/Grenadines	AMPS	12/90			BOATPHONE		Plexsys
Sudan	GSM	1998			Daewoo & Sudatel		Alcatel
Suriname	AMPS-TDMA	1993	2,000	12/95	Telesur		Nortel

Country	Standard	Start-Up	Subscribers	As of	System Operator	U.S. Partners	Vendor
Swaziland	Unknown	1996?			SPTC		Undecided
Sweden	NMT-450	6/81	247,945	5/96	Telia Mobitel AB		Ericsson, Mitsubishi, Mobira
	NMT-450	8/81	0	5/96	Comvik AB	Millicom 20%	Ericsson, E.F. Johnson
	NMT-900	12/86	706,091	5/96	Telia Mobitel AB		Ericsson, Mitsubishi, Mobira
	GSM	11/92	484,300	2/96	Telia Mobitel AB		Ericsson, Nokia
	GSM	9/92	475,800	2/96	Comvik GSM	Millicom 20%	Motorola, Siemens
	GSM	9/92	214,000	5/96	NordicTel Holdings	AirTouch 51%	Nokia
Switzerland	NMT-900	9/87	312,000	5/96	PTT		Ericsson, Philips
	GSM	3/93	212,000	5/96	PTT		Ericsson, Motorola, Philips
Syria	GSM	1996?			PTE		Tendered, withdrawn
Taiwan	AMPS	5/89	600,000	6/95	Directorate General of Telecommunications—LDTA		Ericsson

Country	Standard	Start Date	Subscribers	Date	Operator		Vendors
	GSM	7/95	200,000	6/95	Presubs Directorate General of Telecommunications—LDTA		Nortel
	GSM	1997			Three licenses in each region, bids due 9/96		Undecided
Tanzania	TACS	10/94	2,303	12/95	MIC Tanzania Ltd.	Millicom 51%	Motorola
	GSM	12/95			Tritel		Motorola, Siemens
	GSM	Unknown			May offer four regional licenses		Undecided
Thailand	NMT-450	7/86	50,000	3/96	Telephone Organization of Thailand		Ericsson, Nokia, NovAtel, Philips
	AMPS	3/87	46,000	12/95	Communications Authority of Thailand		Motorola
	NAMPS	9/91	527,263	3/96	Total Access Communication Co. Ltd.	Motorola	Motorola, Nokia
	NMT-900	9/90	540,000	8/95	Advanced Information Services		Ericsson, Nokia
	GSM	10/1/94	32,000	6/95	Advanced Information Services		Ericsson, Motorola, Nokia
Tonga	Unknown	1993?			Unknown		Unknown

Country	Standard	Start-Up	Subscribers	As of	System Operator	U.S. Partners	Vendor
Trinidad and Tobago	AMPS	12/91	3,100	12/94	Telecom. Services of Trinidad & Tobago		Nortel
Tunisia	NMT-450	4/85	7,800	12/95 total	Ministry of Communications		Ericsson
	GSM	End 96			Ministry of Communications		To be tendered
Turkey	NMT-450	10/86	9,500	2/96	Turkcell		Mobira, Nokia
	GSM	3/94	290,000	6/96	Turkcell		Ericsson
	TACS	1993?			Turkcell		Motorola
	GSM	4/94	115,000	2/96	Telsim Mobil		
Turkmenistan	AMPS	1994?			Unknown		Plexsys
Turks and Caicos	Undecided	Unknown			Cable & Wireless		Undecided
Uganda	GSM	5/95	1,747	12/95	Clovergem Celtel Ltd.		Motorola, Siemens
Ukraine	NMT-450	7/93	20,000	6/96	Ukrainian Mobile Communications		Nokia
	GSM	1998			Ukrainian Mobile Communications		Siemens
	GSM	1998			Ukraine Radio Systems		Undecided

Country	System	Date	Subscribers	Date	Operator	Vendor
	DAMPS	End 96		End 96	Telecel	Unknown
	DAMPS	End 96		End 96	Digital Cellular Communications	Ericsson
United Arab Emirates	ETACS	5/89	115,000	6/95 Total	Emirates Telecom Corp.	Ericsson
	GSM	1994			Emirates Telecom Corp.	Ericsson
	GSM	1994			Emirates Telecom Corp.	Alcatel
	GSM	9/94			Emirates Telecom Corp.	Motorola, Siemens
	GSM	4/93			Emirates Telecom Corp.	Lucent
United Kingdom	ETACS	1/85	1,975,200	5/96	Cellnet	Motorola
	ETACS	12/87	1,880,000	5/96	Vodafone	Ericsson
	GSM	7/14/94	459,000	5/96	Cellnet	Ericsson, Motorola, Nokia, Siemens
	GSM	7/92	644,000	5/96	Vodafone	Ericsson, Nokia, Orbitel
	GSM	6/94	400	12/94	Jersey Telecoms	Alcatel
	GSM	3/96			Guernsey Telecoms	Unknown
	GSM	3/96			Manx Telecom	Unknown

Country	Standard	Start-Up	Subscribers	As of	System Operator	U.S. Partners	Vendor
United States	AMPS	10/83	31,000,000	6/95	Multiple		Multiple
	AMPS-TDMA	1993	1,500,000	End 95	Multiple		Multiple
	NAMPS				Multiple		Multiple
Uruguay	AMPS	11/91	20,000	12/95	Abiatar	BellSouth 38%, Motorola	Motorola
	AMPS-TDMA	10/94	19,000	12/95	Antel		Ericsson, NEC
Uzbekistan	NMT-450	10/92	6,500	12/95 total	Uzdunrobita JB	ICG	Unknown
	AMPS-TDMA	6/94			Uzdunrobita JB	ICG	Nortel
	DAMPS	8/96			Uzbekistan Digital Network		Ericsson, Siemens
Venezuela	AMPS-TDMA	10/88	168,000	12/95	Movilnet	GTE, AT&T	Ericsson
	NAMPS	1991	250,331	12/95	Telcel	BellSouth 50%	Motorola
	GSM	1995–6			One license may be awarded after PCS		Undecided
Vietnam	AMPS	5/92	22,000	6/95	Saigon Mobile Telephone Co.		Ericsson
	GSM	4/6/94	20,000	9/95	Vietnam Mobile Telecom Services Co.		Alcatel, Ericsson

Country	System	Date	Subscribers	Launch	Operator	Partner	Equipment
	GSM	6/96			Vietnam Telecom Services Co.		Motorola, Siemens
Virgin Islands (Br.)	AMPS	12/86	120	12/92	Caribbean Cellular Telephone Boatphone	Boston Communications Corp.	Plexsys
West Bank/Gaza	GSM	1997			PalTel		To be tendered 1996
Yemen	TACS	5/92	1,500	Mid-93	Cable & Wireless		Unknown
Zaire	NAMPS	1986	12,000	12/95	TELECEL-Zaire	Telecel International Ltd.	Motorola, Plexsys
Zambia	AMPS	1994			Express Communications		Motorola
	TACS	End 94			Trans Global Telecom		Motorola
	AMPS	1995			Zamtel		Motorola, NEC
	GSM	Unknown			Zamtel and partner		
	CDMA	Late 96			Telecel-Zambia	Telecel International Ltd.	Motorola
Zimbabwe	GSM	Unknown			PTC	Possible	Tendered
	GSM	4/96			Retrofit		Unknown

Index